Extreme Waves and Shock-Excited Processes in Structures and Space Objects

Extreme Waves and Shock-Excited Processes in Structures and Space Objects

Shamil U. Galiev

CRC Press is an imprint of the
Taylor & Francis Group, an **informa** business

First edition published 2020
by CRC Press
6000 Broken Sound Parkway NW, Suite 300, Boca Raton, FL 33487-2742

and by CRC Press
2 Park Square, Milton Park, Abingdon, Oxon, OX14 4RN

© 2020 Taylor & Francis Group, LLC

CRC Press is an imprint of Taylor & Francis Group, LLC

Reasonable efforts have been made to publish reliable data and information, but the author and publisher cannot assume responsibility for the validity of all materials or the consequences of their use. The authors and publishers have attempted to trace the copyright holders of all material reproduced in this publication and apologize to copyright holders if permission to publish in this form has not been obtained. If any copyright material has not been acknowledged please write and let us know so we may rectify in any future reprint.

Except as permitted under U.S. Copyright Law, no part of this book may be reprinted, reproduced, transmitted, or utilized in any form by any electronic, mechanical, or other means, now known or hereafter invented, including photocopying, microfilming, and recording, or in any information storage or retrieval system, without written permission from the publishers.

For permission to photocopy or use material electronically from this work, access www.copyright. com or contact the Copyright Clearance Center, Inc. (CCC), 222 Rosewood Drive, Danvers, MA 01923, 978-750-8400. For works that are not available on CCC please contact mpkbookspermissions@tandf.co.uk

Trademark notice: Product or corporate names may be trademarks or registered trademarks, and are used only for identification and explanation without intent to infringe.

Library of Congress Cataloging-in-Publication Data
[Insert LoC Data here when available]

ISBN: 9780367480653 (hbk)
ISBN: 9781003038498 (ebk)

Typeset in Times
by codeMantra

I dedicate this book to the memory of my parents
- Valitova Ajar Khayrutdinovna and
Galiev Usman Zakirovich

For nothing is *hidden* that *will not become* evident,
nor anything *secret* that *will not be known* and come to light.
Luke 8:17

Contents

Preface xiii
Acknowledgments xv
Author xvii

PART I Basic Models, Equations, and Ideas

Chapter 1 Models of Continuum 3
      1.1   The System of Equations of Mechanics Continuous Medium 3
      1.2   State (Constitutive) Equations for Elastic and
Elastic–Plastic Bodies 8
      1.3   The Equations of Motion and the Wide-Range Equations
of State of an Inviscid Fluid 12
      1.4   Simplest Example of Fracture of Media within
Rarefaction Zones 17
            1.4.1   The State Equation for Bubbly Liquid 17
            1.4.2   Fracture (Cold Boiling) of Water During
Seaquakes 19
            1.4.3   Model of Fracture (Cold Boiling) of Bubbly
Liquid 21
      1.5   Models of Moment and Momentless Shells 24
            1.5.1   Shallow Shells and the Kirchhoff–Love
Hypotheses 24
            1.5.2   The Timoshenko Theory of Thin Shells and
Momentless Shells 28

Chapter 2 The Dynamic Destruction of Some Materials in Tension Waves 33
      2.1   Models of Dynamic Failure of Solid Materials 34
            2.1.1   Phenomenological Approach 34
            2.1.2   Microstructural Approach 36
      2.2   Models of Interacting Voids (Bubbles, Pores) 39
      2.3   Pore on Porous Materials 43
      2.4   Mathematical Model of Materials Containing Pores 46

Chapter 3 Models of Dynamic Failure of Weakly Cohesived Media (WCM) 51
      3.1   Introduction 51
            3.1.1   Examples of Gassy Material Properties 53
            3.1.2   Behavior of Weakly Cohesive Geomaterials
within Extreme Waves 54

3.2 Modeling of Gassy Media 57
    3.2.1 State Equation for Condensed Matter–Gas Mixture 58
    3.2.2 Strongly Nonlinear Model of the State Equation for Gassy Media 60
    3.2.3 The Tait-Like Form of the State Equation 62
    3.2.4 Wave Equations for Gassy Materials 65
3.3 Effects of Bubble Oscillations on the One-Dimensional Governing Equations 66
    3.3.1 Differential Form of the State Equation 66
    3.3.2 The Strongly Nonlinear Wave Equation for Bubbly Media 67
3.4 Linear Acoustics of Bubbly Media 69
    3.4.1 Three-Speed Wave Equations 70
    3.4.2 Two-Speed Wave Equations 71
    3.4.3 One-Speed Wave Equations 72
    3.4.4 Influence of Viscous Properties on the Sound Speed of Magma-like Media 73
3.5 Examples of Observable Extreme Waves of WCM 75
    3.5.1 Mount St Helens Eruption 75
    3.5.2 The Volcano Santiaguito Eruptions 77
3.6 Nonlinear Acoustic of Bubble Media 79
    3.6.1 Low-Frequency Waves: Boussinesq and Long-Wave Equations 80
    3.6.2 High-Frequency Waves: Klein–Gordon and Schrödinger Equations 81
3.7 Strongly Nonlinear Airy-Type Equations and Remarks to Chapters 1–3 82
References 84

PART II   *Extreme Waves and Structural Elements*

Chapter 4 Extreme Effects and Waves in Impact-Loaded Hydrodeformable Systems 91

4.1 Introduction 91
4.2 Underwater Explosions and Extreme Waves of the Cavitation: Experiments 93
4.3 Experimental Studies of Formation and Propagation of the Cavitation Waves 96
    4.3.1 Elastic Plate–Underwater Wave Interaction 97
    4.3.2 Elastoplastic Plate–Underwater Wave Interaction 98
4.4 Extreme Underwater Wave and Plate Interaction 101
    4.4.1 Effects of Deformability 101

Contents ix

		4.4.2	Effects of Cavitation on the Plate Surface	103
		4.4.3	Effects of Cavitation in the Liquid Volume on the Plate–Liquid Interaction	106
		4.4.4	Effects of Plasticity	114
	4.5	Modeling of Extreme Wave Cavitation and Cool Boiling in Tanks		121
		4.5.1	Impact Loading of a Tank	122
		4.5.2	Impact Loading of Liquid in a Tank	124

Chapter 5 Shells and Cavitation (Cool Boiling) Waves 127

    5.1 Interaction of a Cylindrical Shell with Underwater Shock Wave in Liquid 127
    5.2 Extreme Waves in Cylindrical Elastic Container 129
        5.2.1 Effects of Cavitation and Cool Boiling on the Interaction of Shells 130
        5.2.2 Features of Bubble Dynamics and Their Effect on Shells 133
    5.3 Extreme Wave Phenomena in the Hydro-Gas-Elastic System 135
    5.4 Effects of Boiling of Liquids Within Rarefaction Waves on the Transient Deformation of Hydroelastic Systems 139
    5.5 A Method of Solving Transient Three-Dimensional Problems of Hydroelasticity for Cavitating and Boiling Liquids 143
        5.5.1 Governing Equations 143
        5.5.2 Numerical Method 145
        5.5.3 Results and Discussion 147

Chapter 6 Interaction of Extreme Underwater Waves with Structures 155

    6.1 Fracture and Cavitation Waves in Thin-Plate/Underwater Explosion System 155
    6.2 Fracture and Cavitation Waves in Plate/Underwater Explosion System 159
    6.3 Generation of Cavitation Waves after Tank Bottom Buckling 163
    6.4 Transient Interaction of a Stiffened Spherical Dome with Underwater Shock Waves 166
        6.4.1 The Problem and Method of Solution 166
        6.4.2 Numeric Method of Problem Solution 169
        6.4.3 Results of Calculations 170
    6.5 Extreme Amplification of Waves at Vicinity of the Stiffening Rib 174

References 177

PART III Counterintuitive Behavior of Structural Elements after Impact Loads

Chapter 7 Experimental Data 183
        7.1   Introduction and Method of Impact Loading 183
        7.2   CIB of Circular Plates: Results and Discussion 187
        7.3   CIB of Rectangular Plates and Shallow Caps 193
                7.3.1   Discussion of CIB of Shallow Caps 194
                7.3.2   Cap/Permeable Membrane System 200
                7.3.3   CIB of Panels 203

Chapter 8 CIB of Plates and Shallow Shells: Theory and Calculations 207
        8.1   Distinctive Features of CIB of Plates and Shallow Shells 207
                8.1.1   Investigation Techniques 207
                8.1.2   Results and Discussion: Plates, Spherical Caps, and Cylindrical Panels 210
        8.2   Influences of Atmosphere and Cavitation on CIB 217
                8.2.1   Theoretical Models 219
                8.2.2   Calculation Details 224
                8.2.3   Results and Discussion 225
        References 230

PART IV Extreme Waves Excited by Impact of Heat, Radiation, or Mass

Chapter 9 Forming and Amplifying of Heat Waves 235
        9.1   Linear Analysis – Influence of Hyperbolicity 235
        9.2   Forming and Amplifying Nonlinear Heat Waves 238
        9.3   Strong Nonlinearity of Thermodynamic Function as a Cause of Formation of Cooling Shock Wave 242
        Conclusions 248

Chapter 10 Extreme Waves Excited by Radiation Impact 249
        10.1   Impulsive Deformation and Destruction of Bodies at Temperatures below the Melting Point 250
                10.1.1 Thermoelastic Waves Excited by Long-Wave Radiation 250
                10.1.2 Thermoelastic Waves Excited by Short-Wave Radiation 250

		10.1.3	Stress and Fracture Waves in Metals During Rapid Bulk Heating	253
		10.1.4	Optimization of the Outer Laser-Induced Spalling	256
	10.2	Effects of Melting of Material under Impulse Loading		260
		10.2.1	Mathematical Model of Fracture under Thermal Force Loading	260
		10.2.2	Algorithm and Results	263
	10.3	Modeling of Fracture, Melting, Vaporization, and Phase Transition		267
		10.3.1	Calculations: Effects of Temperature	270
		10.3.2	Calculations: Effects of Vaporization	273
		10.3.3	Calculations: Effect of Vaporization on Spalling	277
	10.4	Two-Dimensional Fracture and Evaporation		279
	10.5	Fracture of Solid by Radiation Pulses as a Method of Ensuring Safety in Space		282
		10.5.1	Introduction	282
		10.5.2	Mathematical Formulation of the Problem	285
		10.5.3	Calculation Results and Comparison with Experiments	287
		10.5.4	Special Features of Fracture by Spalling	290
		10.5.5	Efficiency of Laser Fracture	293
		10.5.6	Discussion and Conclusion	295
	Conclusion			296
	Reference			296

Chapter 11 Melting Waves in Front of a Massive Perforator ... 297

	11.1	Experimental Investigation	297
	11.2	Numerical Modeling	300
	11.3	Results of the Calculation and Discussion	301
	References		302

Index ... 307

Preface

Humanity always operates in conditions of incomplete information.
-Valentin Valentinovich Novozhilov

This book focuses on the study of extreme waves. Their formation can significantly change properties of the medium within the wave; that is, a phase state of the medium may be changed. Very often these waves threaten the strength of the medium or the existence of technical systems.

The presented here theory is based on conservation and state equations of continuum media. Cases are considered when these equations may be simplified to the dynamic theory of shells or to strongly nonlinear wave equations in gas, liquid, and solid media, bubbly liquids, and some gassy materials. Thus, the presented theory demonstrates the interdisciplinary nature of the waves under the investigation. In particular, we want to show that the emergence and evolution of many extreme waves are determined by similar physical mechanisms.

This book examines the reaction of structural elements and space objects to the dynamic actions of different durations and intensities. The load is carried out by pulses of pressure, heat, radiation, or mass.

Much attention is paid to cases of shock loading of shells containing water or liquefied gas. The effects of cavitation in the water and cold boiling of cryogenic liquids on the formation of extreme waves inside the shell are studied. It is shown that both cavitation and cold boiling can decisively affect both the strength of structures and the possibility of their beneficial functioning. In particular, the reaction of plates and shells contacting with a liquid to an explosion is studied. The main attention is paid to studying the influence of cavitation, geometric and physical nonlinearity of shells on their strength, and the formation of extreme waves in the liquids. The influence of plastic deformations, large displacements, and reinforcing elements on the wave deformation of the structural elements is studied also. In particular, the effect of the elastoplastic properties of the material on the dynamic behavior of such structures, in particular on the possibility of the so-called counterintuitive behavior (CIB) of plates and shallow shells, is studied.

Then, the extreme waves excited by very short impact of heat, radiation, or mass are studied. In these cases, thermomechanical behavior of the material during the action of the impact loading, in many respects, is determined by the time and the action amplitude. These parameters determine the temperature and pressure arising in the material (matter) and, consequently, the phase state of the matter within extreme waves. Processes of melting and evaporation, the heat propagation, and the emergence of fracture (spalling) are considered independently when the influence of them on extreme waves is studied.

Using the indicated approximations, this book examined a wide range of problems from extreme heat waves in cryogenic liquids and in metal targets perforated by projectiles or laser beams to the use of lasers to ensure life on Earth. In particular, recommendations are given for optimizing the pulsed laser action on dangerous space objects to ensure the fragmentation of them or the change their orbits.

Developed theoretical approaches are supplemented by experimental data. Much attention is paid to the experimental study of CIB of structural elements and to waves of melting in front of a massive perforator. The formation and amplification of nonlinear heat waves in liquid He II is illustrated also by our calculations.

This book is devoted to a very relevant and rapidly developing field of science, in which the problems of mechanics and physics of high energies and pressures are closely related. The field is so vast that its full consideration in one book is impossible. By virtue of this, only some problems from their vast spectrum are considered which the author was directly interested with. The choice of material reflects the level of knowledge of the author and his personal tastes. The author apologizes to those scientists whose results are not reflected in this book. The author thanks for the contribution of all those outstanding scientists and authors of scientific articles and books, results of which have been used in this book.

This book was written for Master's and PhD students as well as for researchers and engineers in the fields of thermal fluids, aerospace, nuclear engineering, and nonlinear wave studies.

Acknowledgments

I thank the University of Auckland and the Department of Mechanical Engineering that provided me with conditions for independent scientific work. The results of this 20-year work are presented in this book.

I thank my scientific teacher Professor Ilgamov Marat Aksanovich, as well as Professors Ganiev R.F. and Nigmatullin R.I. (the Academy of Science of Russia). They have largely determined the research directions presented in this book. I want to recall with the deep gratitude of Professors Novozhilov V.V. (the former Academy of Science of USSR), Pisarenko G.S. (the former Academy of Science of USSR), and my friend Sakhabutdinov Zh.M. (the former Academy of Science of USSR). They supported all my scientific beginnings.

It is my great pleasure to acknowledge my friends—Professors Fakirov S. (the University of Auckland, New Zealand), Lazarev V.A. (Kuban State University, Russia), and Akhmediev N. (the Australian National University). The support and attention of them were very important for both beginning and completing the writing of this book.

This book was written using some results of my students and colleagues—Drs. Abdirashidov A., Astanin V.V., Ivashchenko K.B., Karshiev A.B., Nechitailo N.V., Romashchenko V.A., Skurlatov E.D., and Zhurakhovskii V.S. (the former Academy of Science of USSR). I thank all of them.

Finally, I am grateful for the contribution of all those outstanding scientists whose results were used in the book. I apologize to those whose valuable results are not reflected in this book. This is due to the limited volume of the book and the fact that the author tried to present a well-defined, unified vision of the problem under consideration.

This project could hardly be realized without the generous permission of the following publishers for using material from their publications:

Center of Modern Education (Moscow, Russia).

Ufa Center of the Russian Academy of Sciences (UFIC RAS) (Ufa, Russia).

Springer, Heidelberg, New York, Dordrecht, London.

Naukova Dumka Publishing House (Kiev, Ukraine).

To all of them—my sincere thanks!

It is hard for a writer to even dream of a better publishing team than the CRC Press team. Many thanks Glenon C. Butler, Jr. Production Editor, Jonathan W. Plant, Executive Editor, Kyra Lindholm, Editor, Bhavna Saxena, Editorial Assistant, and Prachi Mishra, Editorial Assistant (CRC Press | Taylor & Francis Group—Mechanical, Aerospace, Nuclear & Energy Engineering) for their professional service and turning my manuscript into this book. Finally, I would like to thank Aswini Kumar, Production Manager (codeMantra Pvt Ltd), whose keen eye, good taste, and sense of structure improved this book in many ways.

Author

Shamil U. Galiev obtained his PhD degree in mathematics and physics from Leningrad University in 1971 and, later, a full doctorate (DSc) in engineering mechanics from the Academy of Science of Ukraine in 1978. He worked in the Academy of Science of former Soviet Union as a researcher, senior researcher, and department chair from 1965 to 1995. From 1984 to 1989, he served as a professor of theoretical mechanics in the Kiev Technical University, Ukraine. Since 1996, he has served as professor, honorary academic of the University of Auckland, New Zealand. He has published approximately 90 scientific publications and is the author of seven books on different complex wave phenomena. From 1965 to 2014, he studied different engineering problems connected with dynamics and strength of submarines, rocket systems, and target/projectile (laser beam) systems. Some of these results were published in books and papers. During 1998–2017, he did extensive research and publication in the area of strongly nonlinear effects connected with catastrophic earthquakes, giant ocean waves, and waves in nonlinear scalar fields. Overall, his research has covered many areas of engineering, mechanics, physics, and mathematics.

Part I

Basic Models, Equations, and Ideas

Fear of the possibility of error should not turn us away from the search for truth.

Claude-Adrien Helvetius

Mechanics of continuous matter is a section of mechanics, devoted to the study of the motion and equilibrium of gases, liquids, plasmas, and deformable solids; it is subdivided into hydro-air-mechanics, gasdynamics, elasticity theory, plasticity theory, etc. The basic assumption of continuum mechanics is that matter can be considered as continuous medium, neglecting its molecular (atomic) structure. The distribution of all characteristics (density, stresses, particle velocities, etc.) of matter is also continuous. These assumptions make it possible to apply in mechanics of a continuous medium, a well-developed theory of continuous functions and corresponding differential equations. Indeed, the dimensions of the molecules are negligibly small in comparison with the particle sizes that are considered in theoretical and experimental studies in mechanics of continuous medium.

The mechanics of a continuous medium is based on equations of motion, from the basic laws of mechanics, on the equation of the continuity of the medium, which is the consequence of law of conservation of mass, and on the equation of conservation of energy.

The characteristics of each particular medium are taken into account by the so-called state equations. They describe for a given medium dependences from stresses, pressure, temperature, and other physical and chemical parameters. The mechanics of a continuous medium is used in various fields of physics and technology.

1 Models of Continuum

In this chapter, certain relationships of the mechanics of continuum, solid deformed body, and fluid as well as the theory of plates and shells are given. Equations of state are introduced, which are valid over a wide range of changes in the thermodynamic parameters of different media.

1.1 THE SYSTEM OF EQUATIONS OF MECHANICS CONTINUOUS MEDIUM

Equations of the continuum include the conservation equations, kinematic (geometric) equations, and state (closing) equations. We present these equations in their general form, using tensor and vector notations. For definiteness, we assume that the motion of the medium is studied in Euler's coordinates. The designations used basically correspond to the books [1,2] where the equations are written in Einstein notation.

Equations of conservation. The system of equations of classical models of continuum includes the equation of conservation of mass

$$\frac{d\rho}{dt} + \rho \nabla \overline{v} = 0, \tag{1.1}$$

momentum

$$\rho \frac{dv_i}{dt} = \nabla_j \sigma_{ij} + \rho F^i, \tag{1.2}$$

energy (the first law of thermodynamics)

$$dE = \rho^{-1} \sigma_{ij} e_{ij} dt + dW + dQ, \tag{1.3}$$

as well as the second law of thermodynamics

$$Tds = dW + dQ^*, \tag{1.4}$$

where ρ is the density, $\overline{v}$ is the velocity vector, v_i is the component of $\overline{v}$, σ_{ij} are the components of the stress tensor ($\sigma_{ij} = \sigma_{ji}$), E is the specific internal energy, e_{ij} are the components of the strain rate tensor, dW is the heat coming from the outside, dQ is the heat inflow due to sources inside the medium, T is the temperature, s is the entropy, and dQ^* is a unfeigned heat. We assume that dQ is given or determined from the additional equation. In addition, the mass forces F^i that characterize the impact of external forces on the medium should be determined.

Equations of conservation of energy (1.2) and (1.3) can be written in the another form, which is often used in applications, that is, expressing σ_{ij} using the pressure p and the deviator of the stress tensor s_{ij}:

$$\sigma_{ij} = s_{ij} - p\delta_{ij}, \quad p = -\tfrac{1}{3}\sigma_{ll}, \qquad (1.5)$$

where δ_{ij} is the Kronecker delta. We assume that i, j, and l may take the value of 1, 2, or 3. In particular, $\sigma_{ll} = \sigma_{11} + \sigma_{22} + \sigma_{33}$. Relations (1.1)–(1.5) hold for any continuous medium. The same generality has a connection of dW with the vector of heat flux $\overline{W}$:

$$dW = -\rho^{-1}\nabla\overline{W}dt. \qquad (1.6)$$

Equation of conservation of energy (1.3) is transformed using (1.6) to the following form:

$$dE = \rho^{-1}\left(\sigma_{ij}e_{ij} - \nabla\overline{W}\right)dt + dQ. \qquad (1.7)$$

Kinematic relations. The unknowns e_{ij} in (1.7) are expressed in terms of v_i by means of the relations following purely geometric reasoning:

$$e_{ij} = \tfrac{1}{2}\left(\nabla_i v^j + \nabla_j v^i\right). \qquad (1.8)$$

Equations (1.1), (1.2), and (1.7) are not included while determining the displacement $\overline{w}$ of the elementary volume of the medium. The displacement of the medium for an infinitesimal interval of the time Δt can be found from the following form:

$$\overline{w} = \overline{v}\Delta t. \qquad (1.9)$$

Five equations (1.1), (1.2), (1.7) contain, if we take into account (1.5), (1.8) and $\sigma_{ij} = \sigma_{ji}$, 15 unknowns: ρ, $\overline{v}$, p, s_{ij}, E, $\overline{W}$. Nonconformity of the number of these equations and the unknowns is natural since conservation laws describe the most fundamental properties of the medium. Individual features of specific materials, fluids, and gases are described by additional differential and algebraic equations [2–8].

State equations [1,2]. We deal with cases when properties of a small-volume medium are completely determined by expressions p, s_{ij}, E, and $\overline{W}$ in terms of the components of the strain rate tensor $e_{\alpha\beta}$, the components of the metric tensor $g^{\alpha\beta}$, the temperature T, and some additional parameters $x_1, x_2, \ldots, x_n$ varying from problem to problem:

$$\begin{aligned}
s_{ij} &= s_{ij}\left(e_{\alpha\beta}, g^{\alpha\beta}, \rho, T, x_1, x_2, \ldots, x_n\right), \\
\overline{W} &= \overline{W}\left(e_{\alpha\beta}, g^{\alpha\beta}, \rho, T, x_1, x_2, \ldots, x_n\right), \\
p &= p\left(\rho, T, x_1, x_2, \ldots, x_n\right), \\
E &= E\left(\rho, T, x_1, x_2, \ldots, x_n\right),
\end{aligned} \qquad (1.10)$$

where p and E are the thermodynamic variables. In the solid mechanics, it is usually assumed that s_{ij} and $\overline{W}$ depend on the components of the deformation tensor $\varepsilon_{\alpha\beta}$, but not on $e_{\alpha\beta}$. The parameters $x_1, x_2, \ldots, x_n$ are different for each state equation (1.10). Values $e_{\alpha\beta}$ transform into $\varepsilon_{\alpha\beta}$ according to (1.10)

Models of Continuum

$$\varepsilon_{\alpha\beta} = e_{\alpha\beta}\Delta t, \quad (1.11)$$

where $\varepsilon_{\alpha\beta}$ are the components of infinitesimal deformations corresponding to the displacement during time Δt. Thus, the five equations of conservation (1.1), (1.2), and (1.7) are supplemented by 14 equations (1.10). A closed system of equations is obtained for finding 19 unknowns: $\rho, \bar{v}, p, s_{ij}, E, \bar{W}$. The system is simplified slightly, if we assume $s_{ij} = s_{ji}$. The components $g^{\alpha\beta}$ and the parameters x_n are considered as known values.

The second law of thermodynamics. Relations (1.10) cannot be arbitrary. They must not contradict the second thermodynamics law, thus ensuring production or at least conservation of the entropy. In the vast majority of important cases, Eqs. (1.10) automatically satisfy this condition. Therefore, we will not discuss Eq. (1.4) additionally.

Examples of the state equations. We give examples of the state equations for $\bar{W}$, p, and E, which are valid for continuous media regardless of whether in solid, liquid, or gaseous states they are.

1. **The law of Fourier heat conductivity.** The following version of this equation is often used in practice:

   $$\bar{W} = -\kappa \nabla T, \quad (1.12)$$

   where κ is the thermal conductivity. This coefficient is constant or depends on temperature and density; $\kappa > 0$.

2. **Modified Fourier law.** As it is known, Eq. (1.12) yields the infinite speed of heat propagation if $\kappa = \text{const}$. This is the fundamentally wrong consequence from (1.12), which does not manifest itself in the solution of many static and dynamic problems of mechanics. However, the finite heat transfer rate must be taken into account in the case of very short heat loadings [9,10] when thermal waves can be formed.

   Approaches based on the dependence of the coefficient of thermal conductivity on T may be used to describe thermal dynamics. Including in the classical Fourier equation a term of the form $\tau \partial \bar{W}/\partial t$, where τ is the thermal relaxation coefficient [9,10], gives the same result. The two approaches can be combined, which, apparently, will allow uniformity in their areas of applicability:

   $$\tau \partial \bar{W}/\partial t + \bar{W} = -\kappa(T)\nabla T. \quad (1.13)$$

   The coefficient τ is of the order of 10^{-11} seconds for metals and 10^{-8} seconds for gases.

3. **Caloric and thermal equations.** Equations for p and E (1.10) are usually valid for one-phase state of matter. At the same time, much attention has recently been paid to wide-range equations for E (caloric equation) and p (thermal equation) that are valid for multiphase state of matter. In the last case, the following thermal and caloric equations can be used if we assume that R is the gas constant, T_c is the critical temperature, and $\bar{\mu}$ is the chemical potential [3–8]:

$$\begin{aligned} p &= p_\rho(\rho) + p_T(\rho,T) + p_e(\rho,T), \\ E &= E_\rho(\rho) + E_T(\rho,T) + E_e(\rho,T) + E_i(\rho,T), \end{aligned} \quad (1.14)$$

where p_ρ and E_ρ are the elastic components that equal to the pressure and the internal energy of matter at zero degrees Kelvin; p_T and E_T are the thermal components determined by the vibrations of lattice atoms; p_e and E_e are the electronic components characterizing the contribution of the electron gas in p and E; E_u determines the contribution of the ionization component energy. When writing (1.14), we assumed independence of interatomic elasticity from lattice vibrations and motion of electrons [3]. Consequently, the changes in p_ρ and E_ρ are not connected with heating of the matter, while the changes in remaining terms in (1.14) are associated with an increase in temperature. Thus, the same pressure can exist in a non-heated deformed body and in a heated, but undeformed matter. Therefore, the specific form for the terms in (1.14) must be established during force and thermal experiments.

Following [4], we write the expressions for the terms in (1.14):

$$p_\rho = KA \frac{\rho_0 mn}{3\rho_c(m-n)} \left[\left(\frac{\rho}{\rho_0}\right)^{m/3+1} - \left(\frac{\rho}{\rho_0}\right)^{n/3+1} \right], \tag{1.15}$$

$$p_T = K \frac{\rho}{\rho_c} \frac{\gamma + z/3}{1 + z/2} E_T, \tag{1.16}$$

$$p_e = \frac{2}{3} \frac{\rho}{\rho_c} KE_e, \tag{1.17}$$

$$E_\rho = A \left\{ 1 + \frac{1}{m-n} \left[n\left(\frac{\rho}{\rho_0}\right)^{m/3} - m\left(\frac{\rho}{\rho_0}\right)^{n/3} \right] \right\}, \tag{1.18}$$

$$E_T = \frac{(2+z)T}{(1+z)T_c}, \quad z = lTT_c^{-1}\varpi^{-k}, \quad \varpi^k = \rho_0^k \rho_c^{-k}, \tag{1.19}$$

$$E_e = \beta_c^{-1} \left(\frac{\rho}{\rho_c}\right)^{2/3} Z^2 \ln\cosh\left[\frac{T\beta_c}{ZT_c}\left(\frac{\rho_c}{\rho}\right)^{2/3}\right], \tag{1.20}$$

$$E_i = \tfrac{2}{3} l N_A R^{-1} T_c^{-1} Z_c. \tag{1.21}$$

Here

$$Z = \left[Z_c \rho_c + \rho \exp\left(a\rho\rho_c^{-1} - a\right) \right] \left[\rho_c + \rho \exp\left(a\rho\rho_c^{-1} - a\right) \right]^{-1}, \tag{1.22}$$

where Z_c is the equilibrium ionization calculated according to the Saha equation; at high densities, $Z_c \to 1$ The index c refers to the critical point. The constants in Eqs. (1.15)–(1.22) are determined by the formulas:

Models of Continuum

$$A = \tfrac{2}{3}\Lambda\bar{\mu}R^{-1}T_c^{-1}, \quad K = \tfrac{3}{2}\bar{\mu}^{-1}RT_c\rho_c p_c^{-1}, \quad \beta_c = \tfrac{2}{3}\bar{\mu}b_c T_c R^{-1}. \tag{1.23}$$

Here the following notations are used: Λ is the volumetric heat of sublimation, b is the electronic heat capacity calculated for the degenerate gas, I is the ionization potential, N_A is Avogadro's constant, and γ is the Grüneisen parameter. These values for some media are presented in [4–7]. The constants m, n, γ, l, k, and a in (1.15)–(1.23) are determined for each medium separately.

The pressure and the elastic component of the energy presented in forms (1.15) and (1.18) qualitatively correctly describe the form of the zero isotherm. In particular, work of expansion of matter from normal density to zero coincides with the heat of sublimation for all the materials considered in paper [4]. There the comparison of data of experiments and formula (1.15) was made. Satisfactory description of the elastic pressure component (up to double compression) for some metals was found.

The interpolation formula (1.19) of the thermal component of energy provides asymptotic values of the molar heat capacity for an ideal gas $c_v = \tfrac{3}{2}R$ and a condensed state $c_v = 3R$. The multiplier in the formula for the thermal pressure (1.16) has the corresponding asymptotes $\tfrac{3}{2}$ and γ.

The interpolation equation for the electronic component (1.20) at high density and relatively low temperatures transforms into the known formula for the internal energy of degenerate electron gas:

$$E_e = \tfrac{1}{2}T^2(\rho/\rho_c)^{-2/3}. \tag{1.24}$$

At low densities and high temperatures, Formula (1.20) yields the equation of energy of an ideal gas (plasma):

$$E_e = ZT/T_c. \tag{1.25}$$

Thus, Formula (1.20) covers a very wide range of parameters. However, the formula is unsuitable for the case of a low-temperature plasma of low density.

By virtue of their generality, Eqs. (1.15)–(1.23) cannot claim on high accuracy of description of matter properties. Moreover, the most part of the domain described by the equations, in general, is not experimentally investigated. In those parameter ranges, where the properties of metals have been studied, the experimental data are satisfactorily described by the equations obtained. In particular, the discrepancy of experimental and calculated pressure, density, and energy of liquid and gas on binodal does not exceed 10%. We note that the value of parameter γ, which is the Grüneisen coefficient for normal conditions, is close to experimental values for all 13 metals considered in [4].

Let us describe by the thermal equation (1.14) the regularities of the change in the parameters ρ, p, and T. Results of numerical study of this equation for aluminum are presented in Figure 1.1(a). It can be seen that the isotherms do not change when temperature increases from zero to critical value. They are parallel to density axis in the region of the gaseous state of matter testing a break on the line R, which defines the boiling boundary.

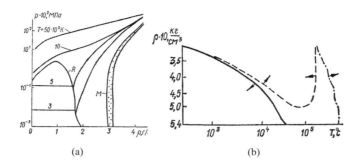

FIGURE 1.1 Phase diagram of aluminum: lines T are isotherms: M and R are the phase boundaries of melting and boiling (evaporation), respectively (a). Areas of predominant influence of the components of the state equation for aluminum [4] (b).

With a further increase in temperature, the break of the isotherms is not observed, and with increasing density, the isotherms approach the phase boundary melting. Note that the resulting phase diagram for aluminum agrees well with those presented in other publications [5].

Let us give certain understanding about zones of predominant influence of the terms of the thermal equation (1.14). The results are shown in Figure 1.1(b) in the plane for ρ and T. The arrows indicate the zones that limit the curves corresponding to not less than 50% of the contribution of one or the other terms to the total pressure. A solid curve defines the zone of predominant influence p_y; dashed and dash-dotted curves define the zones of not less than 50% of the influence p_T and p_e, respectively.

Thus, we briefly discussed the equations for $\bar{W}$, p, and T (1.10). The region of applicability of the first Eq. (1.10) is less extensive. In particular, the contribution of this equation in the behavior of a continuous medium decreases with increasing temperature or pressure. Therefore, this equation is usually determined in less cases of general models of continuum: solid deformable body and fluid.

1.2 STATE (CONSTITUTIVE) EQUATIONS FOR ELASTIC AND ELASTIC–PLASTIC BODIES

Elastic body [8,11]. The equations of the elasticity theory follow from (1.2) if the latter are linearized and $\rho = \rho_0$. The state equations (generalized Hooke's law) are written as follows:

$$\sigma = \tfrac{1}{3} K \varepsilon_{ll}, \quad s_{ij} = 2\mu \left(\varepsilon_{ij} - \tfrac{1}{3} \varepsilon_{ll} \delta_{ij} \right), \tag{1.26}$$

where K is the bulk modulus. Relations (1.26) correspond to the first and third equations of Eqs. (1.10). If we use Eqs. (1.26) and take into account (1.5), (1.8), (1.9), and (1.11), then Eq. (1.2) can be reduced to the Lamé equation:

$$\rho_0 \frac{\partial^2 \bar{w}}{\partial t^2} = (\lambda + \mu)\nabla(\nabla \bar{w}) + \mu \nabla^2 \bar{w}, \tag{1.27}$$

where λ and μ are the Lamé constants. Thus, problems of the elasticity can be solved using Eqs. (1.2), (1.5), and (1.26) if $F^i = 0$ and ρ = const (according to the approach of continuum mechanics), or using Eq. (1.27) written in terms of the displacements. The latter approach usually allows us to solve problems more easily.

Theory of plastic flow [8,12–15]. Hooke's law is valid for relatively small deformations (elastic deformations). During a further loading, many structural materials exhibit plastic properties. Plastic deformation takes place only after the elastic limit is exceeded. It is easy to determine a value of the stress when plastic properties begin to appear, in the case of simple loads, for example, from experiments with stretching of cylindrical specimens. The magnitude of the stress when plastic properties begin to appear is denoted as σ_0 (Figure 1.2).

Stress–strain tensile diagram consists of three parts (Figure 1.2(a)). In the first part, $\sigma \le \sigma_0$. It is an elastic part that is characterized by the angle E (Young's modulus). In the second (elastic–plastic) part, $\sigma_0 < \sigma < \sigma_N$. This part is characterized by the angle E_1 and the flow curve (see curve 2 in Figure 1.2(b)). The pores begin to form in ductile materials if σ is near σ_N. Pores coalesce if the strain continues to grow. Thus, the cracks are formed. This process corresponds to the third part of the stress–strain tensile diagram. As a result of growth of the cracks, the materials can be fragmented (see, also, Figure 2.1).

Of course, the transition of an elastic state to a plastic state is complicated in cases of a complex stress state. During the nineteenth century and the beginning of the twentieth century, many experimental studies of the plastic behavior of the body during complex loadings were made and a number of criteria for the transition of the elastic state to the plastic state were formulated [13,14]. The criterion for such a transition should be based on the stress state, which may be obtained from the simplest experiments. In this case, this approach can be accepted for engineering purposes.

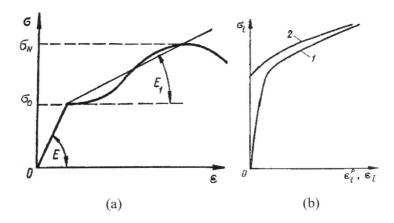

FIGURE 1.2 (a) Simplified stress–strain diagram of ductile metals (thick solid curve) and its ideal scheme (two straight lines that are determined by angles E and E_1). (b) Elastoplastic curve $g(\varepsilon_i, \sigma_i) = 0$ (1.33) (see curve 1) was determined from stretching experiments of cylindrical specimens. Curve 2 corresponds to curve 1, but describes only pure plastic deformation. This flow curve corresponds to the equation $G(\varepsilon_i^p, \sigma_i) = 0$ (see the text).

One of the most common criteria of the transition of material from elastic to plastic state is associated with a name of Mises (von Mises criteria):

$$\sigma_i = \sigma_0, \tag{1.28}$$

where σ_i is the intensity of the stresses characterizing the resistance of a material to the shear stress, $\sigma_i = \sqrt{\frac{3}{2} s_{kj} s_{kj}}$. Consequently, the appearance of plastic flow is primarily associated with the deviator of the stress tensor. It is determined by the first equation of Eqs. (1.10). The value of σ_0 depends on the temperature; that is, it will be high in high-temperature region. With increasing T, this value falls and the effect of strength properties (elasticity, plasticity) on behavior material is reduced. In many cases, σ_0 is also dependent on the attained stress state, the loading rate, and sometimes the pressure p.

Thus, there is the condition (1.28) determining approximately the boundary of plasticity zones in a deformable material. Hooke's equations become inapplicable there. It is necessary to correct Hooke's equations within these zones.

There are many different theories having different accuracy and generality describing the behavior of plastic material. Let us describe two simple models of the plastic body, fairly widespread in the practice of dynamic calculations. They are valid, in general, for small elastic and plastic deformations.

Ideal plasticity. This model is determined in Figure 1.2(a) by the constant E and dashed line $\sigma = \sigma_0$ (flow rule). It is assumed that in the Mises condition (1.28), the value σ_0 does not depend on the value of the attained stress state. The value σ_0 is the constant of material. Plastic deformations begin to grow without limit when σ_i increases till σ_0. The material is elastic when $\sigma_i < \sigma_0$.

Flow plasticity theory. In contrast with the ideal model, this model takes into account a hardening law for the material.

It is assumed that the total strain in a body can be decomposed into an elastic part and a plastic part. The elastic part of the strain can be computed according to the elastic model. However, the plastic part of the strain is needed a special law.

The theory of plastic flow establishes a connection between infinitesimal increments of deformations $d\varepsilon_{kj}$ and stresses $d\sigma_{kj}$, the stress–strain state, and some parameters. The theory allows us to take into account the possibility of unloading, secondary plastic deformations, and loading complexity.

The initial assumptions of the following version of the theory are as follows: the body is isotropic; the relative change in volume is insignificant, and an elastic deformation is proportional to the pressure $\varepsilon_{ii} = -3Kp_e$; full increments of the components of $d\varepsilon_{kj}$ are a sum of the increments of the components of the elastic deformation $d\varepsilon_{kj}^e$ and plastic deformation $d\varepsilon_{kj}^p$:

$$d\varepsilon_{kj} = d\varepsilon_{kj}^e + d\varepsilon_{kj}^p. \tag{1.29}$$

Let the deviatoric components of the plastic deformation increments are proportional to the components of the stress deviator. Since the volume change is the elastic process, we have

$$d\varepsilon_{kj}^p = d\lambda s_{kj}, \tag{1.30}$$

Models of Continuum

where $d\lambda$ is an infinitesimal scalar value (a hardening parameter). Obviously, during loading and plastic deformation, $d\lambda > 0$, while during unloading and elastic deformation, $d\lambda = 0$. Let us express $d\lambda$ in terms of quantities, characterizing the stress–strain state of the material.

For this, we substitute (1.30) into the expression for the intensity of plastic deformation increments:

$$d\varepsilon_i^p = \sqrt{2\left(d\varepsilon_{kj}^p - \delta_{kj}d\varepsilon_{ll}^p/3\right)\left(d\varepsilon_{kj}^p - \delta_{kj}d\varepsilon_{ll}^p/3\right)/3}. \tag{1.31}$$

Since $d\varepsilon_{ll}^p = 0$ in (1.31) and taking into account (1.30), we have

$$d\lambda = \tfrac{3}{2} d\varepsilon_i^p \sigma_i^{-1}. \tag{1.32}$$

Since $\sigma_i = \sqrt{\tfrac{3}{2} s_{kj} s_{kj}}$. The value of equality (1.32) is that its right-hand side includes the quantities $d\varepsilon_i^p$ and σ_i. The relationship between them can be experimentally determined at each step of loading. For example, we can obtain $g(\varepsilon_i, \sigma_i)$ from tensile tests of cylindrical samples (Figure 1.2(b)):

$$g(\varepsilon_i, \sigma_i) = 0, \tag{1.33}$$

where ε_i is the strain intensity. Extracting from this expression the elastic component of the strain intensity, we obtain $G(\varepsilon_i^p, \sigma_i) = 0$ (Figure 1.2(b)). The relations of the flow plasticity theory contain both finite and infinitesimals values and are valid only on small intervals of plastic deformation. In view of this, sufficiently precise solutions of complex elastic–plastic problems are possible only with the help of the step-by-step procedure [15].

We divide the loading time by J small steps. The increments of plastic deformation in expression (1.29) at the step n ($n = 1, 2,\ldots, N,\ldots J$), we denote by the index n. Let us consider the loading step N.

Summing the increments in (1.29), we obtain $\varepsilon_{kj} = \varepsilon_{kj}^e + \sum_{n=1}^{N} \Delta_n \varepsilon_{kj}^p$. From this, $\varepsilon_{kj}^e = \varepsilon_{kj} - \sum_{n=1}^{N} \Delta_n \varepsilon_{kj}^p$. The components of elastic deformation are associated with the pressure and the deviator components of the stress tensor according to Hooke's law (1.26). Therefore, we have

$$p = p_e = -\tfrac{1}{3} K \varepsilon_{ll}, \quad s_{kj} = 2\mu \left(\varepsilon_{kj} - \tfrac{1}{3} \varepsilon_{ll} \delta_{kj} - \sum_{n=1}^{N} \Delta_n \varepsilon_{kj}^p \right) \tag{1.34}$$

State equations (1.34) differ from those written for elastic body (1.26) by terms that take into account the accumulated plastic deformation. After their determination, the problem is analogous to the problem of elastic theory. The main problem is to establish these terms at each step of solving the equations. These terms can be determined using the iteration method (Figure 1.3).

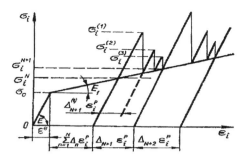

FIGURE 1.3 Scheme of the stress–strain diagram and the algorithm of calculation of ε_{kj} and for ductile metals [15].

Let us dwell briefly on one of the possible schemes determining $\Delta_n \varepsilon_{kj}^p$ at an arbitrary loading step $N+1$. We assume that the accumulated plastic deformation and all the remaining quantities at the Nth step are known.

At the beginning, according to (1.34), where the term $\Delta_{N+1}\varepsilon_{kj}^p$ is not taken into account, the first approximation is sought for σ_i^{N+1}. From Figure 1.3, we assume for this case $\sigma_i^{(1)} = \sigma_i^{N+1}$. Then, we use the intensity of plastic strain, which is presented as $\varepsilon_i^p = \sum_{n=1}^{N} \Delta_n \varepsilon_i^p + \Delta_{N+1}\varepsilon_i^p$. $\sigma_i^{(1)}$ and ε_i^p are substituted in $G(\varepsilon_i^p, \sigma_i^{(1)} = 0)$ (see Figures 1.2(b) and 1.3). Further, we find the first approximation for $\Delta_{N+1}\varepsilon_i^p$, namely, $\Delta_{N+1}^{(1)} \varepsilon_i^p = F\left(\sum_{n=1}^{N} \Delta_n \varepsilon_i^p, \sigma_i^{(1)}\right)$. If $\Delta_{N+1}^{(1)} \varepsilon_i^p = 0$, then the process on the step $N+1$ is elastic. On the contrary, we use Expressions (1.30) and (1.32). As a result, we have the first approximation [15]:

$$\Delta_{N+1}^{(1)} \varepsilon_{kj}^p = \tfrac{3}{2} \Delta_{N+1}^{(1)} \varepsilon_i^p s_{kj} / \sigma_i^{(1)} \tag{1.35}$$

The obtained expression allows us to find the refined value for s_{kj} and σ_i due to the first iteration on the step $N+1$. By repeating the above calculation sequence, we can find the second correction $\Delta_{N+1}^{(2)} s_{kj}^p$. Continuing the calculations, one can obtain the solution at the $N+1$ step with the required accuracy.

From the expressions obtained above for s_{kj}, Eqs. (1.26) and (1.34) can be considered as the simplest state equations describing elastic and plastic deformation of the material.

Thus, in the last two sections, we gave the basic equations of the continuum and a number of variants of the state equations.

1.3 THE EQUATIONS OF MOTION AND THE WIDE-RANGE EQUATIONS OF STATE OF AN INVISCID FLUID

In the case of a viscous liquid, the components of the stress deviator (1.10) are usually determined by the Navier–Stokes law. The model of inviscid (ideal) fluid is based on the assumption that in (1.10), $s_{ij} = 0$. Different versions of model of the

Models of Continuum

ideal fluid differ only in the state equations for p and E (1.10). A simple example of the state equation for ideal fluid is Tait's equation:

$$(p+B)(p_0+B)^{-1} = (\rho/\rho_0)^k, \qquad (1.36)$$

where p_0 and ρ_0 are the pressure and density of undisturbed fluid, respectively; k is the adiabatic exponent for liquid; and B is a constant. Equations (1.34) are valid for water at pressures lower than 3.10^3 MPa.

For the normal temperature, $k = 7.15$ and $B = 304.5$ MPa. It is emphasized that Eq. (1.36) does not depend on the temperature. In this case, equations of hydrodynamics reduce to a system of Eqs. (1.1), (1.2), (1.5), and (1.36). We emphasize that Eq. (1.36) is also valid for certain gas if $B = 0$. Thus, (1.34) is an example of the simple equation describing states of two media.

Ways of simplifying the equations. Formulas (1.26), (1.34), and (1.36) provide the important way of simplifying the mathematical formulation of problems that is connected with assumptions about the form of the state equations (1.10).

1. Further simplification of the equations of hydrodynamics is possible while introducing an assumption about the nature of fluid flow. Let us consider the irrotational flow. This assumption allows us to introduce the velocity potential ψ [15–17]:

$$\bar{v} = \nabla \psi. \qquad (1.37)$$

We introduce (1.37) into the equation of motion (1.2), which is rewritten taking into account (1.5) and $F^i = 0$. Eliminating the operation ∇, we obtain the Cauchy–Lagrange integral:

$$\frac{\partial \psi}{\partial t} + \tfrac{1}{2}(\nabla \psi)^2 + \int_{p_0}^{p} \rho^{-1} dp = f(t). \qquad (1.38)$$

In what follows, the liquid at infinity will be assumed to be at rest: $p_\infty = p_0$, $\rho_\infty = \rho_0$, and $\psi_\infty = 0$. Taking advantage of this, we find that the function $f(t) = 0$. The last term on the left-hand side of expression (1.38) integrates using the equation (1.36), thus resulting in the Cauchy–Lagrange integral:

$$p = -B + (p_0+B)\left\{1 - (k-1)k^{-1}\rho_0(p_0+B)^{-1}\left[\psi_t + \tfrac{1}{2}(\nabla\psi)^2\right]\right\}^{k/(k-1)}. \qquad (1.39)$$

Here, $\psi_t = \partial \psi / \partial t$. Then, we determine the density from Eq. (1.36)

$$\rho = \rho_0 \left\{1 - (k-1)k^{-1}\rho_0(p_0+B)^{-1}\left[\psi_t + \tfrac{1}{2}(\nabla\psi)^2\right]\right\}^{1/(k-1)}. \qquad (1.40)$$

Thus, the assumption about an irrotational motion allowed us to determine the velocity (1.37), pressure (1.39), and density (1.40) through the single function ψ. The equation for ψ is obtained from Eqs. (1.1), (1.37), and (1.40) [15–17]:

$$c^2\nabla^2\psi = \psi_{tt} + \left[(\nabla\psi)^2\right]_t + \tfrac{1}{2}(\nabla\psi\nabla)(\nabla\psi)^2,$$
$$c^2 = c_0^2 - (k-1)\left[\psi_t + \tfrac{1}{2}(\nabla\psi)^2\right]. \tag{1.41}$$

where $\psi_{tt} = \partial^2\psi/\partial t^2$, ∇^2 is the Laplace operator (Laplacian), $c_0^2 = k(B+p_0)\rho_0^{-1}$, and $c^2 = dp/d\rho$ is the sound speed in the liquid. In (1.41), c_0^2 is the unperturbed sound speed in the liquid.

Thus, using particularities of the flow, one can essentially simplify the mathematical formulation of the hydrodynamic problem. In the derivation of (1.41), no assumption was made about the magnitude of the perturbations in the liquid. Consequently, the application of Eqs. (1.41) is limited only by Tait's equation (1.36), which is valid up to $p \leq 3 \times 10^3$ MPa.

2. An introduction of the assumption about small enough perturbations opens the new way for the simplification of the equations. We will use it. Density and pressure are represented in the form of a sum: $\rho = \rho_0 + \rho_1$ and $p = p_0 + p_1$, where ρ_1 and p_1 are the small deviations from ρ_0 and $B + p_0$, respectively. We further consider the perturbations of the initial state as small values of the same of order μ (μ is a small parameter): $\rho_1\rho_0^{-1} \approx \mu$ and $p_1(B+p_0)^{-1} \approx \mu$. In addition, we require that the ratio of the convection acceleration to the local acceleration in Equation (1.2) has the order μ, that is, $(\partial v_i/\partial t)^{-1} v_k \nabla_k v_i \approx \mu$. In this case, the processes in the liquid have mainly a wave (acoustic) nature, and the hydrodynamics equations can be substantially simplified to the following form:

$$\psi_{tt} - c_0^2\nabla^2\psi = -\left[(\nabla\psi)^2 + \tfrac{1}{2}(k-1)c_0^{-2}(\psi_t)^2\right]_t. \tag{1.42}$$

Here, the values were eliminated, which were smaller than the second order of smallness. The pressure is determined by the formula:

$$p = p_0 - \rho_0\left[\psi_t + \tfrac{1}{2}(\nabla\psi)^2 - \tfrac{1}{2}c_0^{-2}\psi_t^2\right]. \tag{1.43}$$

Equations (1.42) and (1.43) allow us to consider waves with the amplitude up to several hundred megapascal. By ignoring the terms of second order of smallness in (1.42), we obtain

$$\psi_{tt} - c_0^2\nabla^2\psi = 0. \tag{1.44}$$

The Cauchy–Lagrange integral (1.43) gives

$$p = p_0 - \rho_0\psi_t. \tag{1.45}$$

Models of Continuum

Despite the limitations in amplitude, the equations of acoustics allow us to consider the propagation, reflection, and interaction of waves of sufficiently high intensity.

It is recommended in [18–20] to use the acoustic approximation up to 35, 50, and 110 Mpa, correspondingly. Our calculations of nonstationary interaction of liquid with deformable boundaries show that acoustic approach can ensure a good accuracy of the calculations for more high pressures (up to 300 MPa), if acoustic conditions are not met only in individual small volumes of the liquid during small time intervals [21]. Thus, approaches based on Tait's equation (1.36) allow us to calculate pressure waves of the considerable amplitude. Let us consider the disadvantages of Eq. (1.36). They are determined by the fact that (1.36) does not describe the behavior of the liquid at relatively low pressures, when the liquid approaches to its two-phase (liquid + gas) state (the rarefaction zones).

Below, we will consider rarefaction zones, the state equations for them, and mathematical modeling of waves in the two-phase zones.

Wide-range equations of state. We emphasized above that Eqs. (1.14) are valid in a very wide range of parameters of a continuous medium, including solid, liquid, and gaseous states. However, this generality of the equations also determines their relatively small local accuracy. It is impossible to study on their basis the details of phase changes of any matter and the transition of one phase of it into another.

Therefore, there are wide-range equations of state that are valid only for liquid and gas. The simple version of similar equation is Tait's equation (1.36). Below equations describing the transition of liquid into its two-phase state are presented.

1. The first interpolation equation, approximately describing the transition of the liquid state of matter into gaseous state in a wide temperature range, was proposed by van der Waals. This equation takes into account the possibility of the existence of a metastable state. In this state, negative pressures are realized and the liquid acts on the confining surface with a force directed inside it [22–25]. For the molar volume V_*, this equation has the following form:

$$p = RT(V_* - B)^{-1} - AV_*^{-2}, \qquad (1.46)$$

where A and B are the constants whose values vary from liquid to liquid. Apparently, two constants in the state equation are insufficient for describing the transformation of liquid into its two-phase state. Therefore, Eq. (1.46) corresponds only qualitatively to the behavior of real liquids. However, beyond doubt, it more correctly describes the behavior of real liquids in low-pressure regions than Tait's equation or similar to it [22–25].

2. Let us consider other interpolation equations of state. One of the most useful empirical equations of state is equation [23]:

$$p = RT(V_* - \beta)^{-1} - \gamma(V_* - \alpha)(V_* - \delta)^{-1}. \tag{1.47}$$

Thanks to four constants α, β, γ, and δ, this equation provides a good description of any isotherm of real liquid in areas of the stability of its thermodynamic state. For example, it was shown in [23] that for water at $T = 555.22$ K, Eq. (1.47) has the following form:

$$p = 3.864 p_c V_c (V - 0.2848 V_c)^{-1} - 13.548 p_c V_c^2$$
$$\times \left[(V - 0.2317 V_c)(V + 2.144 V_c)\right]^{-1}, \tag{1.48}$$

where V is the volume, and p_c and V_c are the pressure and volume in the thermodynamic critical point. Equation (1.48) takes into account the fact that during high speed of dropping the pressure, it can drop significantly below the saturation pressure p_H. There is the question about the action of similar rarefaction zones on elements of constructions.

3. Let us consider a thermal equation for water, which is valid over a wide range of thermodynamic parameters [25]. The equation consists of three expressions, each of which describes a certain interval of variation ρ. If $1 \leq \rho < 2.3 \, \text{g/cm}^3$, then

$$p = \left(1 - 0.012\bar{\rho}^2 F\right) \frac{3050(\bar{\rho}^{7.3} - 1)}{1 + 0.7(\bar{\rho} - 1)^4} + 4.7\bar{\rho}F(T - 273). \tag{1.49}$$

If $0.8 < \rho < 1 \, \text{g/cm}^3$, then

$$p = \zeta^4 - 470\zeta \bar{\rho} F + 4.7\bar{\rho}F(T - 273),$$
$$\zeta = 10(1 - \bar{\rho}) + 66(1 - \bar{\rho}^2) - 270(1 - \bar{\rho})^3. \tag{1.50}$$

If $0 < \rho < 0.8 \, \text{g/cm}^3$, then

$$p = \zeta^4 - 470\zeta \bar{\rho} F + 4.7\bar{\rho}F(T - 273), \quad \zeta = 6.6(1 - \bar{\rho})^{0.57} \bar{\rho}^{0.25}. \tag{1.51}$$

Here, notations were introduced $\bar{\rho} = \rho/\rho_0$, $\rho_0 = 1 \, \text{g/cm}^3$,

$$F = F(\bar{\rho}) = \left(1 + 3.5\bar{\rho} - 2\bar{\rho}^2 + 7.27\bar{\rho}^6\right)\left(1 + 1.09\bar{\rho}^6\right)^{-1}. \tag{1.52}$$

The presented formulas consist of two terms: the first depends on ρ only, and the second on ρ and T. At $T = 273.16$ K and $\rho > 1 \, \text{g/cm}^3$, the first Formula (1.50) generalizes Tait's equation (1.36) for the cases when pressure reaching 10^4 MPa.

Models of Continuum

The phenomenological equations (1.36), (1.46)–(1.52) overlap the significant area of change of thermodynamic parameters. At the same time, they do not explicitly describe some features of the transformation of liquid into gas (steam).

1.4 SIMPLEST EXAMPLE OF FRACTURE OF MEDIA WITHIN RAREFACTION ZONES

Above, we have considered the cases when the appearance of a new phase is described by wide-range equations of state. Here, an example of a medium (bubble liquid) is given, which transforms into a new phase during the process of destruction (transition into the gas–liquid mixture). This process can be observed in rarefaction zones of extreme waves. It can be described in a very rough approximation as a growth of gas bubbles and the fusion of them up to the formation of the gas–liquid mixture.

In general, the task of this book is not a detailing description of complex processes characterizing the destruction of continuous media. However, during the formation of extreme waves, their propagation and reflection from the boundaries, regions of rarefying often arise where the continuous medium is fractured. Therefore in Chapter 2, a few models of fracture will be described. Then in Chapters 4–6, the effects of cavitation and boiling of liquid within extreme waves on the behavior of elastic containers will be studied.

In this section, we give the simplest model of fracture as an introduction in the subject.

1.4.1 THE STATE EQUATION FOR BUBBLY LIQUID

The case of the transition of a liquid into a gas–liquid state within rarefaction waves (cold boiling) can be considered as the simplest example of the destruction of a medium. The destruction is described as the growth and fusion of gas bubbles initially existing in the liquid.

The Rayleigh equation. The motion of the bubble wall in the approximation of an incompressible fluid is described by the Rayleigh equation [26]:

$$R\ddot{R} + \tfrac{3}{2}\dot{R}^2 = \rho_{0l}^{-1}\left(p_g - p - 2\sigma R^{-1}\right), \tag{1.53}$$

where ρ_{0l} is the unperturbed density of the liquid, p_g is the gas pressure in the bubble, and σ is the coefficient of surface tension. Points denote the time derivatives.

Let us rewrite Eq. (1.53) for the bubble volume $V = 4\pi R^3/3$:

$$aV^{-1/3}\ddot{V} + \tfrac{1}{3}aV^{-4/3}\dot{V}^2 = p_g - 2\sigma(3V/4\pi)^{-1/3} - p, \tag{1.54}$$

where $a = \rho_{0l}(3/4\pi)^{2/3}$. The bubble oscillation depends on the gas properties and a pressure wave in the liquid. The heat transfer can be neglected and use adiabatic law for gas if the wave pressure changes enough rapidly. Thus, we have

$$p_g = p_{0g}\left(\rho_g/\rho_{0g}\right)^\gamma = p_{0g}\left(V_0 V^{-1}\right)^\gamma, \tag{1.55}$$

where p_g is the gas pressure, ρ_g is the gas density, ρ_0 is the unperturbed gas density, γ is the adiabatic exponent for gas (for normal air $\gamma = 1.4$), and V_0 is the initial volume of the bubble. The oscillations can be considered isothermal ($\gamma = 1$ in Eqs. (1.55)) if the pressure in the liquid changes sufficiently slowly. In this case, there is substantial heat exchange between the liquid and the gas.

Equations (1.54) and (1.55) are not enough for the closure of the system of equations of ideal liquid. Let us describe the way of obtaining the missing equation.

The state equation. The motion of a liquid with bubbles of gas or vapor is described in the general case by the equations of mechanics of multiphase (heterogeneous) media [27]. These equations differ from (1.1)–(1.3) since new terms that take into account the transition of mass from one phase to the other are included. As a result, the exchange of momentum and energy takes place between them. Usually, the exact expressions for these new terms are unknown.

However, liquid with gas bubbles can be described within the framework of an ideal fluid if the exchange of mass, momentum, and energy between phases is very small in it [28–33]. This approach will be used many times in this book.

Let the gas volume is much smaller than the volume of the mixture. We assume that the wavelength of the perturbation is much larger than the distance between the bubbles and the radius of them. An interaction of bubbles is absent. They are in conditions uniform all-round pressure. The velocity of bubbles relative to each other is zero. The temperature of the medium is assumed a constant.

Taking into account the noted above, the initial density of the mixture is represented as follows:

$$m = \rho_{0m} U_{0m}, \tag{1.56}$$

where ρ_{0m} and U_{0m} are the initial density and the volume of the mixture, respectively. The change in the mixture volume is determined by changes in liquid and gas components. Since the mass of the same volume does not change, we have

$$m = \rho U_m = \rho(U_l + nV), \tag{1.57}$$

where ρ is the density of the mixture, U_m is the volume of the mixture, U_l is the volume of the pure liquid, n is the number of bubbles, and V is the volume of the single bubble. Now we find the change of the medium when an extreme wave passes through it. We take into account that the medium is a mixture of pure liquid and gas bubbles, that is, $U_m = U_l + nV$. It is emphasized that the compressibility of the medium is primarily determined by the compressibility of the bubbles.

First, we take into account the change in the volume of the liquid. Equation (1.36) is rewritten in the form:

$$(p + B)(p_0 + B)^{-1} = (U_l / U_{l0})^{-k}. \tag{1.58}$$

Here $p = p_0 + p_1$, where p_1 is the pressure perturbation. Equation (1.58) yields approximately

$$U_l = U_{0l}\left[1 - \lambda(p - p_0)\right], \tag{1.59}$$

Models of Continuum 19

where $\lambda = 1/k(p_0 + B)$. The mass of the medium does not change in the extreme wave. Let us equate (1.56) and (1.57). Taking into account (1.59), we obtain

$$\rho_{0m} U_{0m} = \rho \{U_{0l}[1 - \lambda(p - p_0)] + nV\}. \tag{1.60}$$

We rewrite (1.60) as

$$\rho_{0m} U_{0m} = \rho \{(U_{0m} - nV_0)[1 - \lambda(p - p_0)] + nV\}. \tag{1.61}$$

Let $U_{0l} = 1$. In this case, Eq. (1.61) yields

$$\rho = \rho_{0m} \{(1 - nV_0)[1 - \lambda(p - p_0)] + nV\}^{-1}. \tag{1.62}$$

We obtain the equation of state that determines the dynamics of the bubble medium in an extreme wave for cases where the initial content of the gas in the mixture is much less than the initial content of the liquid in it. Eq. (1.62) closes the system of equations of mechanics of inviscid nonthermal bubble liquid: (1.1), (1.2), (1.5), (1.54), (1.55), $s_{ij} = 0$. If $nV = 0$, then (1.62) coincides with the linearized Tait equation (1.36).

The presence of gas bubbles significantly changes the properties of the liquid. The liquid medium becomes strongly nonlinear, preserving nonlinear properties even for low gas concentrations. Let us show this. If the wavelength of the disturbance is much larger than the size of the bubbles, then the volumes of the latter change relatively slowly. In this case, the equation (1.54) yields $p = p_g - 2\sigma(3V/4\pi)^{-1/3}$. Neglecting further the influence of the surface tension and using (1.55), we write Eq. (1.62) in the following form:

$$\rho = \rho_{0m} \{(1 - nV_0)[1 - \lambda(p - p_0)] + nV_0 (p_0/p)^{1/\gamma}\}^{-1}. \tag{1.63}$$

The term $nV_0(p_0/p)^{1/\gamma}$ in this expression is small since $nV_0 \ll 1$. However, it determines the variation in ρ for any finite value of nV_0 if $\rho \to 0$ within the extreme wave. Bubbles grow, and whatever their initial concentration, they begin to affect the mechanical properties of the mixture.

1.4.2 Fracture (Cold Boiling) of Water during Seaquakes

Every day there are earthquakes that cause local oscillations of the ocean surface. If the depth is large, these oscillations can evolve near a coast into a tsunami. However, situations are possible where the amplitude of seabed oscillations may be comparable with the depth, or where the seabed accelerations considerably exceed g, or there is resonance of vertical waves. In these cases, there are strongly nonlinear wave phenomena on the surface and at the depth. The bubbles can appear as a result of local collapses of water within the rarefaction waves (nonstationary cavitation) that take place near the oscillating seabed and the water surface. The explosions accompanied by jets of liquid and gas are connected with bursting on the ocean surface of large volumes of gas (bubbles).

During the time of great geographical discoveries, some Spanish and Portuguese sailors had defined such behavior as "sea shakes", and some Dutch captains reported "awful seas". This phenomenon was described in many publications. In particular, there is following description [33] "a quite smooth surface of the sea suddenly was covered by waves. They promptly grew up to a height approximately 8 meters and then promptly fell down, forming deep craters. People began to run captured by panic fear along a vessel; and some, having lost their minds, began to jump from the vessel. Suddenly the vessel was shaken by a very strong blow, and some persons were thrown overboard. The blows from the bottom followed one after another, and it seemed that the vessel was banged about a rocky bottom. The bangs then stopped instantly".

Similar extreme waves are the result of seabed vertical oscillations. Almost periodical fracture (cold boiling) of the water near oscillating bottom occurs. Let us consider the results of Natanzon's experiments [28] with a water column in order to analyze this periodical fracture of the water.

Oscillograms of the pressure, which is measured near the piston, are presented in Figure 1.4 (see, also, Figure 3.5). The sinusoidal curves show the piston position. During the experiments, the piston acceleration was slowly increased from $0.2\,g$ to $2\,g$ and then was slowly reduced to $0.37\,g$. One can see that in the frequency range extending from 7.7 to 17 Hz, there are large variations in the pressure waves. When the down acceleration increases above g, the harmonic oscillations transform into peaks, which correspond to collisions of the water and the base. The smooth parts of the pressure curves correspond to free flight of the water volume. During these moments, above the base, the rarefaction (cavitation) zone is formed. When the acceleration reduces, a strong hysteretic effect takes place. For this case, discontinuous waves exist in the system up to $0.5\,g$ and then instantly transform into harmonic waves of very small amplitude. Thus, according to the experiments (see Figures 1.4 and 3.5), the strongly nonlinear discontinuous waves may be generated

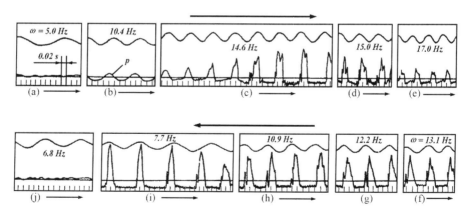

FIGURE 1.4 Hysteretic dynamics of strongly nonlinear oscillations of the vertically excited water column (the exciting amplitude is 0.002 m, and the column length is 7 m (see, also, Figure 3.5)). The vertical acceleration is slowly increased from 0.2 to 2 g (forced frequency from 5 to 17 Hz) and then is slowly reduced to 0.37 g (forced frequency from 13.1 Hz to 6.8 Hz) [33].

Models of Continuum

in the upper-lying material (water) when the down acceleration of the base exceeds the gravity acceleration g. The linear continuous waves are excited if the vertical acceleration is smaller than $0.5\,g$.

Thus, if the forced acceleration increases above some threshold level, then the linear prediction breaks down. That nonlinearity changes the amplitude and form of the forced waves. In particular, the rarefaction (cavitation) wave is periodically generated above the vibrating base. However, as shown in Figure 1.4, the amplitude of the forced waves is bounded and depends on the hysteretic effect. Because of this effect, the maximal amplification takes place when the forced acceleration approximately equals $0.5\,g$.

The experiments show that in waves of small amplitude (5 Hz or 6.7 Hz), the water remains unchanged, although it contains, of course, bubbles. But when the amplitudes of the waves reach extreme values, the water begins to periodically fracture near the piston. Thus, in the water column, periodically zones of fracture and bubbly water are formed. Of course, the motion of the medium in these zones is described by different equations, and certain conditions of fracture are fulfilled on their boundaries.

We will return to the discussion of Natanzon's experiments more than once, and for now, as an introduction to the phenomenon, we consider a simple model of liquid destruction.

1.4.3 Model of Fracture (Cold Boiling) of Bubbly Liquid

It was emphasized that sometimes the multiphase media may be described in terms of known classical models. The inhomogeneity manifests itself only in the state equation. We obtained the state equation (1.62) for the bubbly liquid. Using (1.62), the governing equations are derived for one-dimensional waves traveling within fractured zones and zones of bubbly liquid.

Plane waves. Equations (1.1) and (1.2) were written using Eulerian coordinates. However, plane (one-dimensional) waves are often studied using the Lagrange coordinates. For this case, we transform from the Euler variable x to the Lagrange variable a in Eqs. (1.1) and (1.2). These coordinates are related by the formula $x = a + u$, where u is the longitudinal displacement of medium. Taking into account these formulas, we obtain

$$\frac{\partial F}{\partial x} = \frac{\partial F}{\partial a}\frac{\partial a}{\partial x} = \frac{\partial F}{\partial a}\left(1+\frac{\partial u}{\partial a}\right)^{-1}. \tag{1.64}$$

The equations of motion and continuity in the Lagrange coordinates can be obtained using (1.64), and discarding in (1.2) (in expression for the acceleration), the convective summand:

$$\rho_{0m}u_{tt} + p_a = 0 \quad \text{and} \quad \rho(1+u_a) = \rho_{0m}. \tag{1.65}$$

Equations of continuity (1.65) and (1.63) yield

$$(1+u_a) = (1-\xi_0)\left[1-\lambda(p-p_0)\right]+\xi_0\left(p_0 p^{-1}\right)^{1/\gamma}. \tag{1.66}$$

Here

$$\xi_0 = nV_0. \tag{1.67}$$

We will confine ourselves further to the consideration of long waves and the isothermal process $\gamma = 1$. In this case, Eq. (1.66) has the two roots:

$$p_\pm = \frac{-(u_a + A) \pm \sqrt{(u_a + A)^2 + 4(1 - \xi_0)\lambda \xi_0 p_0}}{2(1 - \xi_0)\lambda}. \tag{1.68}$$

Here

$$A = \xi_0(1 + \lambda p_0) - \lambda p_0. \tag{1.69}$$

The last expression takes into account the effects of both gas and liquid.

Both expressions p_+ and p_- may be regarded as closing (state) equations for system of equations (1.65). We can also assume that the root p_- determines the state equation of the bubbly liquid. On the contrary, the root p_+ determines the state equation of the mixture after its fracture.

Let

$$(u_a + A)^2 \gg 4(1 - \xi_0)\lambda \xi_0 p_0. \tag{1.70}$$

Fractured liquid. For this case, we have from (1.68) and (1.70) that

$$p_+ \approx \xi_0 p_0 (u_a + A)^{-1}. \tag{1.71}$$

The first equation from Eqs. (1.65) yields

$$p_{0m} u_{tt} = \xi_0 p_0 (u_a + A)^{-2} u_{aa}. \tag{1.72}$$

It is reminded that $p_{0m} = p_{0l}(1 - \xi_0) + p_{0g}\xi_0$ in (1.72). The linear version of (1.72) is

$$u_{tt} - c_{0f}^2 u_{aa} = 0. \tag{1.73}$$

Thus, the sound velocity in the fractured medium is determined from

$$c_{0f}^2 = c_{0g}^2 \left[p_{0l} p_{0g}^{-1} - 1(\xi_0^{-1} - 1) + 1 \right]^{-1} \left[\xi_0(1 + \lambda p_0) - \lambda p_0 \right]^{-2}. \tag{1.74}$$

The last expression is valid in the zones of complete destruction of the medium. Let us consider the limit case when $\xi_0 \approx 1$. In this case, (1.74) determines the sound velocity for isothermal gas. c_{0g}^2. It is seen from (1.74) that the presence of water particles in gas + water mixture reduces sound velocity when compared with c_{0g}^2. We emphasize that in the zones of the complete fractured medium, the sound velocity in (1.74) is different from the sound velocity for the bubbly liquid which was found in [29–31] (see, also, (1.78) and Chapter 3).

Models of Continuum

It can be assumed that the gas bubbles have already merged and the medium was transformed from the bubbly liquid to a gas–liquid mixture. Thus, (1.71) and (1.72) describe strongly nonlinear behavior of the mixture in the rarefaction zones even for arbitrarily low concentration of the bubbles.

Bubbly liquid. On the contrary, the root p_- corresponds to zones where there are compression or weak tensions. For the last case,

$$p_- \approx -(u_a + A)(1 - \xi_0)^{-1} \lambda^{-1} - \xi_0 p_0 (u_a + A)^{-1}. \tag{1.75}$$

The first equation from Eqs. (1.65) yields

$$\rho_{0m} u_{tt} - (1 - \xi_0)^{-1} \lambda^{-1} u_{aa} + \xi_0 p_0 (u_a + A)^{-2} u_{aa} = 0. \tag{1.76}$$

The linear version of (1.76) is

$$u_{tt} - c_{0m}^2 u_{aa} = 0. \tag{1.77}$$

Here

$$c_{0m}^2 = \left[(1 - \xi_0) + \xi_0 \rho_{og} \rho_{0l}^{-1} \right]^{-1} (1 - \xi_0)^{-1} c_{0l}^2$$
$$- \left[\rho_{0l} \rho_{og}^{-1} (1 - \xi_0) + \xi_0 \right]^{-1} \xi_0 \left[\xi_0 - (1 - \xi_0) \lambda p_0 \right]^{-2} c_{0g}^2. \tag{1.78}$$

If $\xi_0 = 0$, then (1.78) determines the sound velocity of the pure liquid. $c_{0l}^2 = c_0^2$ Thus, the gas bubbles reduce the sound velocity in the mixture as (1.78) shows. We emphasize that (1.78) corresponds well to the results of [29–31] and the results presented in Chapter 3.

This equation is valid for the bubbly liquid. At the very rough approximation, we can assume that it can be used until the moment of fracture of the mixture. Then, it is necessary to use Eq. (1.72). It is shown that different equations describe the wave propagation in compression and fractured zones of bubbly mixture. Of course, the sound velocity is different within these zones (see Eqs. (1.74) and (1.78)).

Thus, we have two equations that describe the different zones of the medium in question. How is it possible to determine these zones during the calculation process? Indeed, fractured zones arise in the process of propagation of an extreme wave, and initially, the medium did not have the fractured zones.

Fracture and restoration of bubbly liquid. At the initial instant, we have the bubbly liquid. The pressure can be calculated at each point of the mixture according to (1.76) and (1.75). At points where this pressure p_- is less than p_H ($p_- < p_H$), we assume

$$p_- = p_H, \tag{1.79}$$

where p_H is some critical value of fracture. The medium is fractured instantly. At these points, the calculation is then made according to Eqs. (1.72) and (1.71) for the fractured medium. However, if in the fracture zones, according to calculations, the pressure begins to exceed p_H,

$$p_+ > p_H, \qquad (1.80)$$

then we assume that the medium instantly restored the nonfractured state. At the next moment of calculations, we again use Eq. (1.76) for the bubbly liquid.

Of course, it is possible to realize this algorithm for extreme waves only numerically. Numerous examples of the implementation of such algorithms will be presented in this book below. We will study dynamics of structural elements loading by extreme waves and impacts, which is accompanied by the destruction of the media and restoration of their continuity.

Now let us return to Natanzon's experiments. It is easy to see that in Figure 1.4, the straight lines correspond to zones of liquid destruction, which are represented by Eqs. (1.71), (1.72), and (1.79). The peaks correspond to Eqs. (1.76), (1.75), and (1.80).

An algorithm of using (1.71), (1.72), and (1.79) and (1.76), (1.75), and (1.80) may be enough complex. Indeed, boundaries separating the zones are not known and changed in time. These boundaries may be determined during the calculations. We will give many examples of similar calculations in this book.

Remarks. We proposed a few models as the first step of an understanding of the very complex problem of dynamic destruction of liquid and solid media. In particular, we described above the model of certain continuum having initial defects (bubbles), and presented equations of motion media and conditions of appearance and disappearance of the fractural zones are formulated. We will study dynamics of structural elements loading by extreme waves and impacts, which is accompanied by the destruction of the media and restoration of their continuity in Chapters 4–6 and 8.

1.5 MODELS OF MOMENT AND MOMENTLESS SHELLS

Some models of media were briefly described in the previous sections. These models will be used in the study of the propagation of waves and their reflection from the boundaries. The effects of the reflection often depend on the deformability of the boundaries. For example, on rather rigid surfaces, waves of pressure amplify. On the contrary, compression waves are reflected from free or very flexible boundaries, as rarefaction waves, where the medium can lose its continuity. Free boundaries or contact surfaces of different media can be described using the traditional formulations of boundary-value problems in the mechanics of continuum.

However, there is a very important case for applications when a continuous medium borders with a thin shell or plate. This case requires special consideration.

1.5.1 Shallow Shells and the Kirchhoff–Love Hypotheses

A shell is a body whose thickness h is small in comparison with the other two dimensions. It can be called shallow if the curvature of any part of the shell is not large. Let us give the basic relations of geometrically and physically nonlinear theory of

Models of Continuum

shallow shells based on equations of this book [34] and the theory of plastic flow [15]. The lines of curvature of the middle surface (equidistant from both surfaces of the shell) are used as the coordinates x and y.

The z-axis is directed along the normal to the middle surface to the center of its curvature. Generally speaking, shells are three-dimensional bodies. If you want to consider of them as two-dimensional bodies, you should use some hypotheses.

The essence of the classical Kirchhoff–Love hypotheses consists in two statements: (1) lines (fibers) initially perpendicular to the middle surface remain perpendicular to this surface during deformation and the length of them does not change, and (2) the normal stresses on sites parallel to the middle surface are ignorable relative to the normal stresses on sites perpendicular to this surface.

Naturally, the higher the accuracy of these hypotheses, the thinner the shell. The accuracy also depends on the length of waves propagating along the shell or a wave loading on the shell. The hypotheses are not applicable when wavelengths are of the order of thickness h.

Let us give the basic equations of motion of the shell element (Figure 1.5):

$$\frac{\partial N_x}{\partial x} + \frac{\partial T}{\partial y} + p_x = \rho h \frac{\partial^2 u}{\partial t^2}, \quad \frac{\partial T}{\partial x}\frac{\partial N_y}{\partial y} + p_y = \rho h \frac{\partial^2 v}{\partial t^2},$$

$$\frac{\partial^2 M_x}{\partial x^2} + \frac{\partial^2 M_y}{\partial y^2} + 2\frac{\partial^2 H}{\partial x \partial y} + K_x N_x + K_y N_y + \frac{\partial}{\partial x}\left(N_x \frac{\partial w}{\partial x} + T\frac{\partial w}{\partial y}\right)$$

$$+ \frac{\partial}{\partial y}\left(T\frac{\partial w}{\partial x} + N_y \frac{\partial w}{\partial y}\right) + p_z = \rho h \frac{\partial^2 w}{\partial t^2}, \tag{1.81}$$

where N_x, N_y, and T are the normal and tangential forces; p_x, p_y, and p_z are the external loads applied to the element in the directions x, y, and z; u, v, and w are the displacements of the points of the middle surface along x, y, and z, respectively; M_x, M_x, and H are the bending and twisting moments; K_x and K_y are the values inversely related to the main radii of curvature at the point of the middle surface.

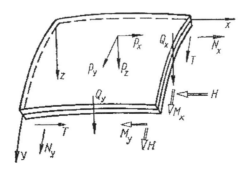

FIGURE 1.5 Forces and moments acting on the middle surface of a shell.

The forces and moments are associated with normal σ_x and σ_y and the tangential σ_{xy} stresses:

$$N_x = \int_{-h/2}^{h/2} \sigma_x\, dz, \quad N_y = \int_{-h/2}^{h/2} \sigma_y\, dz, \quad T = \int_{-h/2}^{h/2} \sigma_{xy}\, dz,$$

$$M_x = \int_{-h/2}^{h/2} \sigma_x z\, dz, \quad M_y = \int_{-h/2}^{h/2} \sigma_y z\, dz, \quad H = \int_{-h/2}^{h/2} \sigma_{xy} z\, dz. \quad (1.82)$$

We divide the loading time of the shell by J small steps (see Section 1.2). At the Nth step, the relationships between the stresses σ_l and the strains $\varepsilon_l = \varepsilon_l^e + \sum_{n=1}^{N} \Delta_n \varepsilon_l^p$ ($l = x, y, xy$) are written taking into account the second part of the Kirchhoff–Love hypotheses. As a result, we have

$$\sigma_x = \frac{E}{1-v^2}\left[\varepsilon_x + v\varepsilon_y - \sum_{n=1}^{N}\left(\Delta_n \varepsilon_x^p + v\Delta_n \varepsilon_y^p\right)\right],$$

$$\sigma_y = \frac{E}{1-v^2}\left[\varepsilon_y + v\varepsilon_x - \sum_{n=1}^{N}\left(\Delta_n \varepsilon_y^p + v\Delta_n \varepsilon_x^p\right)\right], \quad (1.83)$$

$$\sigma_{xy} = \frac{E}{2(1-v)}\left(\varepsilon_{xy} - \sum_{n=1}^{N}\Delta_n \varepsilon_{xy}^p\right),$$

where, as before, $\sum_{n=1}^{N}\Delta_n\varepsilon_l^p$ are the increments of components of the plastic deformation at the Nth step of the loading, E is the Young modulus, and v is the Poisson coefficient.

Expressions for deformations of the elongation ε_x and ε_y and the shift ε_{xy} in the layer of the shell located at a distance z from the middle surface are associated with the displacements by the following geometrical relations:

$$\varepsilon_x = \frac{\partial u}{\partial x} - K_x w + \frac{1}{2}\left(\frac{\partial w}{\partial x}\right)^2 - z\frac{\partial^2 w}{\partial x^2}, \quad \varepsilon_y = \frac{\partial v}{\partial y} - K_y w + \frac{1}{2}\left(\frac{\partial w}{\partial y}\right)^2 - z\frac{\partial^2 w}{\partial y^2},$$

$$\varepsilon_{xy} = \frac{\partial u}{\partial y} + \frac{\partial v}{\partial x} + \frac{\partial w}{\partial x}\frac{\partial w}{\partial y} - 2z\frac{\partial^2 w}{\partial x \partial y}. \quad (1.84)$$

We note that in the derivation of (1.84), we used the first part of the Kirchhoff–Love hypotheses.

Models of Continuum

Let us express the forces and the moments of Eqs. (1.81), through the displacements and increments of the components of plastic deformation using dependences (1.82)–(1.84):

$$T = \frac{hE}{2(1+v)}\left(\frac{\partial u}{\partial y} + \frac{\partial v}{\partial x} + \frac{\partial w}{\partial x}\frac{\partial w}{\partial y} - \sum_{n=1}^{N}\Delta_n^\alpha \varepsilon_{xy}\right),$$

$$N_x = \frac{hE}{(1-v)^2}$$

$$\times \left\{\frac{\partial u}{\partial x} - K_x w + \frac{1}{2}\left(\frac{\partial w}{\partial x}\right)^2 + v\left[\frac{\partial v}{\partial x} - K_y w + \frac{1}{2}\left(\frac{\partial w}{\partial y}\right)^2\right] - \sum_{n=1}^{N}\left(\Delta_n^\alpha \varepsilon_x + v\Delta_n^\alpha \varepsilon_y\right)\right\},$$

$$N_y = \frac{hE}{1-v^2}$$

$$\times \left\{\frac{\partial v}{\partial y} - K_y w + \frac{1}{2}\left(\frac{\partial w}{\partial y}\right)^2 + v\left[\frac{\partial u}{\partial x} - K_x w + \frac{1}{2}\left(\frac{\partial w}{\partial x}\right)^2\right] - \sum_{n=1}^{N}\left(\Delta_n^\alpha \varepsilon_y + v\Delta_n^\alpha \varepsilon_x\right)\right\},$$

$$M_x = \frac{hE}{1-v^2}\left[\frac{1}{12}h^2\left(\frac{\partial^2 w}{\partial x^2} + v\frac{\partial^2 w}{\partial y^2}\right) + \sum_{n=1}^{N}\left(\Delta_n^\beta \varepsilon_x + v\Delta_n^\beta \varepsilon_y\right)\right],$$

$$M_y = \frac{hE}{1-v^2}\left[\frac{1}{12}h^2\left(\frac{\partial^2 w}{\partial y^2} + v\frac{\partial^2 w}{\partial x^2}\right)^2 + \sum_{n=1}^{N}\left(\Delta_n^\beta \varepsilon_y + v\Delta_n^\beta \varepsilon_x\right)\right], \quad (1.85)$$

$$H = -\frac{hE}{2(1+v)}\left(\frac{1}{6}h^2 \frac{\partial^2 w}{\partial x \partial y} + \sum_{n=1}^{N}\Delta_n^\beta \varepsilon_{xy}\right).$$

Here

$$\Delta_n^\alpha \varepsilon_l = \frac{1}{h}\int_{-h/2}^{h/2} \Delta_n \varepsilon_l^p \, dz, \quad \Delta_n^\beta \varepsilon_l = \frac{1}{h}\int_{-h/2}^{h/2} \Delta_n \varepsilon_l^p z \, dz. \quad (1.86)$$

Taking into account (1.85), Eqs. (1.81) are rewritten as

$$\frac{\partial^2 u}{\partial x^2} + \frac{1-v}{2}\frac{\partial^2 u}{\partial y^2} + \frac{1+v}{2}\frac{\partial^2 v}{\partial x \partial y} - (K_x + vK_y) + \frac{\partial w}{\partial x} + \frac{\partial w}{\partial x}\frac{\partial^2 w}{\partial x^2} + \frac{1-v}{2}\frac{\partial w}{\partial x}\frac{\partial^2 w}{\partial y^2} + \frac{1+v}{2}\frac{\partial w}{\partial y}$$

$$\times \frac{\partial^2 w}{\partial x \partial y} - \sum_{n=1}^{N}\frac{\partial}{\partial x}\left(\Delta_n^\alpha \varepsilon_x + v\Delta_n^\alpha \varepsilon_y\right) - \frac{1-v}{2}\sum_{n=1}^{N}\frac{\partial}{\partial y}\Delta_n^\alpha \varepsilon_{xy} = \frac{1-v^2}{hE}\left(\rho h \frac{\partial^2 u}{\partial t^2} - p_x\right);$$

$$\frac{\partial^2 v}{\partial y^2} + \frac{1-v}{2}\frac{\partial^2 v}{\partial x^2} + \frac{1+v}{2}\frac{\partial^2 v}{\partial x \partial y} - (K_y + vK_x)\frac{\partial w}{\partial y} + \frac{\partial w}{\partial y}\frac{\partial^2 w}{\partial y^2} + \frac{1-v}{2}\frac{\partial w}{\partial y}\frac{\partial^2 w}{\partial x^2} + \frac{1+v}{2}\frac{\partial w}{\partial x}$$

$$\times \frac{\partial^2 w}{\partial x \partial y} - \sum_{n=1}^{N} \frac{\partial}{\partial y}\left(\Delta_n^\alpha \varepsilon_y + v\Delta_n^\alpha \varepsilon_x\right) - \frac{1-v}{2}\sum_{n=1}^{N}\frac{\partial}{\partial x}\Delta_n^\alpha \varepsilon_{xy} = \frac{1-v^2}{hE}\left(\rho h \frac{\partial^2 v}{\partial t^2} - p_y\right);$$

$$\frac{h^2}{12}\nabla^2\nabla^2 w - \left(K_x + vK_y\right)\frac{\partial u}{\partial y} - \left(K_y + vK_x\right)\frac{\partial v}{\partial y} + \left(K_x^2 + K_y^2 + 2vK_xK_y\right)w$$

$$-\frac{1}{2}(K_x + vK_y)\left(\frac{\partial w}{\partial x}\right)^2 - \frac{1}{2}(K_y + vK_x)\left(\frac{\partial w}{\partial x}\right)^2$$

$$-\frac{\partial}{\partial x}\left\{\frac{\partial w}{\partial x}\left[\frac{\partial u}{\partial x} + v\frac{\partial v}{\partial y} - (K_x + vK_y)w\right]\frac{1-v}{2}\frac{\partial w}{\partial y}\left(\frac{\partial u}{\partial y} + \frac{\partial u}{\partial x}\right)\right\}$$

$$-\frac{\partial}{\partial y}\left\{\frac{\partial w}{\partial y}\left[\frac{\partial v}{\partial y} + v\frac{\partial u}{\partial x} - (K_y + vK_x)w\right] + \frac{1-v}{2}\frac{\partial w}{\partial x}\left(\frac{\partial u}{\partial y} + \frac{\partial u}{\partial x}\right)\right\}$$

$$+ \sum_{n=1}^{N}\left[\frac{\partial^2}{\partial x^2}\left(\Delta_n^\beta \varepsilon_x + v\Delta_n^\beta \varepsilon_y\right) + \frac{\partial^2}{\partial y^2}\left(\Delta_n^\beta \varepsilon_y + v\Delta_n^\beta \varepsilon_x\right) + (1-v)\frac{\partial^2}{\partial x \partial y}\Delta_n^\beta \varepsilon_{xy}\right]$$

$$+ \sum_{n-1}^{N}\left\{K_x\left(\Delta_n^\alpha \varepsilon_x + v\Delta_n^\alpha \varepsilon_y\right) + K_y\left(\Delta_n^\alpha \varepsilon_y + v\Delta_n^\alpha \varepsilon_x\right)\right.$$

$$+\frac{\partial w}{\partial x}\left[\frac{\partial}{\partial x}\left(\Delta_n^\alpha \varepsilon_x + v\Delta_n^\alpha \varepsilon_y\right) + \frac{1-v}{2}\frac{\partial}{\partial y}\Delta_n^\alpha \varepsilon_{xy}\right]$$

$$+\frac{\partial w}{\partial y}\left[\frac{\partial}{\partial y}\left(\Delta_n^\alpha \varepsilon_y + v\Delta_n^\alpha \varepsilon_x\right) + \frac{1-v}{2}\frac{\partial}{\partial x}\Delta_n^\alpha \varepsilon_{xy}\right]$$

$$+ (1-v)\frac{\partial^2 w}{\partial x \partial y}\Delta_n^\alpha \varepsilon_{xy} + \frac{\partial^2 w}{\partial x^2}\left(\Delta_n^\alpha \varepsilon_x + v\Delta_n^\alpha \varepsilon_y\right) + \frac{\partial^2 w}{\partial y^2}\left(\Delta_n^\alpha \varepsilon_y + v\Delta_n^\alpha \varepsilon_x\right)\right\}$$

$$= \frac{1-v^2}{hE}\left(p_z - \rho h \frac{\partial^2 w}{\partial t^2}\right). \tag{1.87}$$

1.5.2 The Timoshenko Theory of Thin Shells and Momentless Shells

The greatest error of the above equations is determined by the assumptions that $\varepsilon_{xz} = \varepsilon_{yz} = 0$. These deformations may be considerable for local dynamic loading, in places of contact of a shell with a more rigid body, and all cases when significant cutting forces Q_x and Q_y occur. It is known that under action of the underwater wave,

Models of Continuum   29

a plate can be cut off on a rigid boundary [15,35]. Calculations of shell failure cannot be made without taking into account deformations ε_{xz} and ε_{yz} and the forces Q_x and Q_y arising near the different stress concentrations.

The Timoshenko theory. Let the normal to the middle surface changes the direction due to rotations of the shell element as an absolutely rigid body, which can be determined by the quantities $\partial w/\partial x$ and $\partial w/\partial y$. The direction also depends on shear strains ε_{xz} and ε_{yz}. The angles formed by shear strains can be approximated by angles β_x and β_y. Then, the angles of rotation of the normal to the middle surface of the shell are simulated as [34]: $\psi_x = \beta_x - \dfrac{\partial w}{\partial x}$ and $\psi_y = \beta_y - \dfrac{\partial w}{\partial y}$. The unknowns w, ψ_x, and ψ_x are determined from equations

$$\dfrac{\partial Q_x}{\partial x} + \dfrac{\partial Q_y}{\partial y} + K_x N_x + K_y N_y + \dfrac{\partial}{\partial x}\left(N_x \dfrac{\partial w}{\partial x} + T \dfrac{\partial w}{\partial y}\right) + \dfrac{\partial}{\partial y}\left(T \dfrac{\partial w}{\partial x} + N_y \dfrac{\partial w}{\partial y}\right)$$

$$+ p_z = \rho h \dfrac{\partial^2 w}{\partial t^2}, \quad \dfrac{\partial H}{\partial x} + \dfrac{\partial M_y}{\partial y} - Q_y = \dfrac{\rho h^3}{12}\dfrac{\partial^2 \psi_y}{\partial t^2}. \tag{1.88}$$

Equations (1.87) for u and v do not change. If we put $\psi_x = \psi_y = 0$ in (1.88), then we arrive at Eqs. (1.81) of the Kirchhoff–Love model. Let us give the relations that allow expressing the forces and moments in (1.88) through displacements. Geometrical relations are

$$\varepsilon_x = \dfrac{\partial u}{\partial x} - K_x w + \dfrac{1}{2}\left(\dfrac{\partial w}{\partial x}\right)^2 + z\dfrac{\partial \psi_x}{\partial x}, \quad \varepsilon_y = \dfrac{\partial v}{\partial y} - K_y w + \dfrac{1}{2}\left(\dfrac{\partial w}{\partial y}\right)^2 + z\dfrac{\partial \psi_y}{\partial y},$$

$$\varepsilon_{xy} = \dfrac{\partial u}{\partial y} + \dfrac{\partial v}{\partial x} + \dfrac{\partial w}{\partial x}\dfrac{\partial w}{\partial y} + z\left(\dfrac{\partial \psi_x}{\partial y} + \dfrac{\partial \psi_y}{\partial x}\right), \tag{1.89}$$

$$\varepsilon_{xz} = \left(\dfrac{\partial w}{\partial x} + \psi_x\right)f(z), \quad \varepsilon_{yz} = \left(\dfrac{\partial w}{\partial y} + \psi_y\right)f(z).$$

The function $f(z)$ characterizes the stress distributions σ_{xz} and σ_{yz} along the shell thickness. We define it as follows [15,34]: $f(z) = 6\left(\tfrac{1}{4} - z^2 h^{-2}\right)$. Equations (1.83) do not change. The stresses σ_{xz} and σ_{yz} are determined like σ_{xy} (1.83):

$$\sigma_{xz} = \dfrac{E}{2(1+v)}\left(\varepsilon_{xz} - \sum_{n=1}^{N}\Delta_n \varepsilon_{xz}^p\right), \quad \sigma_{yz} = \dfrac{E}{2(1+v)}\left(\varepsilon_{yz} - \sum_{n=1}^{N}\Delta_n \varepsilon_{yz}^p\right). \tag{1.90}$$

Substituting Expressions (1.89) into (1.90) and using (1.82), we obtain the refined expressions for the moments

$$M_x = \frac{hE}{1-v^2}\left[\frac{1}{12}h^2\left(\frac{\partial \psi_x}{\partial x}+v\frac{\partial \psi_y}{\partial y}\right)-\sum_{n=1}^{N}\left(\Delta_n^\beta \varepsilon_x + v\Delta_n^\beta \varepsilon_y\right)\right],$$

$$M_y = \frac{hE}{1-v^2}\left[\frac{1}{12}h^2\left(\frac{\partial \psi_y}{\partial y}+v\frac{\partial \psi_x}{\partial x}\right)-\sum_{n=1}^{N}\left(\Delta_n^\beta \varepsilon_y + v\Delta_n^\beta \varepsilon_x\right)\right], \quad (1.91)$$

$$H = \frac{hE}{2(1-v)}\left[\frac{1}{12}h^2\left(\frac{\partial \psi_x}{\partial y}+\frac{\partial \psi_y}{\partial x}\right)-\sum_{n=1}^{N}\Delta_n^\beta \varepsilon_{xy}\right].$$

Formulas for forces remain the same as (1.88). Cutting forces are defined by the following way:

$$Q_x = \frac{E}{2(1+v)}\left[K^2\left(\frac{\partial w}{\partial x}+\psi_x\right)-\int_{-h/2}^{h/2}f(z)\sum_{n=1}^{N}\Delta_n\varepsilon_{xz}^p dz\right],$$

$$Q_y = \frac{E}{2(1+v)}\left[K^2\left(\frac{\partial w}{\partial y}+\psi_y\right)-\int_{-h/2}^{h/2}f(z)\sum_{n=1}^{N}\Delta_n\varepsilon_{yz}^p dz\right]. \quad (1.92)$$

The constant K^2 can be found from the relation $K^{-2} = h^{-1}\int_{-h/2}^{h/2}f^2(z)dz$. Approximately $K^2 = 5/6$. Substituting the expressions for the forces N_x, N_y, and T; moments M_x, M_y, and H; and forces Q_x and y_0 in the equations of motion, we obtain the equations that determine u, v, w, ψ_x, and ψ_y.

These equations will not be presented here because of their cumbersomeness. For the case of elastic shells, they can be found in the monograph [34].

The equations given above are valid if the maximum displacement of a shell is of the same order as a shell thickness or even is more than h. But the displacement should be much smaller than another dimension of the shell. In this case, the boundary conditions on the boundary forming by liquid and shell can be written on the undeformed middle surface of the shell [34,35]. Therefore, in the future, the linearized boundary conditions are used even for the case of large displacements. In particular, we will assume in (1.87) $P_x = P_y = 0$.

Axisymmetric momentless (flexible) shells. Let us consider thin-walled bodies of rotation that experience an axisymmetric deformation. The change in the stress state along the thickness of the shell does not take into account. Thus, the stress state does not depend on bending moments and shear forces.

This theory can be used when performing the following conditions: the shape of the shell must be smooth, without sharp changes of curvature; loads should be uniform or should change smoothly; the edges of the shell are free from significant cutting and moment forces. It should be noted that the theory is applicable if the listed conditions are not completely observed but a local bend is very local.

Models of Continuum   31

In this case, at a small distance from the zone bending, a stressed state is practically momentless [36,37].

It is important that this theory is valid for large displacements (much more of h) of the shell points.

The equations of motion of the shell are written with respect to the fixed (Eulerian) rectangular coordinates x and y:

$$\frac{1}{x}\left[\frac{\partial}{\partial s}(N_y x \cos\phi) - N_x\right] + p\sin\phi = \rho h \frac{\partial^2 x}{\partial t^2},$$
$$-\frac{1}{x}\frac{\partial}{\partial s}(N_y x \sin\phi) p\cos\phi = \rho h \frac{\partial^2 y}{\partial t^2}, \qquad (1.93)$$

where s is the length of the arc of the shell along the meridian, measured from its pole; N_x and N_y are the forces directed along the lines of meridians and parallels, respectively. The angle φ between the tangent to the middle surface of the shell and the positive direction of the x-axis is determined in the following way:

$$\sin\varphi = -\frac{dy}{ds}, \quad \cos\varphi = \frac{dx}{ds}, \quad ds^2 = dx^2 + dy^2. \qquad (1.94)$$

The forces N_x and N_y are expressed in terms of deformations according to Formulas (1.85) and (1.86). Deformations of ε_x and ε_y along the meridional and latitudinal directions are determined by the current and initial values of coordinates x_0 and y_0

$$\varepsilon_y = \tfrac{1}{2}\left(ds^2 ds_0^{-2} - 1\right), \quad \varepsilon_x = \tfrac{1}{2}\left(x^2 x_0^{-2} - 1\right). \qquad (1.95)$$

Conclusion. The presented dynamical equations of the shallow shells are valid with relatively small displacements when usually the moment terms in the equations of motion (1.87) manifest themselves. With further growth of displacements, the influence of moment forces decreases and the assumption about small curving of the shell can become invalid. Generally speaking for some cases of large deformations, the influence of the moment forces on the final shape of the shell may be small.

This stage of deformation can be studied on the basis of theory of the membrane shortly presented above. Sometimes the entire calculation of a shell can be made using only the membrane theory.

Final remark. The above are presented in compressed and simplified forms of the main approaches to the analysis of the problems considered in this book. Their more extended presentation is given in the remaining chapters of this part.

2 The Dynamic Destruction of Some Materials in Tension Waves

This chapter is devoted to extreme waves of destruction. This subject is very complex and may be treated from different points of view. We will use the continuous damage mechanics approximation [38–41]. This model was used in many researches. For a full enough consideration of this approach, the reader must refer to the special treatises (see, for example, [42–50]). In this book, only some results of different theories are presented. Our goal of this chapter is to give the reader a certain understanding of the features of the dynamic destruction of some materials and basic approaches to the mathematical description of these features. It will allow us to simulate the extreme fracture waves in certain materials.

The process of ductile failure in metals involves the nucleation and growth of voids. Voids nucleate at the weak points in the material when the tensile stress exceeds some critical level. The material around growing voids undergoes a large plastic deformation. As a result of these processes, the initial material obtains the properties of porous materials. The voids link up to form the fracture surface. One aspect of continued interest is the effect of high strain rates (the effects of inertia on pore growth).

Extreme waves are simulated often as a plane-fronted compressive wave. It propagates through the material and reflects off a free surface. The compressive wave becomes tensile (rarefaction wave) when it is reflected from the surface. If the tension exceeds the material strength, fracture occurs. This kind of dynamic fracture is known as spallation. Our attention in this section will be focused on this type of dynamical fracture of ductile materials. This effect may be important in processes such as projectile penetration and impact fragmentation of ductile materials. Extreme waves and dynamic fracture can also be resulted from the action of the enormously powerful lasers on structural elements.

Let us look at the microstructure of a material and its evolution within extreme tensile wave (Figure 2.1). In the micrographs, the different stages of the evolution can be distinguished. At the initial stage, the deformation is elastic or weakly plastic. Then, when the necking formation starts, pores begin to form. This mechanism leads to the formation of macro-cracks, which ultimately result in the fracture of the material.

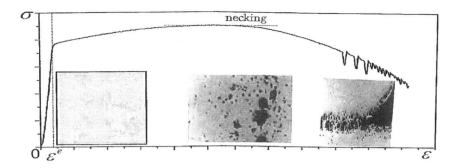

FIGURE 2.1 Typical stress–strain diagram of a ductile metal and micrographs illustrating three parts of the diagram: practically undamaged material (nucleation of pores, left), growth of pores (center), and the formation of crack and fracture of the material (right).

2.1 MODELS OF DYNAMIC FAILURE OF SOLID MATERIALS

Many models have been proposed for dynamic fracture processes. At the same time, there are two main approaches to modeling these processes: phenomenological and microstructural. In the first approach, there are usually the analytical expressions describing an interaction between damage and stresses in material during the loading. Phenomenological models deal mainly with continuum variables such as average stresses, strains, and void volume. The second approach is based on deduction of integral damage characteristics of the material on the basis of the analysis of the behavior of an individual microdefect. Microstructural models take into account often the inertia of voids growth (pores). Here, we discuss a few models to provide a brief sampling in the subject.

High tensile stresses may develop a material within the extreme rarefaction wave; as a result, dynamic failure in the form of an internal cavitation may take place. This type of failure is termed the spall fracture [42].

2.1.1 Phenomenological Approach

In this case, a unified integral characteristic of the process of fracture is introduced. In particular, hypotheses are introduced that predicts fracture as soon as some value reaches a critical number. This value can be determined by stresses, strains, or energy of deformation.

The simplest example of similar hypothesis is the theory of instantaneous fracture [42,50]

$$\sigma \geq \sigma_{cr}. \tag{2.1}$$

The fracture is if (2.1) realizes in some point of material. It is emphasized that we use similar fracture condition earlier when the destruction of bubbly liquid was considered (Section 1.4).

Sometimes the hypotheses take into account time effects

$$\int_0^t \sigma \, dt < \bar{\sigma}\Delta t, \qquad (2.2)$$

where Δt is the time till the fracture under the tension $\bar{\sigma}$. Generally speaking, similar hypotheses do not suggest the change in mechanical properties of material during deformation. This change take into account continuum damage mechanics [38–41]. According to this theory, during the deformation, microdefects begin to grow in material. Let the relative volume of them is ξ,

$$\xi = V_p/V, \qquad (2.3)$$

where V_p is the total volume of voids in the microvolume V. The fracture is characterized by porosity α, $\alpha = V/V_m$, in a number of papers. Here, V_m is the volume of the material of the matrix (volume of the deformed material without voids (pores)). Thus, we have $\xi = 1 - 1/\alpha$. The general enough form of the equation describing the damage evolution is as follows:

$$\partial \xi / \partial t = \Phi(\xi, \sigma), \qquad (2.4)$$

where σ is the maximal principal tensile stress or the pressure in the material. For porosity, the corresponding equation has the form:

$$\partial \alpha / \partial t = \alpha^2 \Phi(1 - 1/\alpha, \sigma). \qquad (2.5)$$

A large group consists of models of dynamic fracture in which the influence of ξ in the right-hand side of Eq. (2.4) is neglected. We list some of them.

The Zhurkov model (modified) is [51]

$$\partial \xi / \partial t = A_1 \left[\exp\left(B_1\left(\sigma - \sigma^*\right)\right) - C_1 \right]. \qquad (2.6)$$

In the classical Zhurkov model [42,51,52], $\sigma^* = C_1 = 0$, but such a formula does not describe well enough the fracture in the microsecond range of durations. If $C_1 = 0$, the Akhmadeev–Nigmatulin model follows from (2.6) [53]. Very popular is the Tuler–Butcher model [54]:

$$\partial \xi / \partial t = A_2 \left(\sigma - \sigma_1^*\right)^n, \quad \sigma > \sigma_1^*. \qquad (2.7)$$

If $n = 2$, $\sigma_1^* = 0$, we arrive at the energy criterion $\partial \xi / \partial t = A_3 \sigma^2$ If $n = 1$, $\sigma_1^* = 0$, we arrive at the impulse criterion $\partial \xi / \partial t = A_4 \sigma$ [44]. For a number of metals, the logarithmic law is used:

$$\partial \xi / \partial t = A_5 \ln\left[B_5\left(\sigma - \sigma_5^*\right) + 1\right]. \qquad (2.8)$$

In Formulas (2.6)–(2.8), the letters A, B, and σ^* with numerical lower indices denote the constants of materials. As indicated in [55,56], the Tuler–Butcher and logarithmic (for iron) formulas describe experimental results most better.

All these theories, however, do not possess universality. They do not take into account a number of important effects, in particular, the change of porosity of material accompanying the destruction. This lack is absent in models in which the right-hand side of (2.4) depends on ξ [56].

The results presented above were obtained primarily during 1960s and 1970s when instrumentation and computational tools were not as developed and as sophisticated as they are today. These tools and the manner in which they are applied to investigate spall problems are well described in many publications (see, for example, [42–61]).

It seems important to have a more in-depth details of the process of dynamical destruction. For this, obviously, it is necessary to attract additional array of experimental data, associated with the emergence of individual microdefects, their growth, interaction, and coalescence. The influence of the size of the defects, inertia of their development, and so on must be investigated.

2.1.2 Microstructural Approach

The phenomenological approach considers the destruction as collective effect, in which whole ensembles of microdamage are involved. It is difficult to study the role of each defect in this process. At the same time, studying the dynamics of a single defect is relatively simple. The difficulty, however, is that the integral characteristics of the destruction are usually observed. However, it is possible by gradually complicating the model taking into account the interaction of voids, their fusion, and unevenness of the initial sizes. Therefore, the necessary stage of the study must be a comparison of the parameters of dynamics microdefects and global characteristics of destruction in time.

Generally speaking, a dynamic fracture model of the microstructural approach includes three systems of equations. First, we should derive equations describing the creation of a single microdefect, its evolution, and merging with other defects. The conclusion must be made on the basis of clearly formulated hypotheses concerning the above phenomena. The second system is based on the statistical methodology. Its most important goal is a transition from local fracture characteristics to averaged values. The third system of equations determines the integral parameters of destruction.

There are serious difficulties connected with realization of microstructural approaches. In this way, empirical data and data of theoretical analysis are usually used. As an example, let us examine a model of fracture of a medium obtained by analyzing data of the high-speed collision of plates [42] and the pore growth:

$$\partial N/\partial t = \dot{N} = N_0 \exp\left[(\sigma - \sigma_{n0})/\sigma_1 - 1\right], \quad \partial R/\partial t = \tfrac{1}{4} \eta^{-1} R(\sigma - \sigma_{g0}), \quad (2.9)$$

where $\dot{N}$ is the rate of void nucleation; R is the void radius; N_0, σ_{n0}, σ_1, and σ_{g0} are the material constants; and η is the viscosity coefficient.

Dynamic Destruction of Some Materials

Now let us examine the increase in void size in accordance with Eqs. (2.9) under the influence of a step-like stress. Let there be the following exponential void distribution at the moment t_0 [61]:

$$P(R_0 > R) = \exp(-c_0 R), \tag{2.10}$$

where c_0 is a constant, and $P(R_0 > R)$ is the probability that a void has a radius greater than R_0.

In accordance with (2.9), void radius changes with time:

$$R = R_0 \exp(\alpha t), \quad \alpha = \tfrac{1}{4}\eta^{-1}(\sigma - \sigma_{g0}). \tag{2.11}$$

At the moment t, the distribution of void radius will be as follows:

$$P(R_0 e^{\alpha t} > R) = \exp(-c_0 e^{-\alpha t} R). \tag{2.12}$$

It follows from this that exponents in the void radius distribution decrease exponentially with time. However, the experiments [56] do not show such a rapid decrease described by the exponents. This discrepancy is due to the fact that Eqs. (2.9) best describe the steady state of void growth and does not consider the acceleration, which can be important during the initial stage of void growth.

On the whole, all of these make it necessary to use a more complex equation for void growth. This equation is obtained by solving the continuum mechanics equations of a viscoplastic medium with a spherically expanding cavity [57] (see also the section 2.3):

$$p = -2\sigma_0 \ln\frac{b}{R} - 4\mu \frac{b^3 - R^3}{Rb^3}\frac{\partial R}{\partial t} + \rho_0\left[\frac{b^4 - R^4}{2b^4}\left(\frac{\partial R}{\partial t}\right)^2 - \frac{b-R}{b}\left(R\frac{\partial^2 R}{\partial t^2}\right) + 2\left(\frac{\partial R}{\partial t}\right)^2\right], \tag{2.13}$$

where b is the radius of the spherical cell embracing the void, b_0 is the unperturbed values of b, ρ_0 is the density of the matrix, and $b^3 - R^3 = b_0^3 - R_0^3$. If $b \gg R$, we have

$$R\ddot{R} + \tfrac{3}{2}\dot{R}^2 + 4\mu(\rho_0 R)^{-1}\dot{R} + \rho_0^{-1}(p + 2\sigma_0 \ln bR^{-1}) = 0. \tag{2.14}$$

The mean density of damaged material and the density of the solid material (matrix) are related by the equation (see Eq. (1.62)):

$$\rho = \rho_0\{(1-\xi_0)[1 - \lambda(p - p_0)] + \xi\}^{-1}, \tag{2.15}$$

where λ is the compressive bulk modulus of the matrix. Here, $\xi = \xi(R)$, since for spherical pores $\xi = \tfrac{4}{3}\pi U^{-1}\sum_{i=1}^{n} R_n^3$. Here, $\tfrac{4}{3}\pi \sum_{i=1}^{n} R_n^3$ is the total volume of n voids in the microvolume U.

Particular cases of (2.14). Let us consider different particular cases of Eq. (2.14).
1. The acceleration can be much larger than the velocity of pore surface at the beginning of the wave action on a pore. In this case, Eq. (2.14) yields

$$R\ddot{R} + 4\mu(\rho_0 R)^{-1} \dot{R} + \rho_0^{-1}\left(p + 2\sigma_0 \ln bR^{-1}\right) = 0. \tag{2.16.1}$$

2. When a pore grows with constant velocity, we have

$$\tfrac{3}{2}\dot{R}^2 + 4\mu(\rho_0 R)^{-1} \dot{R} + \rho_0^{-1}\left(p + 2\sigma_0 \ln bR^{-1}\right) = 0. \tag{2.16.2}$$

3. For strong enough viscosity, Eq. (2.14) yields the expression that resembles (2.9):

$$\dot{R}^2 = -\tfrac{1}{4}\mu^{-1} R\left(p + 2\sigma_0 \ln bR^{-1}\right). \tag{2.16.3}$$

This expression agrees with (2.9). It coincides with a formula presented in [42].

4. Let us examine the case of the absence of viscosity $\mu = 0$ in detail [26]. If the change of the term $\ln(b/R)$ has weak effect, Eq. (2.14) yields

$$\dot{R} = \left[\tfrac{2}{3}\rho_0^{-1}\left(p + 2\sigma_0 \ln(b/R)\right)\left(R_0^3 R^{-3} - 1\right)\right]^{1/2}. \tag{2.17}$$

We have assumed in (2.17) that during the initial stage of fracture, the radius can be represented as a sum of the initial radius R_0 and its perturbation:

$$R = R_0 + \Delta R. \tag{2.18}$$

Inserting (2.18) into (2.17) and omitting small values, we have

$$\partial(\Delta R)/\partial t = \left[-2\rho_0^{-1}\left(p + 2\sigma_0 \ln(b/R)\right)\Delta R/R_0\right]^{1/2}. \tag{2.19}$$

Solving this equation finally gives us

$$R = R_0 - \tfrac{1}{2}\rho_0^{-1} R_0^{-1}\left(p + 2\sigma_0 \ln(b/R)\right) t^2. \tag{2.20}$$

It is seen that the last expression is different from (2.11). We think that this expression describes less rapid growth of the pores than the prediction of Expressions (2.9) and (2.11).

Numerical study. We present in Figure 2.2 the results of the study of the dynamics of isolated pore in a viscoplastic material. Equation for the pore (2.14) was solved by the Runge–Kutta method of fourth order of accuracy.

We study the effects of various terms in Eq. (2.14). The results of the calculations are presented in the summary diagram (Figure 2.2). Along the abscissa the corresponding dimensionless parameter is changed, and along the ordinate axis the dimensionless values of the radius at $t = 1$ μs is presented. Initial (scale) values of

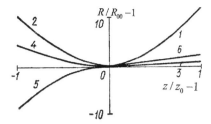

FIGURE 2.2 Dependence of the change in the final radius of the pore on the variation in the material parameters: (1) $z = p$, (2) $z = \sigma_0$, (3) $z = \rho$, (4) $z = \eta$, (5) $z = \ln b$, and (6) $z = R_0$.

parameters used in the calculations are taken for aluminum and denoted by zero indices. In particular, R_{00} is the radius at $t = 1$ μs and for $p = 2 \times 10^9$ Pa. The flow stress σ_0 was compared with the flow stress of aluminum, which was denoted as σ_{00}.

It can be seen that values σ_0 and p have the most significant effect on final radius of the pores. Therefore, they should be defined in experiments with the greatest correction. The rest values weakly affect the process of destruction according to these calculations.

Remark. Thus, a great variety of models have been proposed for fracture process. What is the reason for the presence of the large number of models?

It is explained by a method of comparing these theories with experimental data. Great part of the formulas was obtained after an analysis of results of the high-speed collisions of plates (impactor and target). During much time in the experiments, the available measurements were only the velocity of the free surface target and the time of the separation of the spalled plate from the target. Noted above, phenomenological theories of dynamic failure accurately enough described the speed and separation time for the rear surface although they were substantially different in detail.

2.2 MODELS OF INTERACTING VOIDS (BUBBLES, PORES)

We studied briefly the nucleation and growth of defects (pores, voids, bubbles). In this section, we consider certain models of interaction of these defects. The nucleation in the medium of submicrodefects with a size of 10^{-7} m is not examined. The analysis is extended to the already-formed defects (pores, gas bubbles), their growth, and interaction. The stage of interaction just prior to the start of coalescence of defects (fragmentation of the medium) is not studied. At the same time in accordance with [39–56], it may be assumed that the areas of initiation of cracks which divide the medium into fragments are determined by the zones of concentrations of defects which formed up to the start of coalescence of the defects (see Figure 2.1).

Since the form of voids in the experiments carried out on viscoplastic materials and many liquids is almost spherical, it is assumed that growth and interaction of the voids are determined by the pressure, viscoplastic strain, and the time, but not by the shear stresses [42,8]. It is also assumed that the rate of displacement of the voids in relation to each other is equal to zero.

The study of void dynamics is based on the subdivision of the investigated medium into parts, whose dimensions are much larger than the size of the voids.

At the same time, the dimensions of the parts are small enough when compared with the characteristic scale of a significant change of stresses and pressure within the extreme wave propagating in the media.

Interaction of two pores (two gas bubbles). Equation (1.53) describes the radial oscillations of a spherical bubble. Comparison of (1.53) and Eq. (2.14) for the radial oscillations of a pore in viscoplastic materials shows that these equations are similar. These equations differ only in the right-hand side. Thus, there is an obvious analogy between the equations for bubbles and pores. We will use this analogy when the interaction of two pores and two bubbles is considered (Figure 2.3).

Let us consider two spherical bubbles that perform radial oscillations in liquid. In this case, the kinetic energy of any point of the medium can be calculated by following Lamb's book [58] and [59]:

$$K = 2\pi\rho\left(R_1^3 \dot{R}_1^2 + R_2^3 \dot{R}_2^2 + 2R_1^2 R_2^2 \dot{R}\dot{R}\right)/\Delta. \quad (2.21)$$

Here, $\dot{R} = \partial R/\partial t$ and $\ddot{R} = \partial^2 R/\partial t^2$. This expression is substituted in the Lagrange equation:

$$\frac{d}{dt}\frac{\partial K}{\partial \dot{R}_1} + \frac{d}{dt}\frac{\partial K}{\partial \dot{R}_2} - \frac{\partial K}{\partial \dot{R}_1} - \frac{\partial K}{\partial \dot{R}_2} = Q_1 + Q_2. \quad (2.22)$$

Here, Q_1 and Q_2 are the generalized forces. Using (2.21) and (2.22), one can find equations describing the motion of two bubbles in a liquid. Various cases of these equations can be found in [8,61–64].

We assume that Eqs. (2.21) and (2.22) also describe the radial oscillations of spherical pores in viscoplastic materials.

Using this analogy, we write down equations of interrelated oscillations of two pores as

$$R_1\ddot{R}_1 + \tfrac{3}{2}\dot{R}_1^2 + 2R_2\dot{R}_2^2/\Delta + R_2^2\ddot{R}_2/\Delta = F(R_1),$$
$$F(R_1) = -\left[p + 2\sigma_0 \ln\left(b_1 R_1^{-1}\right) + 4\mu\dot{R}_1 R_1^{-1}\right]/\rho_0. \quad (2.23)$$

In (2.23), R_1 and R_2 are the radii of interacting voids (pores, bubbles), Δ is a distance between centers of the spherical voids. Constants b_1 and b_2 are the radii of spherical cells enclosing the pores, with these values being much larger than the distance

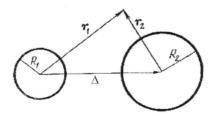

FIGURE 2.3 Two interacting pores (bubbles).

between the pores, but less than the radius of the zone of plasticity around pores. Equations (2.23) are written for R_1. Corresponding equation for R_2 is followed from (2.23) if subscript 1 is replaced by 2 and subscript 2 is replaced by 1. Equations (2.23) differ from those written for gas bubbles [8,64,65] only by right-hand sides. The expressions for the right-hand sides correspond to the equation written for the single pore (2.14).

Calculations show [61] that the mutual effect of the cavities which are taken into account by cubic terms starts to operate at $\Delta \leq (5-6)R$, where R is the radius of the larger cavity. There is no interaction when $\Delta > (5-6)R$. In this case, the material damage dynamics can be described using the equation for an isolated pore. Thus, Eqs. (2.23) are useful only if $\Delta < (5-6)R$. On the other hand, the pores should be located far enough from each other. Thus, it seems that the equations are valid only for sufficiently small amplitudes of pore oscillations and when the condition $3R \leq \Delta \leq (5-6)R$ approximately.

Unidimensional model of interaction of voids. Due to the growth of pores, the initially slightly damaged material can turn into a porous material. Let us examine a linear chain of defects. We separate in the chain an arbitrary sequence of three cavities with the radii R_{i-1}, R_i, and R_{i+1}, where $i = 2, 3...$ (Figure 2.4).

We also assume that the ith cavity is influenced only by the adjacent cavities R_{i-1} and R_{i+1}. The interaction between the defects R_{i-1} and R_{i+1} is neglected assuming that it is a very small quantity. Because of the above assumptions, we can easily derive an equation that generalizes Eqs. (2.23) and takes into account the interaction of three cavities:

$$R_i\ddot{R}_i + \tfrac{3}{2}\dot{R}_i^2 + R_{i-1}\left(2\dot{R}_{i-1}^2 + R_{i-1}\ddot{R}_{i-1}\right)/\Delta_{i-1} + R_{i+1}\left(2\dot{R}_{i+1}^2 + R_{i+1}\ddot{R}_{i+1}\right)/\Delta_{i+1} = F(R_i), \quad (2.24)$$

where $\Delta_{i\pm1}$ are the distances between the voids i and $i+1$, and between the voids i and $i-1$, respectively. While deriving Eq. (2.24), it is assumed that the terms in the round brackets are of the same order of smallness. They are smaller than the first two terms. Consequently, the distances $\Delta_{i\pm1}$ are considerably greater than the radii but this difference is not sufficient to ignore the interaction between the voids (pores and bubbles). Equation (2.24) holds for any void in the chain. Therefore, they take into account the mutual effect of all the voids through each other.

Equation (2.24) can be used directly for calculating the propagation of damage waves in a medium. However, cases are possible in which (2.24) can be conveniently replaced by a single equation in partial derivatives. To obtain this equation, taking into account the smallness of the distance between the defects, we introduced expansions of the following type:

$$R_{i\pm1} = R_i \pm \Delta_{i\pm1}\,\partial R_i/\partial x + \tfrac{1}{2}(\Delta_{i\pm1})^2\,\partial^2 R_i/\partial x^2 + \cdots \quad (2.25)$$

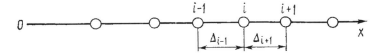

FIGURE 2.4 Chain of pores.

Thus, R is considered as the certain continuum function. Consequently, Eq. (2.24) gives

$$R\ddot{R} + \tfrac{3}{2}\dot{R}^2 + f^-(x)R(2\dot{R}^2 + R\ddot{R}) + f^+(x)\Phi(R, R_x, R_{xx}) = F(R). \qquad (2.26)$$

Here, $f(x)$ is the distribution function of the voids, which determines the distances between their centers $f^-(x) = \Delta_{i-1}^{-1} + \Delta_{i+1}^{-1}$, $f^+(x) = \Delta_{i-1} + \Delta_{i+1}$ and

$$\Phi(R, R_x, R_{xx}) = \ddot{R}R_x^2 + RR_{xx}\ddot{R} + 2RR_x\ddot{R}_x + \tfrac{1}{2}R^2\ddot{R}_{xx}$$

$$+ \tfrac{1}{2}\dot{R}^2 R_{xx} + R\dot{R}_x^2 + 2\dot{R}R_x\dot{R}_x + R\dot{R}\dot{R}_{xx}. \qquad (2.27)$$

Here, the subscript x indicates the derivative with respect to the coordinate. Apparently, the range of application of Eqs. (2.27) is smaller than that of (2.24). This range is discussed in [61].

Case of two spatial coordinates. We will examine a layer of voids distributed in the x and y plane of the Cartesian coordinate system. The plane will be divided by straight lines parallel to the coordinate axes into a nonuniform rectangular grid with the smooth variation of the distance between the nodes, and we will assume that the defects are positioned at the nodes of the grid (Figure 2.5). Let it be that each void i, j ($i, j = 1, 2, 3,\ldots$) is influenced by only the adjacent voids in accordance with the law of the type of Eq. (2.24). Consequently, for the void with the indices i and j, we can write the following equation of the oscillations:

$$R_{i,j}\ddot{R}_{i,j} + \tfrac{3}{2}\dot{R}_{i,j}^2 + \left[R(2\dot{R}^2 + R\ddot{R})/\Delta\right]_{i-1,j} + \left[R(2\dot{R}^2 + R\ddot{R})/\Delta\right]_{i+1,j}$$

$$+ \left[R(2\dot{R}^2 + R\ddot{R})/\Delta\right]_{i,j-1} + \left[R(2\dot{R}^2 + R\ddot{R})/\Delta\right]_{i,j+1} = F(R_{i,j}). \qquad (2.28)$$

Here, the subscripts relate to the quantities included in the square brackets and denote the point at which the void is positioned. In particular, the value $\Delta_{i-1,j}$ is the distance between the point i, the point j, and the point $i - 1, j$. The transition from the system

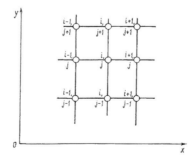

FIGURE 2.5 Distribution of voids in the x and y plane.

Dynamic Destruction of Some Materials

of the ordinary differential Eq. (2.28) to the equations in the partial derivatives can be easily carried out using the same procedure as for the chain of voids:

$$R\ddot{R} + \tfrac{3}{2}\dot{R}^2 + 4\Delta^{-1}R(2\dot{R}^2 + R\ddot{R}) + 2\Delta\left[\Phi(R, R_x, R_{xx}) + \Phi(R, R_y, R_{yy})\right] = F(R). \quad (2.29)$$

Function Φ in Eq. (2.29) was determined above (see Eqs. (2.27)). Equation (2.29) is derived using the assumption of the uniform distribution of the defects, and a distance between them is denoted by Δ. The range of application of Eq. (2.29) is smaller than that of the initial system (2.28). The restrictions on the application of Eq. (2.29) can be determined by the same procedure as that used for the chain of voids [61].

Spatial model. Two-dimensional model described above can be generalized for the three-dimensional case if we assume that all the voids are positioned in the points of intersection of the planes parallel to the coordinate surfaces of the x, y, and z of the Cartesian coordinate system.

Let it be that the spacing between the voids changes smoothly and the void $R_{i,j,k}$ positioned at the points i, j, and k is affected only by the adjacent voids positioned in the planes passing through this point. In this case, the system of equations generalizing Eq. (2.28) is written in the following form:

$$\left(R\ddot{R}\right)_{i,j,k} + \tfrac{3}{2}\dot{R}^2_{i,j,k} + \sum_m \left[R(2\dot{R}^2 + R\ddot{R})/\Delta\right]_m = F(R_{i,j,k}). \quad (2.30)$$

Here, summation is carried out in respect of six values of the indices $i \pm 1$, $j \pm 1$, and $k \pm 1$. The equation in the partial derivatives corresponding to the system (2.30) is obtained using the procedure applied to the unidimensional and two-dimensional cases.

2.3 PORE ON POROUS MATERIALS

We have considered above the cases when the dynamics of the pore under consideration depends on oscillations of surrounding pores. However, it is assumed above that pores do not affect the plastic properties of the material. Let us consider the pore dynamics in the initially porous material.

Pore in a compressible porous material. Damaged materials are considered as continuum. Oscillations of a spherical pore in this continuum are described by equations of conservation of mass and momentum:

$$\frac{\partial \rho}{\partial t} + v\frac{\partial \rho}{\partial r} + \rho\frac{\partial v}{\partial r} + \frac{2}{r}\rho v = 0,$$

$$\rho\left(\frac{\partial v}{\partial t} + v\frac{\partial v}{\partial r}\right) = \frac{\partial \sigma_r}{\partial r} + \frac{2}{r}(\sigma_r - \sigma_\varphi). \quad (2.31)$$

We also introduce the yielding condition of a porous material with the matrix metal that is not subjected to the strain hardening [66,67]. In the case of the spherically symmetrical stress state, the yielding condition is written in the form:

$$(\rho/\rho_0)^{2n} \sigma_0^2 = (\sigma_r - \sigma_\varphi)^2 + 6.25\xi p^2, \tag{2.32}$$

where $\rho = \rho_0(1-\xi)$. In (2.32), ρ_0 is the density of the matrix metal, which is assumed to be incompressible; ξ is the relative volume of the pores (2.32); p is the pressure; σ_0 and n are the constants. We will assume in (2.32) that $n = 2.5$. Let us formulate boundary conditions of the problem. The following kinematic and force conditions are fulfilled on the surface of the pore $r = R$,

$$v = \dot{R}, \tag{2.33}$$

$$\sigma_r = 0. \tag{2.34}$$

Subsequent analysis is carried out on the basis of the assumption that the effect of porosity on the strain state of the medium and the plasticity condition (2.32) is determined by the quantity of the second order of smallness in comparison with the main terms. Therefore, the quantities of the third order of smallness are subsequently ignored. In this case, Eq. (2.32) yields

$$(\sigma_r - \sigma_\varphi) = \sigma_0(1 - 2.5\xi) - 3.125\xi p^2 \sigma_0^{-1}. \tag{2.35}$$

Here, σ_r and σ_φ depend on r and t. Additionally, we take into account the viscous properties of the material. In this case,

$$(\sigma_r - \sigma_\varphi) = \sigma_0(1 - 2.5\xi) + 2\mu\left(\frac{\partial v}{\partial r} - \frac{v}{r}\right) - 3.125\xi p^2 \sigma_0^{-1}. \tag{2.36}$$

This gives for $\xi = 0$ the plasticity condition for the viscous matrix [64]. For spherical symmetrical problem, we have $p = -(\sigma_r + 2\sigma_\varphi)/3$. In this case, Eq. (2.36) yields

$$\sigma_r = -p + \frac{2}{3}\left[\sigma_0 + 2\mu\left(\frac{\partial v}{\partial r} - \frac{v}{r}\right) - 2.5\xi\sigma_0 - 3.125\xi p^2 \sigma_0^{-1}\right]. \tag{2.37}$$

The problem is reduced to the consideration of Eqs. (2.31) and conditions (2.33), (2.34), and (2.37). The equation of motion yields

$$\rho\left(\frac{\partial v}{\partial t} + v\frac{\partial v}{\partial r}\right) = -\frac{\partial p}{\partial r} + \frac{2}{3}\frac{\partial}{\partial r}\left[2\mu\left(\frac{\partial v}{\partial r} - \frac{v}{r}\right) - 3.125\xi p^2 \sigma_0^{-1}\right]$$
$$+ \frac{2}{r}\left[\sigma_0(1 - 2.5\xi) + 2\mu\left(\frac{\partial v}{\partial r} - \frac{v}{r}\right) - 3.125\xi p^2 \sigma_0^{-1}\right]. \tag{2.38}$$

We assumed that the volume content of the pores in the vicinity of the examined pore depends only on time; consequently, the continuity Eqs. (2.31) is written approximately as follows:

Dynamic Destruction of Some Materials

$$-\frac{\partial \xi}{\partial t} + \frac{\partial v}{\partial r} + \frac{2v}{r} = 0. \tag{2.39}$$

We ignored here the term $\xi\left(\frac{\partial v}{\partial r} + \frac{2v}{r}\right)$. The solution of this equation that satisfies the kinematic condition (2.33) has the form:

$$v = r^{-2}\left[R^2\dot{R} + \tfrac{1}{3}(r^3 - R^3)\dot{\xi}\right]. \tag{2.40}$$

Now we can rewrite the equation of the motion (2.38) using (2.40):

$$\rho_0(1-\xi)(1-R^3r^{-3})\left(rR^2\ddot{R} + 2Rr^{-2}\dot{R}^2 + \tfrac{1}{3}r\ddot{\xi} - \tfrac{4}{3}R^2r^{-2}\dot{R}\dot{\xi}\right)$$
$$= -\partial p/\partial r - 2.08\xi\sigma_0^{-1}\left(\partial p^2/\partial r\right) - 6.25\xi r^{-1}\sigma_0^{-1}p^2 + 2\sigma_0 r^{-1}(1-2.5\xi). \tag{2.41}$$

This equation is integrated from $R(r = R(t))$ to $b(R = b)$, where b is the radius of the sphere and $b \gg R$. At the same time, b is many times smaller than the radius of the zone of plasticity around the pore. As a result, we have

$$(1+\xi)[p(b) - p(R)] = -\rho_0\left[\left(R^2\ddot{R} + 2R\dot{R}^2 = \tfrac{1}{3}R^3\ddot{\xi} - \tfrac{4}{3}R^2\dot{R}\dot{\xi}\right)\left(R^{-1} - b^{-1}\right)\right.$$
$$+ \frac{1}{2}R^4\dot{R}\left(\tfrac{2}{3}R\dot{\xi} - \dot{R}\right)\left(R^{-4} - b^{-4}\right) - \tfrac{1}{6}\ddot{\xi}\left(R^2 - b^2\right)\right] + 2\sigma_0(1 - 1.5\xi)\ln(b/R)$$
$$- 2.08\xi\sigma_0^{-1}\left(\partial p^2/\partial r\right)(R-b) + 6.25\xi\sigma_0^{-1}\int_R^b p^2 r^{-1}\,dr. \tag{2.42}$$

We have obtained the equation connecting the pore radius R and the porosity of material ξ. The latter is determined by the distribution of pores located around the pore under consideration. In addition to the indicated unknown quantities, Eq. (2.42) includes the quantity $p(R)$. In accordance with Eqs. (2.34) and (2.37),

$$P(R) = \tfrac{2}{3}\sigma_0 - 4\mu R^{-1}\dot{R}. \tag{2.43}$$

Now Eq. (2.42) is rewritten in the form:

$$R\ddot{R} + \tfrac{3}{2}\dot{R}^2 - \tfrac{1}{3}R^2\ddot{\xi} - \tfrac{4}{3}R\dot{R}\dot{\xi} + \tfrac{1}{3}RR\dot{\xi} + \tfrac{1}{6}b^2\ddot{\xi} = -(1+\xi)\rho_0^{-1}\left[p(b) - \tfrac{2}{3}\sigma_0 + 4\mu R^{-1}\dot{R}\right]$$
$$+ \rho_0^{-1}\sigma_0^{-1}\left[2\sigma_0^2(1-1.5\xi)\ln(b/R) + 6.25\xi\int_R^b p^2 r^{-1}dr + 2.08b\xi\left(\partial p^2/\partial r\right)\right]. \tag{2.44}$$

Let us retain only the most important quantities in Eq. (2.44). Then, we have (2.14)

$$R\ddot{R} + \tfrac{3}{2}\dot{R}^2 + 4\mu\rho_0^{-1}R^{-1}\dot{R} + \rho_0^{-1}\left[p + 2\sigma_0\ln(b/R)\right] = 0. \qquad (2.45)$$

Coalescence of pores. We considered qualitatively the results of fusion of pores [8] using (2.45). A chain of pores are considered as simple model that allows us to study this phenomenon. Interaction of only two pores within the chain is taken into account at each moment of time. First (stage 1), the largest pore i of the chain is chosen and its interaction with the largest next pore ($i + 1$ or $i - 1$) is taken into account. Let the largest from the two is $i + 1$. As a condition of fusion of them, the following condition was applied:

$$k_R(R_i + R_{i+1})/\Delta = 1. \qquad (2.46)$$

Here, k_R is a certain constant, and Δ is a distance between centers of the pores. The radius R of the new pore is found assuming that its volume is $V = V_i + V_{i+1}$. The speed of $\dot{R}$ was found using condition of the conservation momentum. As a result of this stage, we find the new largest pore R in the chain. At the next stage, we repeat the process using new pore R and its neighbors R_{i-1} or R_{i+2} (we chose the largest from R_{i-1} and R_{i+2}). As a result, we have new volume for pore R. This volume depends on all the volumes considered for above pairs of pores. Similar calculation for many stages allows us to simulate growth and coalescence of pores. It was found that the merger, after beginning, continues avalanche-like. As a result, a single pore is formed by absorbing all pores existing before.

It was found that at the initial stage of destruction (tens of nanoseconds), the effect of inertia on pore growth is important. However, after some time from the starting of the interaction of the pores, the speed of their growth reduces. This phenomenon takes effect when the average pore size is several times smaller than Δ in (2.46). At the next stage of the dynamic destruction, when radiuses of the pores approach a distance Δ, the process of merger obtains aggravated nature, leading to the formation of macro-crack.

2.4 MATHEMATICAL MODEL OF MATERIALS CONTAINING PORES

We presented theoretical models for calculations of the change in pore sizes and their volume content in the material. Of course, the dynamic characteristic of damage should be included in certain state equations (1.10). Let us correct the equations of Section (1.1) taking into account the possibility of fracture of media when it is in solid state.

It is assumed that pores are equally distributed in all directions of material. In this case, damaged materials can be often described within the framework of well-known classical models. The space-averaged values of strains, stresses, displacements, pressure, and density are introduced to simplify the mathematical model. The averaging takes place over the volume element, which can contain much pores, but has a small size with respect to the characteristic length of the extreme wave. It is assumed that the number of pores per unit volume does not change. The pores do

Dynamic Destruction of Some Materials

not move within the network of the material. In this case, the damaged material can be approximately considered as a continuum, and its motion may be described by equations of continuum mechanics (1.1)–(1.7) [14,38,39].

Now it is necessary to take into account the change in properties of material in the corresponding equations of state because of the porosity change within an extreme wave. This is a very difficult task. The situation resembles the appearance of plastic deformations. Going from the elastic material model to the plastic one, we changed Hooke's law (1.26) to the form taking into account the accumulation of plastic deformation (1.34)!

In particular, the state equations for solid are usually known for undamaged materials (1.10):

$$\begin{aligned} s_{ij} &= s_{ij}\left(\tilde{e}_{\alpha\beta}, g^{\alpha\beta}, \tilde{\rho}, T, x_1, x_2, \ldots x_n\right), \\ \overline{W} &= \overline{W}\left(\tilde{e}_{\alpha\beta}, g^{\alpha\beta}, \tilde{\rho}, T, x_1, x_2, \ldots x_n\right), \\ p &= p_{\tilde{\rho}}(\tilde{\rho}) + p_T(\tilde{\rho}, T) + p_e(\tilde{\rho}, T), \\ E &= E_{\tilde{\rho}}(\tilde{\rho}) + E_T(\tilde{\rho}, T) + E_e(\tilde{\rho}, T) + E_i(\tilde{\rho}, T). \end{aligned} \quad (2.47)$$

where $\tilde{\rho}$ and $\tilde{e}_{\alpha\beta}$ are the values that are determined for the undamaged material. However, we have from calculations the density and components of the strain rate tensor for the damaged material. Therefore, Relations (2.47) should be rewritten using some damaged values since we use Eqs. (1.1)–(1.7), which are written for the damaged values. Indeed, we usually use in this book the averaged values and the continuum hypothesis [14,38,39].

We assume the smallness of the surface energy of the pores; therefore, the specific internal energy of the porous material is equal to the energy of the solid component substance. In addition, we consider the destruction processes to be independent from heat dynamics [65].

To form full system of equations, we must determine the equations that determine the growth of damage. The effect of ξ on the material density can be determined as

$$\tilde{\rho} = \rho(1-\xi)^{-1}. \quad (2.48)$$

To describe the damage evolution, we will use the particle case of (2.4), which is based on Akhmadeev–Nigmatulin model [65] (see, also, (2.7)). According to the model,

$$\frac{\partial \xi}{\partial t} = \frac{1}{t_k}\left[\exp\left(\frac{\sigma - \sigma_t}{\sigma_k}\right) - 1\right], \quad \sigma < \sigma_t, \quad \xi < \xi^*, \quad (2.49)$$

where ξ^* is the critical level of microdamages; and t_k, σ_k, and σ_t are certain material constants. For noted above case, Eq. (2.47) is rewritten

$$s_{ij} = s_{ij}\left(e_{\alpha\beta}, g^{\alpha\beta}, \rho, T, x_1, x_2, \ldots x_n\right)\left[1 - f(\xi)\right],$$

$$\overline{W} = \overline{W}\left(e_{\alpha\beta}, g^{\alpha\beta}, \rho, T, x_1, x_2, \ldots x_n\right),$$

$$p = p_\rho(\rho)\left[1 - f(\xi)\right] + p_T(\rho, T) + p_e(\rho, T), \quad (2.50)$$

$$E = E_\rho(\rho) + E_T(\rho, T) + E_e(\rho, T) + E_i(\rho, T).$$

where $f(\xi)$ is a function taking into account an influence of ξ. The function $f(\xi)$ is also determined by the noted model:

$$f(\xi) = \sqrt{\xi/\xi^*}. \quad (2.51)$$

On the whole, the growth of the pores (voids, bubbles) leads to an increased complexity of mathematical formulations of considering problems. Therefore, a more lucid understanding of the impact on the behavior of constructive elements of dynamic destruction and plastic deformation materials can be achieved by numerical experiments using different approaches to describing the process of destruction. For example, it is possible to calculate the destruction according to the instantaneous destruction model (2.1), by the kinetic equation (2.4) or the equations for spherical voids. In the last case, we have

$$\xi = \tfrac{4}{3}\pi U^{-1} \sum_{i=1}^{n} R_n^3, \quad (2.52)$$

where $\tfrac{4}{3}\pi \sum_{i=1}^{n} R_n^3$ is the total volume of n voids in the microvolume U.

The noted above shows difficulties in studying extreme waves in destructible media. The study may be realized by numerical methods using step-by-step algorithms.

The growth of pores affects only the state equations for s_{ij} and p. For the calculation of s_{ij} and p, as in the case of an elastoplastic body, it is necessary to break the loading time into several stages. At each stage, there are increments:

$$\Delta e_{\alpha\beta} = \Delta e_{\alpha\beta}^e f_1(\xi) + \Delta e_{\alpha\beta}^p f_2(\xi), \quad (2.53)$$

where $\Delta e_{\alpha\beta}^e$ and $\Delta e_{\alpha\beta}^p$ are the increments of elastic and plastic deformations during the step from moment $t = t_m$ to $t = t_{m+1}$. The functions $f_1(\xi)$ and $f_2(\xi)$ determine the influence of pores. In general case, these functions are different. Apparently, $f_1(\xi)$ is more influenced by the incipient pores, but $f_2(\xi)$ is more influenced by their growth. However, here, for simplicity, we will neglect the generation of pores and take these functions equal. Then, (2.53) yields

$$\Delta e_{\alpha\beta} = \left(\Delta e_{\alpha\beta}^e + \Delta e_{\alpha\beta}^p\right) f(\xi). \quad (2.54)$$

It is assumed that the function $f(\xi)$ is known (2.51). Expressions for $\Delta e_{\alpha\beta}^{e}$ и $\Delta e_{\alpha\beta}^{p}$ can be found at each step of loading according to the theories presented in Chapter 1 (Section 1.2).

Remark. We presented in this chapter the results that were obtained during 1950–1980 years of the last century. In general, the presented dependencies correspond to the theory of "continuum damage mechanics" [38,39,14]. According to this theory, if the effect of plasticity is small, there is a well-established hypothesis of strain equivalence. This hypothesis was presented by Lemaitre and Chaboche as: "The strain associated with a damaged state under the applied stress is equivalent to the strain with its undamaged state under the effective stress" [41].

We have considered several models describing the dynamic destruction of materials. Similar models will be used in this book for studying the propagation of extreme waves in various liquid and solid fracturable media.

3 Models of Dynamic Failure of Weakly Cohesived Media (WCM)

We have considered some models of the destruction of solids in Chapter 2. On the other hand, certain mechanism of liquid fracture was considered in Section 1.4. Some of the results presented above are applicable to weakly cohesived media (WCM) such as mixtures of gas (gas phase), and solid and liquid particles (condensed phase).

In particular, the model of an instant destruction (2.1) is widely used when the effect of the destruction of liquid media on the dynamic behavior of engineering structures is considered. On the other hand, liquid and WCM have the capacity to withstand tensile stresses. This ability can manifest itself under the action of short-time tensile stresses [68–70]. In the latter case, the strength of such mixtures can be estimated approximately according to (2.2).

An enormous literature is devoted to the investigation of dynamic properties of these mixtures. We do not have the opportunity to review even the main results obtained in this field. Therefore, here we will outline only some results that enable the reader to get acquainted with the problem. For us, these approaches are especially important as they will be used in the relevant parts of this book.

The principal differences between approaches developed below and considered in Chapter 2 are that here we consider weakly-cohesived media (the initially strongly damaged media) and we do not take into account the plastic properties of the mixture [71–80].

Before considering mathematical models, we give examples of considering media and their behavior in rarefaction waves and during vibrations.

3.1 INTRODUCTION

Granular materials, weakly cohesive deposits, and benthonic dirt can be considered as some analogues of bubbly liquids. However, in rest, these deposits resemble solid materials since they can resist tension action [33,75,80] (Figure 3.1(a)).

But if you shake this material vigorously, forcing particles move relative to each other, and the properties of the material change from "solid" to "gas" states. In particular, standard experiments with the granulated materials show an appearance of bubbles of gas in them under vibrations [33,76,80] (Figure 3.1(b)).

It is possible to talk about "melting" of WCM, and the degree of shaking will be related to an "effective temperature" (fluctuations of particle velocity or fluctuations of particle acceleration) of the mixture. In this state, the WCM are not in equilibrium. It needs a constant input of vibrational energy to retain its new state.

52 Extreme Waves and Shock-Excited Processes

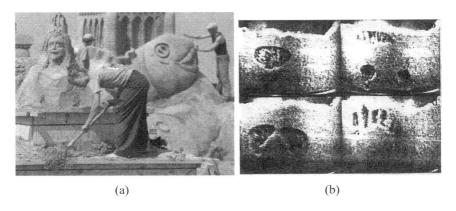

(a) (b)

FIGURE 3.1 Sand can be sculpted into solid forms (a) [33]. In material consisting of small-diameter grains that are intensively shaken, bubbling of air voids can occur (b) [33,76,80].

If the amplitude of the vibrations is reduced, then the mixture (granular liquid) should reach a point, at which it "freezes" into a solid [33].

Another example of the described process is illustrated in Figure 3.2. A tube containing gas, solid particles, and a piston is represented, where the gas gets strongly compressed by the piston. Three positions of the piston are shown. When the piston lifts very quickly, the expanding gas moves the particles into new positions, forming new skeletons having a lesser and lesser strength. Similar phenomena can take place within extreme rarefaction waves. In particular, this sort of the spalling can take place during a reflection of a seismic wave of compression from a free surface of a sediment layer. The WCM can transform into their gas-like state in which the material volume increases strongly. As a result, the uplift of a land surface can occur.

FIGURE 3.2 Scheme of the uplift of initially compressed granular material during sharp decompression [33,80].

Dynamic Failure of WCM

Apparently, this uplift of the land surface cannot take place during a long time, and the earthquake-induced vertical acceleration should be sufficiently large, greater than g.

3.1.1 EXAMPLES OF GASSY MATERIAL PROPERTIES

The presence of gas bubbles can have a significant effect on the behavior of many natural materials. A quarter of the land surface of the northern hemisphere contains the permafrost and about 100 billion tonnes of carbon. The properties of the permafrost strongly depend on the temperature. An increase in the carbon content yields carbon dioxide or methane. Methane is bubbling up out and transforms the land surface into the gassy soil (see Figure 3.3(a)). Another example of natural gassy material is offshore sediments frequently containing bubbles of gas, which is normally methane produced by biothermal process and another process.

The behavior of the weakly cohesive geomaterials depends on their history. In particular, the properties of the materials before and after the earthquake may be quite different. Due to strong earthquake-induced vibrations, there are multiple displacements, cracks, tensional features between oscillating material blocks, and tensional features between the surface deposit and the solid base. The material is loosened and transforms into liquefied (or gas like) state since the strength and stiffness of the material drop to zero.

The paper [78] explicitly describes these phenomena. A change in the characteristics of the soil of "Treasure Island" during the 1989 Loma Prieta earthquake was described. The initial speed of a shear wave was equal about 160 m/s. After the earthquake, it was less than 10 m/s (Figure 3.3(b)). Figure 3.4 shows the change in soil properties caused by earthquake. The variation in stress–strain diagram within an interval between 6 and 30 seconds after the beginning of the strong oscillations is shown in this figure.

One can see progressive reductions in the shear modulus during the earthquake. The measured response of soil of Treasure Island may be considered as a reasonable

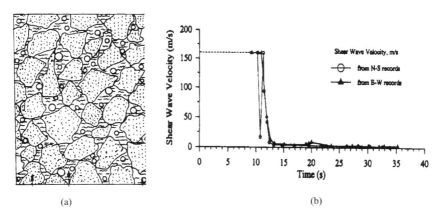

FIGURE 3.3 Schematic presentation of gassy soil (a). Shear-wave velocity history of liquefied deposit at Treasure Island (b) [33,78].

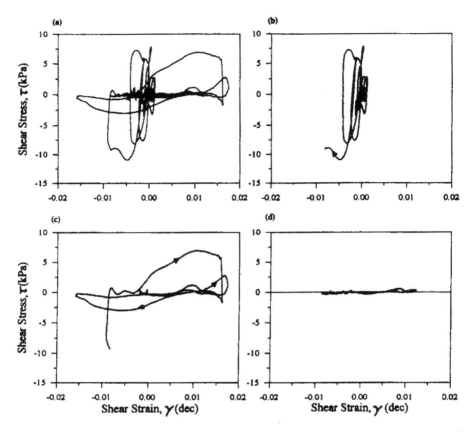

FIGURE 3.4 East–west shear stress–strain history of liquefied deposit at Treasure Island: (a) Full record; (b) 8–14 seconds; (c) 14–22 seconds; (d) 22–32 seconds [33,78].

analogue of the earthquake-induced change in properties of different deposits. The similar plots were recorded for soil of a site the Wildlife [78]. There, the soil density is higher than that of sand of Treasure Island. This difference in the densities is reflected in Table 1 [78]. The peak residual moduli show a reduction 15–35 times of the initial modulus, and the low-strain residual modulus shows a reduction 75–500 times of the initial modulus.

The presented data are typical for many deposits. The particles of the surface layer can begin to lose their elastic properties if the vertical acceleration is large enough.

3.1.2 Behavior of Weakly Cohesive Geomaterials within Extreme Waves

Our interest is behavior of geomaterials in case of the extreme waves, in particular, arising at earthquakes. The behaviors draw the general attention, and at the same time, it was quite poorly studied. It is connected with difficulty in their theoretical and experimental modeling. As we already emphasized, natural media can change

Dynamic Failure of WCM   55

the mechanical properties during earthquakes. In particular, strong earthquakes can significantly change the dynamic properties of water in oceans and magma in volcanoes. Since the mathematical description of the change in these properties is an aim of this chapter, we will yield some results from observations and experimental researches of behavior of WCM during weak and strong vibrations.

At first, let us consider seaquakes to illustrate certain results of this behavior using the results of Natanzon's experiments (Figures 1.4 and 3.5) [28,33,77]. Natanzon studied unexpected cavitation behavior of a water column excited by a piston. We assume that the seabed oscillations during underwater earthquakes may be modeled as the oscillating piston in Figure 3.5(a), where the upper surface of the column was free. Thus, we consider Natanzon's experiments as the simplest scheme of a seaquake. We will use Natanzon's data to understand qualitatively a large-scale phenomenon, which can occur during severe seaquakes.

Figure 3.5(a) sketches the experimental apparatus which Natanzon used. Experiments were conducted using a vertical metal tube having an inner diameter of 240 mm and a thickness of 2.5 mm. The piston produced vertical oscillations of the water. The frequency of harmonic vibrations of the piston could be smoothly changed from 0.5 to 50 Hz, and the amplitude l from 0 to 15 mm. The pressure on the free surface of the water column was atmospheric pressure, and the speed of sound in the tube was 750 m/s. The length of the water column varied from 4 to 7.5 m. Therefore, the fundamental frequency of the water column varied from 300 to 160 Hz approximately. During the experiments, the piston displacement and the pressure near the piston were measured. Experiments showed that large-amplitude steep-front waves (shock waves) can be generated in the tube as a result of the piston–water collision when the piston acceleration exceeded g.

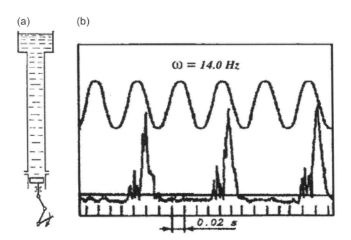

FIGURE 3.5 Sketch of Natanzon's experimental apparatus (a). Strongly nonlinear oscillations of the water column (the exciting amplitude is $l = 0.004$ m, the column length is 6 m, and the forcing frequency is 14 Hz). The flight time is so large that subharmonic oscillations of the water column are excited (b) [28,33,77].

In Figure 3.5(b), the oscillograms of the water pressure are presented, and the sinusoidal curves show the piston position. The smooth parts of the oscillograms may be explained by the periodical appearances of the gap (bubbly or cavitation zone) in the water near the piston. The distance between the peaks determines the free flight time of the column. The flight time of the water column is determined by the amplitude and the forcing frequency. The pressure peaks are the result of collapse of the gap and the column–piston collision.

According to simple calculations, if $l = 0.002$ m, then a cavitation zone forms when the angular frequency of the piston is 11 Hz. If $l = 0.0011$ m, then it forms when the angular frequency of the piston is 15 Hz. In both cases, the piston acceleration slightly exceeds g. As a result of periodic collisions of the water column by the piston, shock-like waves were excited near the piston rather than sinusoidal acoustic waves (Figure 1.4) [28].

Another interesting example of the formation of bubbles (cavitation zones) in the vertical tubes was presented in [79]. The mechanics of forming bubbles was investigated by using glass beads lying on a vertical vibrating base of the tube. The aim of the experiments was to determine whether bubbling is possible in the granular beds that are subjected to vertical vibrations. Figure 3.6 (a) shows an image of the bed in which the first bubble is created near the base, the second bubble is located in the middle of the bed, and the third bubble is approaching the free surface of the system. As the vertical acceleration increases, the bubbles become larger as shown in Figure 3.6(b). Bubbles close to each other may coalesce and form a larger bubble, and as a result, the bed grows in height. A further increase in the acceleration results in elongated bubbles, and the bed grows further in height.

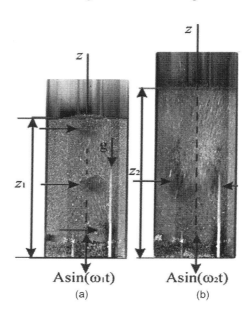

FIGURE 3.6 Results of experiments with vertically excited WCM. Bubbles are produced in granular materials [79]. Number and dimension of the bubbles depend on frequency ω and amplitude A of the vertical excitation.

Dynamic Failure of WCM

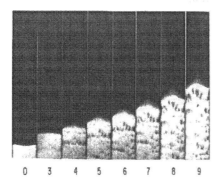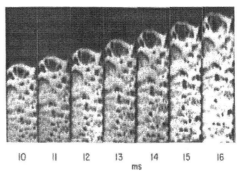

FIGURE 3.7 Wave phenomena during fast depressurization of a high-pressure pipe containing gas and granulated material. Expansion and rising of the surface of the layer of 0.25 mm glass balls. The initial layer thickness is 17 cm [33,80].

We think that Figure 3.6 shows the bubbling that takes place in Natanzon's experiments. On the other hand, Figure 3.6 illustrates the processes near the sea bottom and in the bottom mud which accompanied the seaquake. The waves, as shown in Figures 3.5 and 3.6, can be formed in different surface layers during great earthquakes and seaquakes.

In contrast with Figures 3.5 and 3.6, the waves and bubbles presented in Figure 3.7 can occur in volcanic conduit magma during earthquakes.

A high-pressure pipe was used in [80]. The pipe models a conduit of a volcano and contains gas and solid particles (Figure 3.7). When the pipe is rapidly opened, an expansion wave is formed, which moves downward. Within this wave, the material expands like magma at the top of the conduit before a strong eruption. Figure 3.7 shows the formation of big bubbles within the expansion wave. Similar fast growth of bubbles in magma qualitatively explains the explosive eruptions of volcanoes.

For this book, it is important that the results presented in Figures 3.6 and 3.7 can be interpreted as the destruction or consequences of the destruction of the initial media under the action of vibrations or impulse load. However, in contrast to the classical destruction (see Chapter 2), the media considered in the experiments restore their initial state after the termination of loading.

For us, it is important that in the media under consideration, there is initially a certain amount of gas. This quantity determines the nonlinear properties of these media. Therefore, the idea arises that their behavior can be described on the basis of a unified mathematical model in which the gas component of the mixture plays an important role.

3.2 MODELING OF GASSY MEDIA

In this book, we already described the phenomena arising in rarefaction waves, using data of experiments and simple mathematical models. These phenomena are now well studied. Fundamental books devoted to the transient cavitation, rarefaction waves, and boiling of different liquids were published. Probably, similar problems were most carefully studied by Nigmatulin [27]. However, existence of fundamental results and equations does not exclude the development of simplified approaches based on analogies, one-dimensional models, and using simplest relations of continuous media. They allow us to qualitatively understand and even analyze various complex processes. In particular, it is possible to consider weakly cohesive deposits and benthonic dirt as some analogues of bubbly liquids. Though, in rest, these deposits resemble solid materials since they can resist the tension action (Figure 3.1).

A mixture of liquid (typically water) and gas (typically air or methane) occupies often pore spaces of many soils (gassy soils). Solid particles of the gassy soil form, though, a skeleton, and they can be considered as a continuum (the dense phase). Because of the high compressibility, the mechanical properties of gassy soils resemble more the properties of bubbly liquids or gas, than those of a solid body. Therefore, we should take into account the high compressibility of the gassy soils. The compressibility usually determines the equation of state. This equation describes an interaction between pressure p and density ρ (or p, ρ, and temperature). The high nonlinearity determines this equation. In this chapter, we will derive the state equations for grain–water–gas mixture and the wave equations for corresponding nonlinear waves.

It is well known that the motion of matter containing grains, gas, and water is described by the equations of the mechanics of multiphase media [1,2,27]. These equations contain terms that describe the transformation of one phase into another and the exchange of momentum and energy between them. The exact expressions for these terms are, usually, unknown, and, therefore, the behavior of multiphase materials is often described within the framework of well-known classical models [3,27]. Therefore, to simplify an analysis of motion of weakly cohesive deposits, we introduce the space-averaged values: strains, stresses, displacements, pressure, and density. The averaging takes place over a volume element of the mixture, which contains much gas, but has a small size with respect to the characteristic length of the seismic wave. It is assumed that the number of gas inclusions per unit volume does not change. The gas does not move within the contact network of grains of the material. In this case, the material (mixture) can be approximately considered as a continuum, and its motion may be described by equations of continuum mechanics. The continuum should take into account the most important property of WCMs. Within the depression wave, the properties of WCMs resemble the properties of gas. However, within the compression zones, the properties of WCM resemble the properties of solid materials.

3.2.1 State Equation for Condensed Matter–Gas Mixture

Gassy media are considered as a homogeneous mixture of the condensed phase (solid particles, soft particles, and water) and gas. The composite bulk modulus λ of the condensed phase (compressibility the condensed phase) is obtained from

$$\lambda^{-1} = K_s^{-1}\phi_s + K_m^{-1}\phi_m + K_l^{-1}\phi_l, \qquad (3.1)$$

where K_s, K_m, and K_l are the bulk moduli of the solid and soft components and the liquid, respectively. Values ϕ_s, ϕ_m, and ϕ_l are the fractions of unit volume of mixture

Dynamic Failure of WCM

occupied by the solid and soft components and the liquid, respectively. Equation (3.1) is known as Wood's equation [29–33].

An undisturbed density ρ_c of the condensed phase of the deposit is given by

$$\rho_{0c} = \rho_s \phi_s + \rho_m \phi_m + \rho_l \phi_l, \qquad (3.2)$$

where ρ_s, ρ_m, and ρ_l are the densities of the solid particles, the mild (soft) particles, and the liquid, respectively.

Then, we write the equation of state for the condensed phase:

$$\rho_c = \rho_{0c}\left[1 + \lambda(p_c - p_{0c})\right], \qquad (3.3)$$

where ρ_c is the averaged density of the condensed phase and ρ_{0c} is the corresponding undisturbed value; p_c and p_{0c} are the pressure and its undisturbed value, correspondingly. The equation of state for the gas is (1.55)

$$p_g = p_{0g}\left(\frac{\rho_g}{\rho_{0g}}\right)^\gamma = p_{0g}\left(\frac{\phi_0}{\phi}\right)^\gamma, \qquad (3.4)$$

where ρ_g is the density of the gas, ρ_{0g} is the corresponding undisturbed value, and γ is the adiabatic exponent of the gas. In the case of isothermal behavior of the gas, $\gamma = 1$. Value ϕ is the gas volume (porosity). It is the fraction of unit volume of mixture occupied by the gas, and ϕ_0 is the undisturbed gas value (undisturbed porosity).

The undisturbed density ρ_0 of the mixture (gassy media) is given by

$$\rho_0 = mU^{-1} = \rho_{0c}(1 - \phi_0) + \rho_{0g}\phi_0, \qquad (3.5)$$

where m is the mass of a small element of the mixture, and U is the volume of it. Now, we note that the volume U is the sum of the gas volume and the condensed phase volume:

$$U = m\rho^{-1} = (U - \phi_0 U)\left[1 - \lambda(p_c - p_{0c})\right] + \phi U, \qquad (3.6)$$

where ρ is the average density of the mixture (see, also, Section 1.4). Using (3.5), we find the equation of state in the form:

$$\rho = \rho_0 \left\{ (1-\phi_0)\left[1 - \lambda(p_c - p_{0c})\right] + \phi_0 \left(\frac{p_{0g}}{p_g}\right)^{1/\gamma} \right\}^{-1}. \qquad (3.7)$$

(see Section 1.4 and (1.63) too). Equation (3.7) is rewritten approximately as:

$$\rho = \rho_0 \left[(1-\phi_0)\left(\frac{p_{0c} + B}{p_c + B}\right)^{1/k} + \phi_0\left(\frac{p_{0g}}{p_g}\right)^{1/\gamma} \right]^{-1}, \qquad (3.8)$$

where $k = 1/\lambda(p_{0c} + B)$ and B is a constant. It is always possible to choose k and B corresponding to some known λ. For example, for normal state of water, $k = 7.15$ and $B = 304.5$ MPa, $\lambda = 1/2100$ MPa (Section 1.3).

Let the pressure in the gas equal the pressure in the condensed phase, $p_c = p_g = p$. In this case, Eq. (3.8) yields

$$\rho = \rho_0 \left[(1-\phi_0)\left(\frac{p_0 + B}{p + B}\right)^{1/k} + \phi_0 \left(\frac{p_0}{p}\right)^{1/\gamma} \right]^{-1}. \tag{3.9}$$

Differentiating (3.9) with respect to p, we find the sound speed in the mixture:

$$c = c(p) = \sqrt{\frac{dp}{d\rho}} = \rho_0^{-0.5} \left[(1-\phi_0)\left(\frac{p_0 + B}{p + B}\right)^{1/k} + \phi_0 \left(\frac{p_0}{p}\right)^{1/\gamma} \right]$$

$$\times \left[\frac{1-\phi_0}{k(p+B)}\left(\frac{p_0 + B}{p + B}\right)^{1/k} + \frac{\phi_0}{\gamma p}\left(\frac{p_0}{p}\right)^{1/\gamma} \right]^{-0.5}. \tag{3.10}$$

Let us consider the undisturbed mixture. In this case using (3.5), we obtain

$$c_0 = \left[(1-\phi_0)^2 c_{0c}^{-2} + \phi_0^2 c_{0g}^{-2} + \phi_0(1-\phi_0)\rho_{0g}^{-1}\rho_{0c}^{-1}\left(\rho_{0c}^2 c_{0g}^{-2} + \rho_{0g}^2 c_{0c}^{-2}\right) \right]^{-0.5}, \tag{3.11}$$

where

$$c_{0c}^2 = k\rho_{0c}^{-1}(p_0 + B) \quad \text{and} \quad c_{0g}^2 = \gamma \rho_{0g}^{-1} p_0. \tag{3.12}$$

are the sound speeds for the condensed phase and the gas, respectively. We emphasize that the low-frequency mixture sound velocity c_0 coincides with Wood's result [29].

Remark. If the volumes of the solid and soft components are zero, then (3.11) and (3.12) determine the sound speed in bubbly liquids. In this case, Eq. (3.12) agrees with Eq. (2.3) in [30] if we identify ϕ_0 with α. Equations (3.12) transform into Eq. (2.4) (from [31]) in the appropriate limit. Finally, Eq. (3.11) can be approximated by

$$c_0^2 = \gamma p_0 \left[\rho_{0c} \phi_0 (1-\phi_0) \right]^{-1} \tag{3.13}$$

if ϕ_0 is far from either zero or unity and $\gamma = 1$. Let us note also if ϕ_0 is not very close to zero or unity, then c_0 is much less than c_{0l} or c_{0g} (3.12).

3.2.2 Strongly Nonlinear Model of the State Equation for Gassy Media

We have previously considered the model of highly nonlinear behavior of a bubble liquid in extreme waves (Section 1.4.3). Here, we generalize the results given in Section 1.4.3 for the case of gassy media.

Since we assume that $p_c = p_g = p$, Eq. (3.7) is rewritten in the form:

$$\rho_0/\rho = (1-\phi_0)[1-\lambda(p-p_0)] + \phi_0 (p_0/p)^{1/\gamma}. \tag{3.14}$$

The appearance of a high-negative pressure is, generally, impossible in the WCMs, according to (3.14). Let us show this analytically. The process of the gas oscillations in the mixture may be considered as isothermal ($\gamma = 1$). In this case, it follows from (3.14) that

$$p = 0.5\lambda^{-1}(1-\phi_0)^{-1}\{b - b[1+4\phi_0\lambda(1-\phi_0)b^{-2}p_0]^{0.5}\}, \tag{3.15}$$

where

$$b = (1-\phi_0)(1+\lambda p_0) - \rho_0 \rho^{-1}. \tag{3.16}$$

Two last equations show that the pressure in the mixture is always positive even when $\rho \to 0$. For example, let $\rho_0\rho^{-1} \gg (1-\phi_0)(1+\lambda p_0)$. Then, $p \approx -\phi_0 b^{-1} p_0 \approx \phi_0 \rho \rho_0^{-1} p_0$. If $\rho_0\rho^{-1} \approx \infty$, then $p \approx 0$.

Equation (3.15) approximately describes in a simple way the real properties of WCMs. The tensile strength of the deposit is approximately determined by the initial (hydrostatic) pressure p_0. The material is being fragmented if the pressure drops below p_0 within the rarefaction wave.

As example of the non-cohesive material, we consider water–gas mixture. Figure 3.8 shows the curves corresponding to a few equations of state for the discussed model of the mixture. Curves 3, 4, and 5 are calculated according to the equation of state (3.15) for $\phi_0 = 10^{-3}$, 10^{-4}, and 10^{-5}, respectively. It is seen that the nonlinearity of the mixture depends on the gas concentration.

Curve 1 corresponds to the classical equation of state for water: $\rho = \rho_0[(p_c+B)/(p_{0c}+B)]^{-1/k}$, $k = 7.15$ and $B = 304.5\,\text{MPa}$. However, this equation is not fair for rarefaction zones. Curve 2 corresponds to a wide-range equation of state for water (see Section 1.3). It is visible that this curve corrects a little unacceptable situation with the accuracy of the classical equation in rarefaction zones.

Only Eq. (3.15) truly describes the behavior of water in rarefaction zones. The nonlinearity increases when the gas concentration decreases. The properties of the mixture begin to resemble the properties of gas within rarefaction zones. In the last case, the main property of the bubbly liquid can be described by two segments (the so-called two-line model): one segment is almost perpendicular to density axis, and the other segment is almost parallel to this axis.

Thus, we have shown that even highly simplified forms of Eq. (3.9) (see Eqs. (3.14) and (3.15)) describe the real properties of water better than the Tait equation and some wide-range equations.

However, Eq. (3.9) is too complex in compare with the equation Tait. It is difficult to use it in compare with using the equation Tait. Therefore, it makes sense to rewrite (397) in the form of the Tait equation.

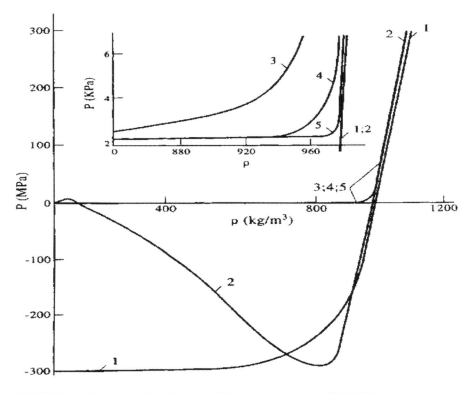

FIGURE 3.8 The curves that illustrate different laws $\rho = \rho(p)$ [33,81,82].

3.2.3 THE TAIT-LIKE FORM OF THE STATE EQUATION

It was shown in Section 1.4 that the influence of a gas on acoustic wave propagation is important. It agrees with well-known results of many investigations [83–86].

1. To simplify the analysis of deposit behavior, state equation (3.9) is presented in the form of the Tait equation. Following [33] and neglecting small (cubic nonlinear) terms, we can write Eq. (3.9) in the form:

$$\frac{p+y}{p_0+y} = \left(\frac{\rho}{\rho_0}\right)^X, \qquad (3.17)$$

where

$$y = \frac{\lambda(1-\phi_0)\gamma^2 p_0^2 + \phi_0 \gamma p_0}{(k+1)(1-\phi_0)\lambda^2\gamma^2 p_0^2 + \phi_0(\gamma+1) - [\lambda\gamma p_0(1-\phi_0) + \phi_0]^2} - p_0,$$

$$X = \left[\lambda(1-\phi_0) + \phi/\gamma p_0\right]^{-1}(y+p_0)^{-1}.$$

Dynamic Failure of WCM

Now, using (3.17), we can find the sound speed

$$c^2 = X(p+y)\rho^{-1}. \tag{3.18}$$

Our calculations have shown that Eq. (3.17) is a good approximation of Eq. (3.9) if $(p - p_0) \leq 0.4 p_0$. In particular, (3.18) is the good approximation of (3.10). Some results of the comparison are shown in Figures 3.9 and 3.10.

It is seen from Figure 3.9 that the pressure curves are quite similar to each other, especially for very small values of ϕ_0 and $\phi_0 \approx 1$.

Figure 3.10 shows the variation in the sound speed, calculated according to Eqs. (3.10) and (3.18) in the interval $0 \leq \phi_0 \leq 1$. Here, $nV_0 = \phi_0$. On the whole, the curves are similar to each other. We do note that $c = 1450$ m/s at $\phi_0 = 0$ for all of the curves of Figure 3.10. At the same time, the curves

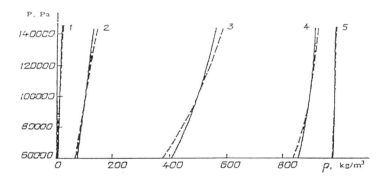

FIGURE 3.9 Curves $p = p(\rho)$ calculated for different values of ρ and for $p_0 = 0.1$ MPa. Dashed lines correspond to Eq. (3.9), while solid lines correspond to Eq. (3.17). Curves 1, 2, 3, 4, and 5 correspond to $\phi_0 = 0.99, 0.9, 0.5, 0.1,$ and 0.01, respectively [33,87].

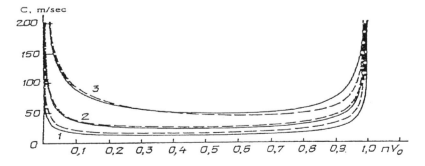

FIGURE 3.10 Curves show the change in the sound speed with increasing gas component. Curves are calculated according to Eqs. (3.10) (dashed lines) and (3.18) (solid lines) for different values of p and p_0: $p = 0.06$ MPa, $p_0 = 0.1$ MPa (curve 1); $p = 0.16$ MPa, $p_0 = 0.2$ MPa (curve 2); $p = 0.3$ MPa, $p_0 = 0.5$ MPa (curve 3). Here, $nV_0 = \phi_0$ [33,87].

in Figure 3.10 reach different values of c at $\phi_0 = 1$, because these values are determined by corresponding values of p and p_0.

2. Curves of Figure 3.11 illustrate the influence of gas on properties of water-based mud located on depths up to 5 km [85].

The sound velocity of the mud can be reduced strongly if the gas phase volume is slightly larger than 0 or slightly smaller than 1. It is interesting that the curves of Figure 3.11 also qualitatively describe the dependence of the sound velocity of the bubbly liquid from the gas component (Figure 3.12).

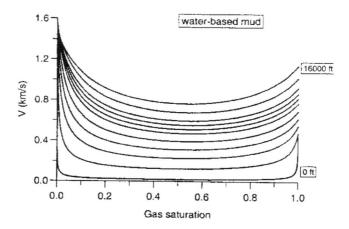

FIGURE 3.11 Sound velocity of water-based drilling mud versus gas saturation. The depth varies, and the frequency is 25 Hz [33,85].

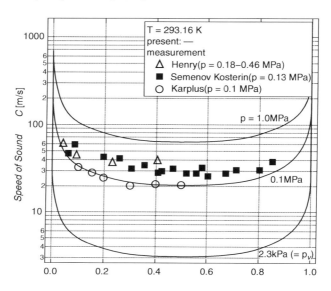

FIGURE 3.12 Sound speed of bubbly liquid under the isothermal condition [33,86].

Dynamic Failure of WCM

Therefore, we assume that many results and models of the theory of bubbly liquids are valid for modeling the weakly cohesive gassy materials.

Equation (3.17) opens the possibility of greatly simplifying the equations of gassy media by introducing the velocity potential.

3.2.4 Wave Equations for Gassy Materials

Here, we assume that the following classical equations of motion and continuity of hydrodynamics are fair for average parameters of weakly cohesive sediments:

$$\frac{d\bar{v}}{dt} = -\frac{1}{\rho}\nabla p, \qquad (3.19)$$

$$\frac{d\rho}{dt} = -\frac{1}{\rho}\mathrm{div}\,\bar{v}, \qquad (3.20)$$

where $\bar{v}$ is the velocity vector. Equations (3.19) and (3.20) are presented in the Eulerian coordinates. The equation of state is the Tait-type equation (3.17). Then, the velocity potential is introduced (see Section 1.3)

$$\bar{v} = \nabla\psi. \qquad (3.21)$$

Substituting (3.21) into Eq. (3.19) and eliminating the operator ∇, we derive

$$\frac{\partial\psi}{\partial t} + \tfrac{1}{2}(\nabla\psi)^2 + \int_{p_0}^{p}\rho^{-1}\,dp = f(t), \qquad (3.22)$$

where $f(t)$ is an arbitrary function. We will hereafter consider the fluid to be at rest at the infinity; using this, we find that $f(t) = 0$. In this case, the last term at the left-hand side of (3.22) is integrated using (3.17). Then, we find that

$$p = -y + (p_0 + y)\left\{1 - (X-1)\rho_0 X^{-1}(p_0 + X)^{-1}\left[\psi_t + \tfrac{1}{2}(\nabla\psi)^2\right]\right\}^{X/(X-1)}. \qquad (3.23)$$

The density is determined as

$$\rho = \rho_0\left\{1 - (X-1)\rho_0 X^{-1}(p_0 + y)^{-1}\left[\psi_t + \tfrac{1}{2}(\nabla\psi)^2\right]\right\}^{1/(X-1)}. \qquad (3.24)$$

The assumption of an irrotational motion has allowed us to determine the velocity, the pressure, and the density using a single arbitrary function ψ. We can obtain a strongly nonlinear equation for ψ by satisfying the continuity equation (3.20). Substituting (3.24) into (3.20), and using (3.21) and the equation for the sound speed $c^2 = dp/d\rho$, we find the equation:

$$c^2\nabla^2\psi = \psi_{tt} + \left[(\nabla\psi)^2\right]_t + (\nabla\psi)^2\nabla^2\psi, \qquad (3.25)$$

where ∇^2 is the Laplacian operator and

$$c^2 = c_0^2 - (X-1)\left[\psi_t + \tfrac{1}{2}(\nabla\psi)^2\right], \tag{3.26}$$

where $c_0^2 = X(y + p_0)\rho_0^{-1}$.

1. **Spherical waves**. In the case of pure spherical quadratically nonlinear waves, Eqs. (3.25) and (3.26) yield

$$\left[c_0^2 - (X-1)\psi_t\right]\left(\psi_{rr} + 2r^{-1}\psi_r\right) = \psi_{tt} + 2\psi_r\psi_{rt}, \tag{3.27}$$

   where r is the radial Eulerian coordinate.
2. **Longitudinal waves**. In the case of pure longitudinal quadratically nonlinear waves, Eqs. (3.25) and (3.26) yield

$$\left[c_0^2 - (X-1)\psi_t\right]\psi_{xx} = \psi_{tt} + 2\psi_x\psi_{xt}, \tag{3.28}$$

   where x is the Eulerian coordinate.

We have developed a simplified theory of wave processes in WCM based on the model of a continuous medium. It is emphasized again that the most important properties of the medium under consideration are determined by the compressibility of the gas. Below, we will pay special attention to the effects of the dynamic properties of the gas on the governing equations. This effect is modeled on the basis of effect of the gas bubble oscillations. We will show that there is a connection between the theory and the classical wave equations.

3.3 EFFECTS OF BUBBLE OSCILLATIONS ON THE ONE-DIMENSIONAL GOVERNING EQUATIONS

Here, we consider WCM under sharp decompression as a mixture of the condensed phase and the gas bubbles [88–90].

The equation of state for WCM, containing oscillating gas bubbles, is derived. Then, for this mixture, the governing nonlinear wave equation is written. It is shown that in the low-frequency limit, and for weakly nonlinear waves, the theory yields the Boussinesq equation. In the case of the high-frequency limit, the Klein–Gordon equation with quadratic nonlinearity is obtained.

3.3.1 DIFFERENTIAL FORM OF THE STATE EQUATION

We modify Eq. (3.7) taking into account oscillations of gas bubbles. A change in the bubble radius R can be defined by the Rayleigh–Plesset equation [91] with the surface tension:

$$\rho_{0c}\left(RR_{tt} + \tfrac{3}{2}R_t^2\right) + 4\mu_g R_t/R + 2\sigma/R = p_g - p_c, \tag{3.29}$$

Dynamic Failure of WCM

where μ_g is the coefficient of the gas viscosity. For the undisturbed mixture, Eq. (3.29) yields

$$p_{0g} = p_0 + 2\sigma R_0^{-1}. \tag{3.30}$$

Now, using (3.29) and (3.30), we can rewrite the equation of state (3.7) as

$$\rho_0 \rho^{-1} = (1-\phi_0)\left[1 - \lambda(p_c - p_{0c})\right] + \phi_0 \left[\frac{p_{0c} + 2\sigma R_0^{-1}}{p_{0c}\left(RR_{tt} + \frac{3}{2}R_t^2\right) 4\mu_g R_t R^{-1} + 2\sigma R^{-1} p}\right]^{1/\gamma}. \tag{3.31}$$

Let us consider the state equation (3.31) attentively. The equation takes into account an inertia of the bubble oscillations. Because of the inertia, a reduction in the density can continue, although an increase in the pressure begins. In other words, a bubble attains its equilibrium volume not instantly after loading, but because of the inertia and the nonlinearity of the bubble oscillations only after a certain period of relaxation. Therefore, the mechanical properties of the mixture, for example, the viscosity, can change with time.

Linear bubble oscillations. The main nonlinearity of the mixture is determined by its high compressibility. Therefore, we can ignore the nonlinearity of the bubble oscillations. For simplicity, we also assume that $\mu_g = 0$ and $\sigma = 0$. The bubble radius is the sum of R_0 and the disturbance R_*:

$$R = R_0 + R_*. \tag{3.32}$$

In this case, we can rewrite (3.31) in the form:

$$\rho_0 \rho^{-1} = (1-\phi_0)\left[1 - \lambda(p_c - p_{0c})\right] + \phi_0 \left(\frac{p_{0c}}{p_{0c} R_0 R_{tt} + p}\right)^{1/\gamma}. \tag{3.33}$$

One can see from the last expression that the elementary volume of the mixture can expand, owing to the inertia of bubbles, although an increase in the pressure begins. On the other hand, a considerable negative pressure can appear in the mixture, owing to the bubble inertia, if the incident decompression wave was very short. For example, the high-negative pressures can appear in liquid, which are not breaking its continuity if the negative pressures exist for very short time, and the initial volume of bubbles is extremely small. Certainly, an appearance of such situations in daily practice is almost impossible. Therefore, such possibility is completely ignored by the models of the "hull" cavitation considered in Chapter 4.

3.3.2 THE STRONGLY NONLINEAR WAVE EQUATION FOR BUBBLY MEDIA

Let $p_c = p$, $p_{0c} = p_0$, but $p_{0g} \neq p_0$. Following [29–33], we will use the averaged parameters of the mixture. A simple homogeneous model is used. In this case, the motion and continuity equations do not differ from the well-known equations of the viscous liquid:

$$\rho u_{tt}(1+u_a) = -p_a + \left(\eta_* + \tfrac{4}{3}\mu_*\right)u_{aat}, \tag{3.34}$$

and

$$\rho(1+u_a) = \rho_0, \tag{3.35}$$

where u is the averaged displacement of the mixture, η_* is the coefficient of second viscosity, and μ_* is the coefficient of internal (kinematic) viscosity. Further, we will use the value which takes into account the specified viscosities: $\mu_1 = \eta_* + \tfrac{4}{3}\mu_*$, μ_1 is the effective viscosity coefficient of the mixture.

Physical properties of the mixture are described by Eq. (3.7). This state equation is rewritten in the form:

$$\rho = \rho_0 \left\{ (1-\varphi_0)\left[1 - \lambda(p - p_0)\right] + \varphi \right\}^{-1}. \tag{3.36}$$

We will use also Eq. (3.29). At first, we introduce the displacement potential, $\Phi(a,t)$:

$$u = \Phi_a. \tag{3.37}$$

Then, (3.34) and (3.35) yield

$$p = p_0 - \rho_0 \Phi_{tt} + \mu_1 \Phi_{aat}. \tag{3.38}$$

Using (3.35) and (3.36), we get

$$\phi = 1 + \Phi_{aa} - (1-\phi_0)\left[(p+B)(p_0+B)^{-1}\right]^{-1/k}. \tag{3.39}$$

Substituting (3.38) into (3.39), we find that

$$\phi = 1 + \Phi_{aa} - (1-\phi_0)\left[1 - (p_0+B)^{-1}(\rho_0 \Phi_{tt} - \mu_1 \Phi_{taa})\right]^{-1/k}. \tag{3.40}$$

On the other hand, the expression for ϕ follows from Eq. (3.29). Taking into account (3.4), we rewrite this equation in the form:

$$\phi = \phi p_{0g}^{1/\gamma}\left[\rho_0\left(RR_{tt} + \tfrac{3}{2}R_t^2\right) + p + 4\mu R_t R^{-1} + 2\sigma R^{-1}\right]^{-1/\gamma}. \tag{3.41}$$

Following Section 1.4.2, we denote the volume occupied by the gas in a unit volume of the undisturbed mixture as nV_0 ($\phi_0 = nV_0$), where n is the number of similar bubbles in a unit volume of the mixture. In this case, if V is the volume of a spherical bubble of radius R, we have $\phi = nV$ and $\phi = \tfrac{4}{3}n\pi R^3$. Using (3.40), we obtain the expression for R in (3.41):

$$R = \left(\tfrac{3}{4}n^{-1}\pi^{-1}\right)^{1/3}\left\{1 + \Phi_{aa} - (1-\phi_0)\left[1 - (B+p_0)^{-1}(\rho_0\Phi_{tt} - \mu_1\Phi_{taa})\right]^{-1/k}\right\}^{1/3}. \tag{3.42}$$

Dynamic Failure of WCM   69

Thus, two expressions for ϕ, which are presented through the function Φ, are found. Expression (3.41) equals (3.40). Therefore, using (3.38), we derive the following equation:

$$\phi_0 p_{0g}^{1/\gamma} \left[\rho_{0c} \left(RR_{tt} + \tfrac{3}{2} R_t^2 \right) + p_0 - \rho_0 \Phi_{tt} + \mu_1 \Phi_{aat} + 4\mu_g R_t R^{-1} + 2\sigma R^{-1} \right]^{-1/\gamma}$$

$$= 1 + \Phi_{aa} - (1 - \phi_0) \left[1 - (p_0 + B)^{-1} (\rho_0 \Phi_{tt} - \mu_1 \Phi_{taa}) \right]^{-1/k}, \qquad (3.43)$$

where R is determined by (3.42). Thus, Eq. (3.43) is strongly nonlinear equation for considered model of the mixture.

Remark. At first glance, the effect of gas on the form of Eq. (3.43) may be very small. If the amplitude of the oscillations of the bubbles is small, then this influence is completely ignorable. But this is not the case since the average density of the mixture ρ_0 depends on the gas density in the bubbles (see Eq. (3.5)).

3.4 LINEAR ACOUSTICS OF BUBBLY MEDIA

The linear versions of (3.43) are considered. The linear terms in (3.43) and (3.42) yield

$$\phi_0 + \Phi_{aa} - \left[(1 - \phi_0) \lambda + \phi_0 p_{0g}^{1/\gamma} p_0^{-1/\gamma - 1} \gamma^{-1} \right] (\rho_0 \Phi_{tt} - \mu_1 \Phi_{aat})$$

$$= \phi_0 p_{0g}^{1/\gamma} p_0^{-1/\gamma} \left[1 - p_0^{-1} \gamma^{-1} \left(\rho_0 R_0 R_{tt} + 4\mu_g R_0^{-1} R_t + 2\sigma R^{-1} \right) \right]. \qquad (3.44)$$

We approximate R (3.42) as

$$R \approx R_0 \left[1 + \tfrac{1}{3} \varphi_0^{-1} \Phi_{aa} - \tfrac{1}{3} \lambda \left(\varphi_0^{-1} - 1 \right) \left(\rho_0 \Phi_{tt} - \mu_1 \Phi_{taa} \right) \right]. \qquad (3.45)$$

Substituting (3.45) into (3.44), we have after long but simple calculations

$$\left(1 - \tfrac{2}{3} \sigma \gamma^{-1} p_{0g}^{1/\gamma} p_0^{-1/\gamma - 1} R_0^{-1} \right) \left(\Phi_{aa} - c_0^{-2} \Phi_{tt} \right) - \tfrac{1}{3} (1 - \phi_0) \lambda \gamma^{-1} p_0 p_{0g}^{1/\gamma} p_0^{-1/\gamma - 1}$$

$$\times \left[4\mu_g \left(\Phi_{tt} - c_1^2 \Phi_{aa} \right)_t + 3\gamma p_0 \omega_0^{-2} \left(\Phi_{tt} - c_2^2 \Phi_{aa} \right)_{tt} \right]$$

$$+ (1 - \phi_0) \mu_1 \lambda \gamma^{-1} \omega_0^{-2} p_{0g}^{1/\gamma} p_0^{-1/\gamma} \Phi_{tttaa} = C_*, \qquad (3.46)$$

where ω_0^{-2} is the bubble resonance frequency

$$\omega_0 = \left(3\gamma p_0 \rho_{0c}^{-1} R_0^{-2} \right)^{0.5} \qquad (3.47)$$

and

$$C_* = -\phi_0 \left(1 - p_{0g}^{1/\gamma} p_0^{-1/\gamma} \right) - 2\sigma \phi_0 \gamma^{-1} p_{0g}^{1/\gamma} p_0^{-1/\gamma - 1} R_0^{-1}, \qquad (3.48)$$

$$c_0^{-2} = \left(1 - \tfrac{2}{3} p_{0g}^{1/\gamma} p_0^{-1/\gamma} \gamma^{-1} p_0^{-1} \sigma R_0^{-1}\right)^{-1} \left[(1-\phi_0)^2 c_{0c}^{-2} + \phi_0^2 p_{0g}^{1/\gamma} p_0^{-1/\gamma} c_{0g}^{-2} \right.$$
$$+ (1-\phi_0)\phi_0 \left(c_{0c}^{-2} \rho_{0c}^{-1} \rho_{0g} + p_{0g}^{1/\gamma} p_0^{-1/\gamma} c_{0g}^{-2} \rho_{0g}^{-1} \rho_{0c} \right)$$
$$\left. - \tfrac{2}{3}(1-\phi_0)\sigma\lambda\gamma^{-1} \rho_0 p_{0g}^{1/\gamma} p_0^{-1/\gamma-1} R_0^{-1} \right], \tag{3.49}$$

$$c_1^2 = \tfrac{3}{4} \mu_1 \mu_g^{-1} \rho_0^{-1} \left[p_{0g}^{-1/\gamma} p_0^{-1/\gamma+1} \gamma + \phi_0 (1-\phi_0)^{-1} \lambda^{-1} - \tfrac{2}{3}\sigma R_0^{-1} \right]$$
$$+ \lambda^{-1} \rho_0^{-1} (1-\phi_0)^{-1}, \tag{3.50}$$

$$c_2^2 = \rho_0^{-1} \left[(1-\phi_0)^{-1} \lambda^{-1} + \tfrac{4}{3} \mu_g \mu_1 \omega_0^2 p_0^{-1} \right]. \tag{3.51}$$

We derived the wave equation (3.46) with three wave speeds: "low-frequency" speed c_0 is defined by Expression (3.49), "high-frequency" speed c_2 is defined by Expression (3.51), and the speed c_1 (3.50) depends essentially on viscous properties of mixture and the compressibility of the condensed phase. We emphasize that the speeds c_1 and c_2 weakly depend on gas concentration in comparison with c_0.

Below, a few special cases of Eq. (3.46) are considered.

3.4.1 Three-Speed Wave Equations

1. **No surface tension.** In this case, $\sigma = 0$ and $p_{0g} = p_0$ (3.30). Since also $p_{0g}^{1/\gamma} = p_0^{-1/\gamma}$, Eq. (3.46) yields

$$\left(\Phi_{aa} - c_0^{-2}\Phi_{tt}\right) - \tfrac{1}{3}(1-\phi_0)\lambda\gamma^{-1}\rho_0 p_0^{-1} \left[4\mu_g \left(\Phi_{tt} - c_1^2 \Phi_{aa}\right)_t \right.$$
$$\left. + 3 p_0 \omega_0^{-2} \left(\Phi_{tt} - c_2^2 \Phi_{aa}\right)_{tt} \right] + (1-\phi_0)\mu_1 \lambda\gamma^{-1} \omega_0^{-2} \Phi_{tttaa} = 0. \tag{3.52}$$

Here

$$c_0^{-2} = \left[(1-\phi_0)^2 c_{0c}^{-2} + \phi_0^2 c_{0g}^{-2} + (1-\phi_0)\phi_0 \left(c_{0c}^{-2} \rho_{0c}^{-1} \rho_{0g} + c_{0g}^{-2} \rho_{0g}^{-1} \rho_{0c} \right) \right], \tag{3.53}$$

$$c_1^2 = \tfrac{3}{4} \mu_1 \mu_g^{-1} \rho_0^{-1} \left[p_0 \gamma + \varphi_0 (1-\varphi_0)^{-1} \lambda^{-1} \right] + \lambda^{-1} \rho_0^{-1} (1-\varphi_0)^{-1}. \tag{3.54}$$

Value c_2^2 is determined by (3.51). We emphasize that (3.53) coincides with Wood's results [29] (see, also, (3.11) and (1.78)).

1.1. Let additionally $\gamma = 1$. In this isothermal case, we should put $\gamma = 1$ in (3.52)–(3.54) and c_2^2.

Thus, the excitation of waves propagating with different velocities is possible according to the model of the bubble medium.

3.4.2 TWO-SPEED WAVE EQUATIONS

Let us consider separately the cases when waves having two sound velocities appear in the mixture.

1. **Very viscous gas in bubbles and an inviscid condensed phase.** In this case, we have

$$\left(1 - \tfrac{2}{3}\sigma\gamma^{-1} p_{0g}^{1/\gamma} p_0^{-1/\gamma-1} R_0^{-1}\right)\left(\Phi_{aa} - c_0^{-2}\Phi_{tt}\right) - \tfrac{1}{3}(1-\phi_0)\lambda\gamma^{-1}\rho_0 p_{0g}^{1/\gamma} p_0^{-1/\gamma-1}$$

$$\times \left[4\mu_g\left(\Phi_{tt} - c_1^2\Phi_{aa}\right)_t + 3p_0\omega_0^{-2}\left(\Phi_{tt} - c_2^2\Phi_{aa}\right)_{tt}\right] = C_*. \quad (3.55)$$

In (3.55), the values C_* and c_0^{-2} are determined by (3.48) and (3.49) correspondingly, and

$$c_1^2 = c_2^2 = \lambda^{-1}\rho_0^{-1}(1-\phi_0)^{-1}. \quad (3.56)$$

2. **Let us consider very viscous** condensed phase and $\mu_g = 0$. In this case, we have

$$\left(1 - \tfrac{2}{3}\sigma\gamma^{-1} p_{0g}^{1/\gamma} p_0^{-1/\gamma-1} R_0^{-1}\right)\left(\Phi_{aa} - c_0^{-2}\Phi_{tt}\right) - \tfrac{1}{3}(1-\phi_0)\lambda\gamma^{-1}\rho_0 p_{0g}^{1/\gamma} p_0^{-1/\gamma-1}$$

$$\times \left[4c_1^2\Phi_{aat} + 3p_0\omega_0^{-2}\left(\Phi_{tt} - c_2^2\Phi_{aa}\right)_{tt}\right] + (1-\phi_0)\mu_1\lambda\gamma^{-1}\omega_0^{-2} p_{0g}^{1/\gamma} p_0^{-1/\gamma}\Phi_{tttaa} = C_* \quad (3.57)$$

In (3.57), the values C_*, c_0^{-2}, and c_2^2 are determined by (3.48), (3.49), and (3.56), correspondingly, and

$$c_1^2 = \tfrac{3}{4}\mu_1\rho_0^{-1}\left[p_{0g}^{-1/\gamma} p_0^{1/\gamma+1}\gamma + \phi_0(1-\phi_0)^{-1}\lambda^{-1} - \tfrac{2}{3}\sigma R_0^{-1}\right]. \quad (3.58)$$

3. **Let us consider** very special case when

$$1 - \tfrac{2}{3}\sigma\gamma^{-1} p_{0g}^{1/\gamma} p_0^{-1/\gamma-1} R_0^{-1} = 0. \quad (3.59)$$

Using (3.59), we rewrite (3.46), (3.50), and (3.51) in the form:

$$4\mu_g\left(\Phi_{tt} - c_1^2\Phi_{aa}\right)_t + 3\gamma p_0\omega_0^{-2}\left(\Phi_{tt} - c_2^2\Phi_{aa}\right)_{tt}$$

$$-3\mu_1\omega_0^{-2}\left(\rho_0 p_0^{-1}\right)^{-1}\Phi_{tttaa} = -3C_*(1-\phi_0)^{-1}\left(\lambda\gamma^{-1}\rho_0 p_{0g}^{1/\gamma} p_0^{-1/\gamma-1}\right)^{-1}. \quad (3.60)$$

Here

$$c_1^2 = \tfrac{3}{4}\mu_1\mu_g^{-1}\rho_0^{-1}\left[p_{0g}^{-1/\gamma} p_0^{1/\gamma+1}\gamma + \phi_0(1-\phi_0)^{-1}\lambda^{-1} - \tfrac{2}{3}\sigma R_0^{-1}\right] + \lambda^{-1}\rho_0^{-1}(1-\phi_0)^{-1}, \quad (3.61)$$

$$c_2^2 = \rho_0^{-1}\left[(1-\phi_0)^{-1}\lambda^{-1} + \tfrac{4}{3}\mu_g\mu_1\omega_0^2 p_0^{-1}\right]. \tag{3.62}$$

Apparently, the formulas for c_1^2 and c_2^2 given are not all ways applicable. Indeed, the whole analysis is based on the assumption that the wavelength is much larger than the size of the bubbles. Therefore, the theory does not cover frequencies close to the resonance frequency of bubbles. The value c_2^2 can serve only as a guide for experiments. On the other hand, we are not sure of the applicability of the models used in the case of very viscous liquids. This case can serve also only as a guide for experiments.

3.4.3 One-Speed Wave Equations

Of course, we do not expect that all the above expressions for wave velocities describe reality. But perhaps they determine some critical values or tendencies that arise when one or another parameter changes.

It is very important to compare the above-presented results with experiments. It is possible when waves have one speed.

1. **An incompressible condensed phase**. As an example of such a tendency, we consider the case of an incompressible condensed phase when $\lambda = 0$. In this case, Eq. (3.46) yields

$$\left(1 - \tfrac{2}{3}\sigma\gamma^{-1}p_{0g}^{1/\gamma}p_0^{-1/\gamma-1}R_0^{-1}\right)\left(\Phi_{aa} - c_0^{-2}\Phi_{tt}\right)$$
$$+ \tfrac{1}{3}\gamma^{-1}p_{0g}^{1/\gamma}p_0^{-1/\gamma-1}\left[\left(3\mu_1\varphi_0 + 4\mu_g\right)\Phi_{aat} + 3p_0\omega_0^{-2}\Phi_{aatt}\right] = C_*. \tag{3.63}$$

In (3.63), the value C_* is determined by (3.48) and

$$c_0^{-2} = \left(1 - \tfrac{1}{3}A_1\right)^{-1}\left[\varphi_0^2 p_{0g}^{1/\gamma}p_0^{-1/\gamma}c_{0g}^{-2}\right.$$
$$\left. + (1-\varphi_0)\varphi_0\left(p_{0g}^{1/\gamma}p_0^{-1/\gamma}p_{0g}^{-1}p_{0c}c_{0g}^{-2}\right)\right]. \tag{3.64}$$

2. **Case $\gamma = 1$ and $\sigma = 0$**. As a result, Eq. (3.63) yields

$$\Phi_{tt} = \tilde{c}^2\Phi_{aa} + \tfrac{1}{3}\varphi_0^{-1}(1-\varphi_0)^{-1}\left(4\mu_g\rho_{0c}^{-1}\Phi_{aat} + R_0^2\Phi_{aatt}\right). \tag{3.65}$$

This equation coincides with that derived in [30].

Equation (3.65) was used for the simulation of experimental data. Namely, it was assumed that $\Phi = \exp i(ka - \omega t)$. The last expression was substituted into (3.65), and the corresponding dispersion equation was obtained. The similar dispersion equation was also obtained earlier in [31]. The dispersion equation was used to describe the results of experiments. The results of the comparison are shown in Figure 3.13.

Dynamic Failure of WCM

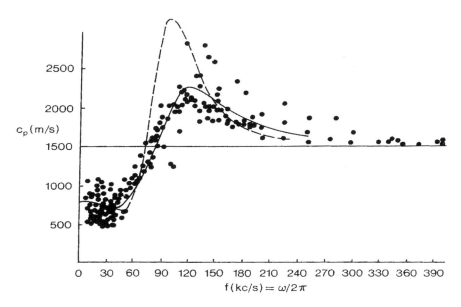

FIGURE 3.13 Sound speed in gas bubble–water mixture as a function of frequency. The solid curve represents the theoretical results for the actual bubble distribution. The broken curve results if all the bubbles have the same size $R_0 = 1.1 \times 10^{-4}$ m. The dots represent the measured values [30,31].

The solid curve in Figure 3.13 represents the theory for the actual experimental distribution, whereas the broken line represents the theoretical result if all the bubbles were of the same radius $R_0 = 1.1 \times 10^{-4}$ m. The difference between these curves clearly demonstrates the influence of a small spread on bubble sizes. With regard to wave speed, the agreement with theory is reasonable. Note in particular that above resonance, the speed of propagation of a wave reaches the values larger than the sound speed in pure water.

3.4.4 Influence of Viscous Properties on the Sound Speed of Magma-like Media

The magma is a mixture of liquid, bubbles, and solid particles [92]. Of course, many features of volcanic activities are ascribed to the presence of bubbles in magma. According to many publications, bubbles can significantly reduce the sound velocity of magma. The motion of huge bubbles in the volcanic conduit can generate seismic waves and acoustic waves in the air. However, these results were obtained for low-viscosity liquid, and their applicability has not been tested. Therefore in [93], experiments were conducted with high-viscosity silicone oil and a commercial syrup. It was investigated the responses of a bubble and liquid–bubble mixture to weak shock waves. It was obtained the following important result for this section: bubbles do not decrease the velocity of pressure waves when both rigidity and viscosity of the liquid are large. Because magma has a large rigidity, bubbles do not decrease the sound velocity of viscous magma.

Thus, a wave speed of some media with a large viscosity [84,93] depends on the viscosity, and the bubbles do not decrease the wave speed of media with a large viscosity, which is determined according to Eq. (3.43).

Let the viscosity μ_1 is very large. Considering this case, we will write (3.43) taking into account the most important viscous terms.

Let $1 - \frac{2}{3}\sigma\gamma^{-1} p_{0g}^{1/\gamma} p_0^{-1/\gamma-1} R_0^{-1} = 0$ and $\omega_0^2 \to \infty$. Then, Eq. (3.46) yields

$$4\mu_g \left(\Phi_{tt} - c_1^2 \Phi_{aa}\right)_t = -3C_* \left(1-\phi_0\right)^{-1} \left(\lambda\gamma^{-1} \rho_0 p_{0g}^{1/\gamma} p_0^{-1/\gamma-1}\right)^{-1}. \qquad (3.66)$$

Here, the value c_1^2 is determined by (3.50).

Let $p_{0g} \approx p_0$. Then the critical condition $1 - \frac{2}{3}\sigma\gamma^{-1} p_{0g}^{1/\gamma} p_0^{-1/\gamma-1} R_0^{-1} = 0$ yields

$$\sigma = \tfrac{3}{2}\gamma p_0 R_0 \qquad (3.67)$$

In this case for large μ_1 the sound speed of the mixture can be mainly determined by viscosity (see Eq. (3.50)). It is known that many magma-type melts have a high value of surface tension, which depends strongly on temperature and varies widely. For example, for tin at 400°C, the surface tension is large and about 520×10^{-3} N/m. Approximately the same value of surface tension has mercury. Of course, these values strongly depend on temperature. Therefore, for some values of magma temperature, its surface tension can be equal to (3.67).

Let the pressure of the medium and the size of the bubbles are such that the critical condition (3.67) is satisfied. Then, the sound velocity in the mixture can be greatly increased, is determined by the viscosity of the medium, and does not depend on the volume of the bubbles. However, if the viscosity or pressure in the medium is changed, so that the critical condition ceases to be fulfilled, the influence of the bubbles begins to appear and the velocity of the sound can be greatly reduced.

In general, this result agrees with experiment [93]. Experiments with silicone oil and syrup containing bubbles showed that:

1. When the viscosity is small, the pressure wave has smooth front followed by an oscillation, which is a typical feature of a shock wave in a bubbly liquid with low viscosity. When the viscosity is large, the wave front is as sharp as the incident shock wave.
2. As viscosity becomes smaller, the propagation velocity approaches the shock wave velocity, which was about 188 m/s. It approaches approximately 3580 m/s as the viscosity becomes large.

In [94–96], it is emphasized that the experiments give an insight about the dynamics of viscoelastic bubbly liquids, which is important for the developing mathematical and experimental methods to investigate seismoacoustic phenomena in volcanoes. Indeed, when the magma moves upward, the pressure temperature in it decreases. All these lead to a change in the surface tension of the bubbles. As a result, the situation arises when the critical condition (3.67) is satisfied and the speed of sound in the magma can also be determined by the viscosity.

Remark. We have considered small oscillations of WCM. We found that even with this restriction, the basic relation (3. 43) determines an interesting specter of equations that describe weakly or completely unexplored features of wave propagation in these media.

Of course, it is one thing to predict a certain phenomenon using mathematical model, and another case, usually more complicated, is to observe this phenomenon in Nature or discover it in an experiment! In particular, we do not know the work in which the propagation of three-speed waves in bubble media is described, and the influence of viscosity on the propagation of waves in bubble media is poorly studied. We associate this with the complexity of observations and the experimental study of these physical effects. Finally, it is possible that such effects are not essential for very strong extreme waves, which are of primary interest for this book. Examples of highly nonlinear behavior of the media under consideration are shown in Figures 3.1–3.7, 3.9, and 3.10.

Of course, such a complex behavior of media can be studied only on the basis of nonlinear wave equations.

3.5 EXAMPLES OF OBSERVABLE EXTREME WAVES OF WCM

Let us consider some nonlinear waves observed in Nature. We remind that WCM are considered [97,98]. If you shake the media vigorously, the properties of them change from "solid" to "gassy material" states.

3.5.1 MOUNT ST HELENS ERUPTION

It is possible to talk about "melting" of WCM during vibrations. In particular, certain eruptions of volcanos may be connected with this phenomenon. Let us consider the catastrophic eruption of Mount St. Helens.

Geologists call Mount St. Helens a composite volcano (or stratovolcano), which is a term for a cone-like volcano having steep, often symmetrical sides. Composite volcanoes are constructed by alternating layers of lava flows, ash, and other volcanic deposits. Those volcanoes tend to erupt explosively, and Mount St. Helens exploded on May 18, 1980, in the worst volcanic disaster in the recorded history of the United States. That was after 2 months of earthquakes and intermittent, relatively weak eruptions.

Intense earthquake activity persisted at the volcano during and between visible eruptive activities. Seismographs began recording occasional spasms of volcanic tremors, a type of continuous, rhythmic ground shaking different from the discrete sharp jolts characteristic of earthquakes. The triggering earthquake was of slightly greater magnitude than any of the shocks recorded earlier at the volcano. For example, shocks of magnitude 3.2 or greater occurred at a slightly increasing rate during April and May with five earthquakes of magnitude 4 or above per day in early April, and 8 per day in the week before May 18th. Initially, there was no direct sign of any eruption, but small earthquake-induced avalanches of snow and ice were reported by aerial observers. Thus, the volcano material became more and more weaker because of the earthquake-induced vibrations (Figure 3.14).

About 8 O'clock in the morning, there was an earthquake of magnitude 5.1, with epicenter at a depth of 2 km under the volcano. As a result, the volcano material fully transformed into gassy state. Edges of the crater of the volcano started to collapse because of a landslip. The gases and magma began to erupt into the atmosphere. What happened next was described by the geologists Keith and Dorothy Stoffel, who were then in a small plane over the volcano's summit.

Among the events they witnessed, they "noticed landsliding of rock and ice debris inward into the crater…the south-facing wall of the north side of the main crater was especially active. Within a matter of seconds, perhaps 15 seconds, the whole north

FIGURE 3.14 Dynamics of the volcano Mount St. Helens on May 18, 1980. Photos by Gary Rosenquist.

Dynamic Failure of WCM   77

side of the summit crater began to move instantaneously.... The nature of movement was eerie... The entire mass began to ripple and churn up, without moving laterally. Then the entire north side of the summit began sliding to the north along a deep-seated slide plane. I (Keith Stoffel) was amazed and excited with the realization, that we were watching this landslide of unbelievable proportions.... We took pictures of this slide sequence occurring, but before we could snap off more than a few pictures, a huge explosion blasted out of the detachment plane. We neither felt nor heard a thing, even though we were just east of the summit at this time". Realizing their dangerous situation, the pilot put the plane into a steep dive to gain speed, and thus, he was able to outrun the rapid eruption cloud that threatened to engulf them. The collapse of the north flank produced the largest landslide-debris avalanche ever recorded in historic time.

A volcano system (vent, conduit, and magma chamber) can be quite stable and single shock cannot transform all volcano materials into weakly cohesived state. However, the focusing of the shock energy in the top can transform the material there in the weakly cohesived state. As a result, some conditions are formed for a short-time eruption [33,77] since the shock can open the conduit vent. In the last case, the pressure drops down and the volcano magma can begin to boil.

To discuss these phenomena, we will consider the dome (or plug) as a cover closing the conduit. When the magma pressure in the conduit exceeds some critical level, the cover lifts and a short eruption occurs. Then, gas pressure in the conduit reduces, and the cover drops down. As a result, converging radial waves and diverging radial waves of the surface deformation are being formed.

Similar processes were observed in the volcano Santiaguito, Guatemala (see Figure 3.15), where the local peak of the vertical acceleration of the crater surface was up to 3 g [33,77].

3.5.2 THE VOLCANO SANTIAGUITO ERUPTIONS

A ring-like eruption of the volcano Santiaguito is shown in Figure 3.15((a), top). The eruption is a result of the dome uplift. Then, the converging wave propagating in atmosphere is formed, and it focuses on the dome center. Figure 3.15((a), bottom) shows the top of the volcano Santiaguito between the eruptions. Thus, on the vibrating surface of the volcano top, there may be a complex interaction of atmosphere, cracks, and boiling magma. As a result of the interaction, a short-time eruption can take place. The interval between eruptions was 20–40 minutes. Considering this and the remarks presented in the previous subsections, it can be assumed that the average speed of waves in magma of Santiago is of the order of 10 m/s, and the height of the magma conduit is about 10 km.

Thus, when the gas pressure near the volcano vent increases above some critical level, the cover is lifted up and a smoke column erupts. During a few seconds, the 20–80 m thick dome (whose radius is about 200 m) is being lifted by the gas pressure up to several tens of centimeters. These large and rapid uplifts occur 1–2 times per hour, together with explosive ash-rich eruptions. Then, the pressure reduces near the vent and the cover drops down. Thus, the phenomena were excited as a result of the periodical degassing of the conduit magma. Waves in the crater were measured: their

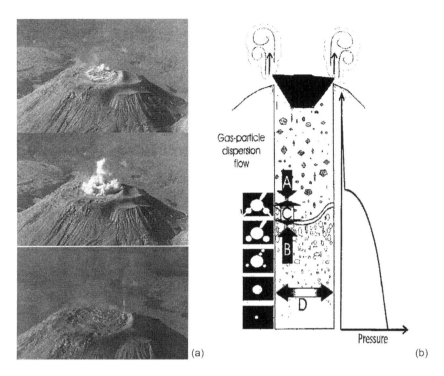

FIGURE 3.15 Focusing of the short-time periodical eruption of the volcano Santiaguito, Guatemala. (After http://www.photovolcanica.com/VolcanoInfo/Santiaguito/Santiaguito.html) (a). A sketch of bubbly magma fragmentation (b). See, also, Figures 3.5–3.7, 3.14, 3.15.

speed was about 30–50 m/s, which is too small for typical elastic waves within solid or fluid magma. The short-duration events and the dome–crater collision generate the periodical earthquakes or trembling of the ground.

The phenomenon is additionally illustrated in Figure 3.15(b) where different areas of erupting magma are discussed. Area D is the region where almost microscopic bubbles exist. It is the high-pressure field. There the wave properties of the medium are apparently described by the linear equations of Section 3.3. Area B is the field of explosive growth of bubbles where pressure is rapidly falling. As a result, the linear approximation is inapplicable and the equations of Section 3.4 must be used. Within the fragmentation layer, the fusion of bubbles begins. Magma is converted into a mixture of condensed particles and vapor. This is a very complex process. Perhaps, this process might be described using continuum approximation and wide-range state equations (1.10) or using some approaches of Chapter 2. As a result of fragmentation, the mixture arises. Its wave behavior can be qualitatively described by equations of the hydrodynamic type presented in Section 3.2.4 if the effects of vortices may be ignored.

Strongly nonlinear vertical waves may be generated in the volcano conduit because of periodical opening of the vent. The simplest scheme of this process is

Dynamic Failure of WCM

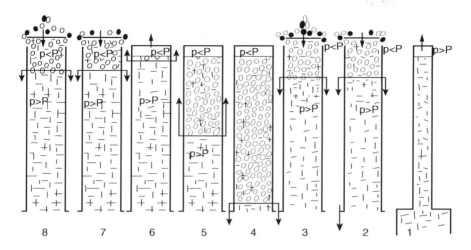

FIGURE 3.16 Simple scheme of the strongly nonlinear wave process, which can be initiated by periodical opening of the volcano conduit. In (2, 3, 7, 8), partly opened conduit when the dome lifts up. In (1, 4–6), closed conduit (see, also, Figure 3.7).

presented in Figure 3.16, where p is the pressure and P is the critical value of the pressure.

The cover (vent filling or dome) of the conduit lifts up when a pressure p reaches the critical value P, and then, magma begins to boil and erupt. As a result, a wave of expansion (depression) is generated at the vent, the pressure drops down, and then, the cover closes the conduit. The process repeats when p under the cover increases above P.

On the whole, the process resembles boiling in a kettle. Its cover rises when the pressure of the steam increases above some critical value, and then, the steam pressure decreases. As a result, the water in the kettle starts to boil more strongly, and steam and foam erupt from the kettle. When the pressure drops below some critical value, the cover closes the kettle and then, the process repeats.

Thus, the volcano oscillations strongly depend on the behavior of magma in rarefaction extreme waves. Therefore, this behavior has been studied in many researches. In particular, a high-pressure pipe is shown in Figure 3.7. The pipe models a conduit of a volcano and contains gas and solid particles. When the pipe is rapidly opened, an expansion wave is formed, which moves downward. Within this wave, the material expands like magma at the top of the conduit before a strong eruption.

3.6 NONLINEAR ACOUSTIC OF BUBBLE MEDIA

We return to the basic equation (3.43) and derive nonlinear equations, by keeping only linear, quadratic, and cubic nonlinear terms. For simplicity, we assume $\mu_g = \sigma = 0$. Therefore, we have that $p_{0g} = p_0$.

3.6.1 Low-Frequency Waves: Boussinesq and Long-Wave Equations

For the low-frequency waves, we assume that $\Phi_a \gg \Phi_t$. Therefore, nonlinear terms containing Φ_t and Φ_{tt} will be ignored.

Boussinesq's equation. For this case, Eq. (3.43) yields

$$\phi_0^{-\gamma} p_{0g}^{-1}\left(\rho_{0c}R_0 R_{tt} + p_0 - \rho_0\Phi_{tt} + \mu_1\Phi_{aat}\right) \approx \phi_0^{-\gamma}\left(1 + \phi_0^{-1}\Phi_{aa}\right)^{-\gamma}. \tag{3.68}$$

According to (3.42), we have in (3.68) that

$$R \approx R_0\left[1 + \tfrac{1}{3}\varphi_0^{-1}\Phi_{aa} - \tfrac{1}{3}\lambda\left(\varphi_0^{-1} - 1\right)\left(\rho_0\Phi_{tt} - \mu_1\Phi_{taa}\right)\right]. \tag{3.69}$$

Substituting (3.69) into (3.68) and discarding the small terms, we obtain

$$\Phi_{tt} - c_{00}^2\phi_0^{-1}\Phi_{aa} - \mu_1\rho_0^{-1}\Phi_{aat} - \tfrac{1}{3}\rho_{0c}\rho_0^{-1}R_0^2\varphi_0^{-1}\Phi_{aatt}$$
$$+ \tfrac{1}{2}(\gamma+1)c_{00}^2\phi_0^{-2}\Phi_{aa}^2 + \tfrac{1}{6}(\gamma+1)(\gamma+2)c_{00}^2\phi_0^{-3}\Phi_{aa}^3 = 0. \tag{3.70}$$

Here

$$c_{00}^2 = \gamma p_0\rho_0^{-1}. \tag{3.71}$$

Let

$$\eta = \Phi_{aa}. \tag{3.72}$$

Using (3.72), we rewrite (3.70) in the form:

$$\eta_{tt} - c_{00}^2\phi_0^{-1}\eta_{aa} - \mu_1\rho_0^{-1}\eta_{aat} - \omega_0^{-2}c_{00}^2\varphi_0^{-1}\eta_{aatt}$$
$$+ \tfrac{1}{2}(\gamma+1)c_{00}^2\phi_0^{-2}\left(\eta^2\right)_{aa} + \tfrac{1}{6}(\gamma+1)(\gamma+2)c_{00}^2\phi_0^{-3}\left(\eta^3\right)_{aa} = 0. \tag{3.73}$$

If the cubic nonlinear term in (3.73) is ignorable and $\mu_1 = 0$, then this equation yields the well-known Boussinesq equation:

$$\eta_{tt} - c_{00}^2\phi_0^{-1}\eta_{aa} - \omega_0^{-2}c_{00}^2\varphi_0^{-1}\eta_{aatt} + \tfrac{1}{2}(\gamma+1)c_{00}^2\phi_0^{-2}\left(\eta^2\right)_{aa} = 0. \tag{3.74}$$

The long-wave equation. Let

$$u = \Phi_a. \tag{3.75}$$

In this case, Eq. (3.70) yields

$$u_{tt} - c_{00}^2\phi_0^{-1}u_{aa} - \mu_1\rho_0^{-1}u_{aat} - \omega_0^{-2}c_{00}^2 u_{aatt}$$
$$+ (\gamma+1)c_{00}^2\phi_0^{-2}u_a u_{aa} + \tfrac{1}{2}(\gamma+1)(\gamma+2)c_{00}^2\phi_0^{-3}u_a^2 u_{aa} = 0. \tag{3.76}$$

Dynamic Failure of WCM   81

This equation and its versions describe many waves of different physical nature. In particular, it is often used for the description of shallow water waves.

3.6.2 High-Frequency Waves: Klein–Gordon and Schrödinger Equations

If the frequencies are high, then $\Phi_{tt} \gg \Phi_{aa}$. We will consider the highly nonstationary oscillations of bubbles under the action of an impact of an extreme wave. In doing so, we have what $RR_{tt} \gg R_t^2$ and $RR_{tt} \approx R_0 R_{tt}$. Therefore, for the case under consideration, it is assumed that $\frac{3}{2} R_t^2 \approx 0$ in (3.43). It is also assumed that $\mu_1 = 0$. The amplitude of the oscillations of the bubble is so small that the nonlinear terms containing the amplitude of the order of R_0^4 are ignored. Using these assumptions, we rewrite Eq. (3.43) in the form:

$$\phi_0 \left(1 + \rho_{0c} p_0^{-1} R_0 R_{tt} - \rho_0 p_0^{-1} \Phi_{tt}\right)^{-1/\gamma} = 1 - (1-\phi_0)\left[1 - (p_0+B)^{-1}(\rho_0 \Phi_{tt})\right]^{-1/k}. \quad (3.77)$$

According to (3.69), the left-hand side of (3.77) gives

$$\left(1 + \rho_{0c} p_0^{-1} R_0 R_{tt} - \rho_0 p_0^{-1} \Phi_{tt}\right)^{-1/\gamma} = 1 - \tfrac{1}{3}\gamma^{-1} p_0^{-1} \rho_{0c} R_0^2$$

$$\times \left[\varphi_0^{-1} \Phi_{aatt} - \left(\varphi_0^{-1} - 1\right) C_{00}^{-2} \Phi_{tttt}\right] + c_{00}^{-2} \Phi_{tt} + \tfrac{1}{2}\gamma\left(\gamma^{-1} + 1\right) c_{00}^{-4} \Phi_{tt}^2$$

$$+ \tfrac{1}{6}\gamma^2 \left(\gamma^{-1} + 1\right)\left(\gamma^{-1} + 2\right) c_{00}^{-6} \Phi_{tt}^3. \quad (3.78)$$

Here

$$C_{00}^{-2} = k^{-1}(B + p_0)^{-1} \rho_0. \quad (3.79)$$

Then, using (3.78), we rewrite Eq. (3.77) in the form:

$$\omega_0^{-2}\left[(1-\phi_0) C_{00}^{-2} \Phi_{tttt} - \Phi_{aatt}\right] + \varphi_0 c_{00}^{-2} \Phi_{tt} + (1-\phi_0) C_{00}^{-2} \Phi_{tt}$$

$$+ \tfrac{1}{2}\left[\gamma(\gamma^{-1}+1)\varphi_0 c_{00}^{-4} - k(k^{-1}+1)(1-\phi_0) C_{00}^{-4}\right]\Phi_{tt}^2$$

$$+ \tfrac{1}{6}\left[\gamma^2(\gamma^{-1}+1)(\gamma^{-1}+2)\varphi_0 c_{00}^{-6} + k^2(k^{-1}+1)(k^{-1}+2)(1-\phi_0) C_{00}^{-6}\right]\Phi_{tt}^3 = 0. \quad (3.80)$$

Let

$$\varphi = \Phi_{tt}. \quad (3.81)$$

In this case, Eq. (3.80) yields

$$\varphi_{tt} - (1-\phi_0)^{-1} C_{00}^2 \varphi_{aa} + \omega_0^2 \left[c_{00}^{-2} C_{00}^2 \phi_0 (1-\phi_0)^{-1} - 1\right]\varphi$$

$$+ \tfrac{1}{2}\omega_0^2 \left[(1+\gamma)c_{00}^{-4} C_{00}^2 \phi_0 (1-\phi_0)^{-1} - (1+k) C_{00}^{-2}\right]\varphi^2$$

$$+ \tfrac{1}{6}\omega_0^2 \left[(1+\gamma)(1+2\gamma)c_{00}^{-6} C_{00}^2 \phi_0 (1-\phi_0)^{-1} + (1+k)(1+2k) C_{00}^{-4}\right]\varphi^3 = 0. \quad (3.82)$$

We have obtained the Klein–Gordon equation for φ.

The Schrödinger equation. We consider the special case of Eq. (3.82) when the term containing φ^2 is absent. We also assume that

$$c^2 = C_{00}^2(1-\varphi_0)^{-1}, \quad \omega_p^2 = \omega_0^2\left[C_0^{-2}C_{00}^2\phi_0(1-\varphi_0)^{-1}+1\right],$$
$$\alpha_0 = \tfrac{1}{6}\omega_0^2\left[(1+\gamma)(1+2\gamma)c_{00}^{-6}C_{00}^2\phi_0(1-\phi_0)^{-1}+(1+k)(1+2k)C_{00}^{-4}\right]. \tag{3.83}$$

As a result, Eq. (3.82) yields

$$\varphi_{tt} - c^2\varphi_{aa} + \omega_p^2\varphi + \alpha_0\varphi^3 = 0. \tag{3.84}$$

A solution of this equation can be written as the quasi-harmonic wave having slowly varying complex amplitude A. The frequency and wave number of this wave are chosen so as to satisfy the linear dispersion relation following from (3.84). In this case, it is possible to derive Equation [99]:

$$i\frac{\partial A}{\partial \tau} - \beta\frac{\partial^2 A}{\partial \xi^2} + \alpha|A|^2 A = 0, \tag{3.85}$$

where β and α are certain constants, and τ and ξ are new variables.

We write one of the most thoroughly investigated equations, usually referred to as the nonlinear Schrödinger equation.

3.7 STRONGLY NONLINEAR AIRY-TYPE EQUATIONS AND REMARKS TO CHAPTERS 1–3

Here, we consider strongly nonlinear models, which are certain limits of Eq. (3.43). We set $\mu_g = \mu_1 = \sigma = 0$.

1. Let $\phi_0 \to 0$. In this case, the right-hand side of (3.43) $\to 0$. We differentiate the left-hand side. Then, we use $u = \Phi_a$ and have

$$u_{tt} - c_{0c}^2(1+u_x)^{-k-1}u_{aa} = 0. \tag{3.86}$$

We obtain the Airy-type equation [26] (see Eqs. (1.72) and (1.76)).

2. Let $\phi_0 \to 1$. In this case, (3.43) becomes

$$1 + \Phi_{aa} = p_{0g}^{1/\gamma}\left(p_{0g} - \rho_{0g}\Phi_{tt}\right)^{-1/\gamma}. \tag{3.87}$$

Using $u = \Phi_a$, we find the following Airy-type equation:

$$u_{tt} - c_{0g}^2(1+u_a)^{-1-\gamma}u_{aa} = 0. \tag{3.88}$$

Dynamic Failure of WCM 83

Remarks. We presented in this chapter a number of linear and nonlinear equations allowing us to investigate the wave processes in suspensions. Some of these equations are well known, and there are solutions for many cases of them. Derivation of these equations was one of our important goals. On the one hand, their conclusion demonstrated the commonality of the theoretical approaches developed. Starting from same mathematical model, we derived a few well-known equations describing the problem. Thus, the relationship of the basic nonlinear wave equations is shown. An idea is that they can have similar solutions. Therefore, the possibility is opened of using known solutions for analyzing the processes of propagation of extreme waves in media that are not very well studied.

Simultaneously with the description of wave fields, some of these equations and solutions determine the change in gas content and porosity of media. The change can characterize dynamic destruction by the growth of pores or bubbles. We remind that the growth of pores and bubbles is described by both wave equation and the corresponding ordinary equations. If the volume of spherical voids reaches a certain critical value, then we can speak of local destruction of the medium in this volume.

We introduced wave fields based on the wide-range equations of state in Chapter 1. However, although this approach is so general that it opens the possibility of taking into account even phase transitions, it should be detailed in the area where the material may be fractured. Therefore, in Chapter 2, we dwelled in more detail on models that make it possible to study the influence of the destruction of a medium on the propagation of waves.

On the other hand, this is determined by the fact that the process of destruction is extremely important for many engineering applications. Therefore, we dedicated Chapters 2 and 3 to it. We give an idea of the dynamic destruction. The fracture of a medium element is complementary to the general picture determined by the wide-range equations of state.

It is emphasized that the detailed investigation of the processes of dynamic destruction is beyond this scope of this book. However, we will touch on some of the most important aspects of this problem taking into account the destruction of media in extreme waves. The theories of similar destruction are very complex, so we have considered the most simple models.

In particular, in Chapter 2 and this chapter, simplified models of media containing spherical defects develop. Namely, equations of the form (2.28) or (3.43) are derived. In particular, equations of Sections 2.2 and 2.4 allow us to directly calculate the porosity dynamics of materials in extreme waves. In contrast, equations following from (3.43) define a function characterizing the growth of pores. Only after determining these functions and using them in the equations of pore dynamics (bubbles) can we calculate the dynamics of material destruction in various parts of a wave traveling in the medium.

For example, different wave conditions can arise when explosive volcanic eruptions occur. There, in a relatively short time, a rapid change in force and temperature fields occurs in a sufficiently large volume of the medium, leading to a colossal growth of initially microscopic bubbles. Apparently different conditions can appear in different parts of the magma at the same time, which can be described by the models of continuous medium presented in Chapters 1–3. In particular from models using wide-range equations of state, it is possible to come to the models reducible to the long-wave equations, the Klein–Gordon and Schrödinger equations.

REFERENCES

1. Sedov LI. *Continuum Mechanics*. Volumes 1–2. Nauka, Moscow (1983) (in Russian).
2. Nigmatulin RI. *Continuum Mechanics*. Nauka, Moscow (2014) (in Russian).
3. Zel'dovich YaB, Ryzer YuP. *Physics of Shock Waves and High-Temperature Hydrodynamic Phenomena*. Nauka, Moscow (1966) (in Russian).
4. Kolgatin SN, Khachaturyats AV. Interpolation equations of the costanding of metals. *High Temp* 20 (3): 447–451 (1982).
5. Anisimov SI, Prokhorov AM, Fortov VE. Application of powerful lasers for investigation of a substance at ultrahigh pressures. *Physics-Uspekhi* 142 (3): 395–434 (1984).
6. Agureikin VA et al. Thermophysical and gas-dynamic problems of antimeteorite protection of the Vega spacecraft. *High Temp* 22 (5): 964–983 (1984).
7. Kanel GI, Razorenov SV, Fortov VE. *Shock-Wave Phenomena and the Properties of Condensed Matter*. Springer Science & Business Media (2013).
8. Galiev ShU. *Nonlinear Waves in Bounded Continua*. Naukova Dumka, Kiev (1988) (in Russian).
9. Galiev ShU, Zhurakhovskii SV, Ivashchenko KB. Influence of the HeII- HeI phase transition on the formation of second sound waves. *Sov J Low Temp Phys*, 17 (8): 1035–1038 (1991) (in Russian).
10. Straughan B. *Heat Waves*. Springer (2011).
11. Novozhilov VV. *Theory of Elasticity*. Pergamon Press, New York (1961).
12. Trusdell K. *The Initial Course of Rational Mechanics of Continuous Media*. The World (1975).
13. Nowacki WK. *Stress Waves in Non-Elastic Solid*. Pergamon Press (1978).
14. Öchsner A. *Continuum Damage and Fracture Mechanics*. Springer (2016).
15. Galiev ShU. *Dynamics of Hydroelastoplastic Systems*. Naukova Dumka, Kiev (1981) (in Russian).
16. Ilgamov MA. *Oscillations of Elastic Shells Containing a Liquid and Gas*. Nauka, Moscow (1969) (in Russian).
17. Kuznetsov VP. Equations of nonlinear acoustics. *Sov Phys (Acoustics)* 16: 467–470 (1971) (in Russian).
18. Galiev ShU. *Dynamics of Structure Element Interaction with a Pressure Wave in a Fluid*. Naukova Dumka, Kiev (1977) (in Russian) (The same: Galiyev ShU. Dynamics of Structure Element Interaction with a Pressure Wave in a Fluid. American Edition, Department of the Navy, Office of Naval Research, Arlington (1980)).
19. Grigalyuk EI, Gorshkov AG. *Nonstationary Hydraelasticity of Shells*. Sudpromgiz, Leningrad (1974) (in Russian).
20. Zakhmlyaev BV, Yakovlev YuS. *Dynamical Loads during Underwater Explosions*. Sudpromgiz, Leningrad (1967) (in Russian).
21. Kedrinskii VK. *Hydrodynamics of Explosion*. Springer (2005).
22. Kutameladze SS, Nakoryakov VE. Heat-Mass Transfer and Waves in Gas-Liquid Systems. Nauka. Novosibirsk1 (984) ((in Russian).
23. Lienhard JN, Karimi A. Homogeneous nucleation and spinodal line. *J Heat Transfer*, 103 (1): 72–76 (1981).
24. Akulichev VA. *Cavitation in Cryogenic and Boiling Liquids*. Nauka, Moscow (1978) (in Russian).

25. Kuzpetsov NM. Equation of state and heat capacity of water in a broad range of thermomechanical parameters. *Journal of Appcied Mechanics and Technical Physics*, 1: 112–118 (1961).
26. Lamb H. *Hydrodynamics*. 6th edn. Dover Publications, New York (1932).
27. Nigmatulin RI. *Dynamics of Multiphase Media; Parts I and II*. Taylor & Francis Group, London (1991).
28. Natanzon MS. *Longitudinal Self-Excited Oscillations of a Liquid-Fuel Rocket*. Mashinostroenie, Moscow (1977) (in Russian).
29. Wood AB. *A Textbook of Sound*. Macmillan, New York (1955).
30. Crighton DG. *Nonlinear acoustics of bubbly liquids*. 1990 (unpublished).
31. Wijngaarden L. van. One-dimensional flow of liquids containing small gas bubbles. *Ann Rev Fluid Mech*, 4 (1972).
32. Naugolnykh K, Ostrovsky L. *Nonlinear Wave Processes in Acoustics*. Cambridge University Press (1998).
33. Galiev ShU. *Charles Darwin's Geophysical Reports as Models of the Theory of Catastrophic Waves*. Centre of Modern Education, Moscow (2011) (in Russian)
34. Valmir AL. *Nonlinear Dynamics of Plates and Shells*. Nauka, Moscow (1972) (in Russian).
35. Pertsev AK, Platonov EG. *Dynamics of Shells and Plastin*. Sudpromgiz, Leningrad, 1987 (in Russian).
36. Galiev ShU, Nechitailo NB. Effect of instantaneous stressed state on dynamic inflation and strength of plates. *Strength Mat* 2: 219–223 (1984).
37. Gilmanov AN, Sahabutdinov ZhM. Problems of dynamics of elastic membranes from an incompressible material. Interaction shells with a liquid. Trudy Seminara. Kazan EK Zavoisky Physical-Technical Institute, 24: 127–145 (1981) (in Russian).
38. Rabotnov Yu N. *Creep Problems in Structural Members*. North-Holland, Amsterdam (1969).
39. Kachanov LM. *Introduction to Continuum Damage Mechanics*. Martinus Nijhoff Publishers, The Netherlands (1986).
40. Lemaitre JA. Continuous damage mechanics model for ductile fracture. *J Eng Mat Tech* 107: 83 (1985).
41. Lemaitre J, Chaboche J. Aspect phenomenologique de la rupture par endommagement. *J Mecanique Appliquee* 2 (3): 317–365 (1978).
42. Tarabay A, Seaman L, Curran DR et al. *Spall Fracture*. Springer (2006).
43. Morozov N, Petrov Y. *Dynamics of Fracture*. Springer (2000).
44. Ramesh RT, Hogan JD, Kimberley J, Stickle A. A review of mechanisms and models for dynamic failure, strength, and fragmentation. *Planet Space Sci* 107: 10–23 (2015).
45. Molinary A, Mercier S, Jacques N. Dynamic failure of ductile materials. *Procedia IUTAM* 10: 201–220 (2014).
46. Rottler J, Robbins MO. Growth, microstructure, and crazes in glassy polymers. *Physical Review E* 68: 011801 (2003).
47. Horgan CO, Polignone DA. Cavitation in nonlinearly elastic solids. *Appl Mech Rev* 48 (8): 471–485 (1995).
48. Huang, Y, Hutchinson JW, Tvergaard, V. Cavitation instability in elastic-plastic solids. *J Mech Phys Solid* 39 (2): 223–241 (1991).
49. Seaman L, Curran DR, Murri WJ. A continuum model for dynamic tensile microfracture and fragmentation. *J Appl Mech*, 52: 593–600 (1985).
50. Barbee TW, Seaman L, Crewdson RC, Curran DR. Dynamic fracture criteria for ductile and brittle metals. *J Mat, JMLSA*, 7: 393–401 (1972).
51. Regel VR, Slutsker AI, Tomashevsky EE. Kinetic the nature of strength of solids. *Physics-Uspekhi*, 15: 45–65 (1972).

52. Fomin VM, Khakimov EM The spalling destruction of the medium in plane rarefaction waves. - Novosibirsk: ITAM SO, The Academy of Sciences of the USSR, Preprint, 1981 (in Russian).
53. Akhmadeev NKh. Dynamic destruction of solids in stress waves. Ufa, BFAN USSR (1988) (in Russian).
54. Tuler FR, Butcher BM. A criterion for the time dependence of dynamic fracture. *Int J Fract Mech* 4 (4): (1968).
55. Seaman L, Development of computational model for microstructural features. In: Shock Waves in Condensed Matter, *Conf. Menlo Park Calif.*, 1981, New York: 118–129 (1982).
56. Seaman L, Curran D, Shockey DA. Comhutational models for ductile and brittle fracure. *J Appl Phys* 47: 4814–4826 (1976).
57. Ruzanov AI. Numerical examination of spallation strength taking into account microdamage. *Izv. Akad Nauk SSSR, Mekh Tverd Tela*, 5: 109–115 (1984) (in Russian).
58. Galiev ShU, Babish UN, Zhurakhovskii SV, Nechitailo NB, Romashchenko VA. *Numerical Modeling of Wave Processes in Bounded Media*. Naukova Dumka, Kiev (1989) (in Russian).
59. Kanel GI, Razorenov SV, Fortov VE. *Shock-Wave Phenomena and the Properties of Condensed Matter*. Springer, New York (2004).
60. Kanel GI et al. Unusual plasticity and strength of metals at ultrashort load durations. *Physics-Uspekhi* 187 (5): 525–545 (2017).
61. Galiev ShU, Zhurakhovskii SV. Fracture of Multilayered Plates in Thermo-Viscoelasto-Plastic Waves. *Strength Mat* 11: 1542–1549 (1984).
62. Golubev VK. On the expansion of pores in plastic metals. *J App Mech Tech Phys* 6: 159–165 (1983).
63. Kuznetsov GN, Shekin IE. Interaction of pulsating bubbles in a viscous fluid. *J App Mech Tech Phys* 18 (4): 565–570 (1962).
64. Voinov OV, Golovin AM. Lagrange equations for a system of bubbles from-J varying radii in a liquid of low viscosity. Izv. AN SSSR. Ser. *Fluid Gas* 3: 117–123 (1970).
65. Galiev ShU. A Model of cavitation dynamic failure of solid viscoplastic and liquid media. *Strength Mat* 10: 3–7 (1986).
66. Shima S, Oyane M. Plasticity theory for porous metals. *Int J Mech Sci* 18 (6): 285–291 (1976).
67. Swegle JW. Constitutive equation for porous materials with strength. *J Appl Phys* 51(5): 2574 (1980).
68. Zel'dovich YaB. To the theory of the formation of a new phase. *Cavitation. J Exp Theor Phys* 12(11-12): 525–538 (1942) (in Russian).
69. Fisher JC. The fracture of liquids. *J Appl Phys* 19: 1062 (1948).
70. Mason WP, Thurston RN. (Eds.) *Physical Acoustics. XIV*, Academic Press, New York (1965).
71. Nakoryakov VE, Dontsov VE, Pokusaev BG. Pressure waves in a liquid suspension with solid particles and gas bubbles. *Int J Multiphase Flow* 22 (3): 417–429 (1996).
72. Jaeger HM, Nagel SR, Behringer RP. Granular solids, liquids, and gases. *Rev Mod Phys* 68: 1259 (1996).
73. Gubaidullin DA, Teregulova EA. Acoustic waves in multifractional gas mixture with the inclusion of different materials and dimensions without phase transformations. *J Phys*: Conference Series 567: 012019 (2014).
74. Kameda M et al. Advancement of magma fragmentation by inhomogeneous bubble distribution. *Sci Rep* 7: 16755 (2017).
75. Buchanan M. Think outside the sandbox. *Nature* 425: 556–557 (2003).
76. Pak HK, Behringer RP. Bubbling in vertically vibrated granular material. *Nature* 371: 231–233 (1994).

77. Galiev ShU. *Darwin, Geodynamics and Extreme Waves*. Springer (2015).
78. Pease JW, O'Rourke TD. Seismic response of liquefaction sites. *J Geotech Eng ASCE* 123: 37–45 (1997).
79. Zamankhan P. Solid structures in a highly agitated bed of granular materials. *App Math Modell* 36: 414–429 (2012).
80. Anilkumar AV. Experimental studies of high-speed dense dusty gases. Thesis of Doctor of Philosophy, California Institute of Technology, Pasadena, CA (1989).
81. Galiev ShU. The influence of cavitation upon anomalous behavior of a plate/liquid/underwater explosion system. *Int J Impact Eng* 19: 345–359 (1997).
82. Galiev ShU. Strongly-nonlinear wave phenomena in restricted liquids and their mathematical description. In: *New Nonlinear Phenomena Research*. Ed. T Perlidze, pp. 109–300, Nova Science Publishers, New York (2008).
83. Richardson MD et al. The effects of free-methane bubbles on the propagation and scattering of compressional and shear wave energy in muddy sediments. *JASA* 96, 5 (2): 3218 (1994).
84. Brujan EA, Willaims PR. Bubble dynamics and cavitation in non-newtonian liquids. *Rheol Rev* 147–172 (2005).
85. Carcione JM, Poletto F. Sound velocity of drilling mud saturated with reservoir gas. *Geophysics* 65 (2): 646–651 (2000).
86. Iga Y, Nohml M, Goto A, Ikohagi T. Numerical analysis of cavitation instabilities arising in the three-blade cascade. *J Fluids Eng* 126(3): 419 (2004).
87. Galiev ShU, Panova OP. Periodic shock waves in spherical resonators (survey). *Strength Mat* 10: 602–620 (1995).
88. Barcilon V, Richter F. Nonlinear waves in compacting media. *J Fluid Mech* 164: 429–448 (1986).
89. Nakoryakov VE, Kuznetsov VV. Dontsov VE. Pressure waves in saturated porous media. *Int J Multiphase Flow* 15: 857–875 (1989).
90. Levy A, Ben-Dor G, Sorek S. Numerical investigation of the propagation of shock waves in rigid porous materials: development of the computer code and comparison with experimental results. *J Fluid Mech* 324: 163–179 (1996).
91. Prosperetti A. The thermal behaviour of oscillating gas bubbles. *J Fluid Mech* 222: 587–616 (1991).
92. Gonnermann HM, Manga M. The fluid mechanics inside a volcano. *Ann Rev Fluid Mech* 39: 321–356 (2007).
93. Ichihara M, Ohkunitani H, Ida Y, Kameda M. Dynamics of bubble oscillation and wave propagation in viscoelastic liquids. *J Volcanol Geotherm Res* 129: 37–60 (2004).
94. Alidibirov M, Dingwell DB. Magma fragmentation by rapid decompression. *Nature* 380: 146–148 (1996).
95. Kameda M, Kuribara H, Ichihara M. Dominant time scale for brittle fragmentation of vesicular magma by decompression. *Geophys Res Lett* 35: L14302 (2008).
96. Spieler O, Dingwell DB, Alidibrov M. Magma fragmentation speed: an experimental determination. *J Volcanol Geotherm Res* 129: 109–123 (2004).
97. Royer JR, Corwin EI, Flior A, Cordero M-L, Rivers ML, Eng PJ, Jaeger HM. Formation of granular jets observed by high-speed X-ray radiography. *Nature Phys* 1: 164–167 (2005).
98. Ciamara MP, Coniglio A, Nicodemi M. Shear instability in granular mixture. *Phys Rev Lett* 94(18): 188001, 2005.
99. Ostrovsky LA, Potapov AI, *Modulated Waves*. The Johns Hopkins University Press (1999).

Part II

Extreme Waves and Structural Elements

Science is built up with facts, as a house is with stones.

But a collection of facts is no more a science than a heap of stone is a house.

Henri Poincaré

We discussed in Part I the issues related to the destruction of solid and liquid media. Typically, this destruction is associated with the occurrence in the condensed media of extreme tension stresses (rarefaction zones). Such destructions often occur as a result of reflection of extreme waves from free surfaces or passing of waves in a less dense medium.

In this part, we will consider actions shock waves and explosions on structural elements. Under the action of an underwater wave on the structure on its surface and near it the destruction of the liquid can take place. The cavitation waves can occur (see Section 2.2). In this part, we focus on details of the destruction of liquid media in rarefaction waves. Such waves usually arise during the reflection of underwater compression waves from free surfaces, or solid, but deformable (flexible) boundaries.

We emphasize that the term "destruction" is used for liquids only in the case of extremely short waves. In the case of sufficiently long waves, the term "cavitation" (cavitation destruction) or "cool boiling" is commonly used. Here, we study effects of cavitation on transient interaction of underwater compression waves with solid but deformable (flexible) boundaries. Effects of destruction of deformable boundaries are taken into account in Chapter 6.

4 Extreme Effects and Waves in Impact-Loaded Hydrodeformable Systems

The phenomenon of cavitation occurs when a liquid evaporates and gas (vapor) bubbles are generated in zones where the pressure falls below its saturated vapor pressure (or boiling point) [1,2]. Steady cavitation is usually observed in areas of high-speed fluids. For example, this phenomenon takes place behind bodies moving with high velocity in water. Unsteady (transient) cavitation may be generated under certain conditions in the flow near propellers, pumps, and within different hydraulic machines. This phenomenon is usually accompanied by the violent expansion and the collapse of gas bubbles. Transient cavitation may also be generated during the reflection of an underwater shock wave from a free water surface [2,3], or from a plate and shells. It must be emphasized that a considerable negative (i.e., tensile) pressure can appear as a result of the reflection of very short and very strong shock waves. A negative pressure can exist before the rupture of liquid by means of generation and expansion of vapor bubbles.

4.1 INTRODUCTION

In this section, we will consider the strongly nonlinear effects connected with the cavitation in a liquid generated because of the interaction of long underwater wave with plates. In the case cavitation appears if the pressure drops to values of the order of (−0.2)–(−0.35) MPa. This phenomenon is important for explosive hydro-forming technology using thin metal plates and in cases when explosions are used for peaceful or military purposes in liquid media.

Explosive hydro-forming of the metal work pieces. This technology uses the energy generated by an explosive detonation in water. It has been used in a wide variety of applications in the automotive, aerospace, and maritime industries for almost 70 years. This technique is especially efficiently for large parts. Since the pressure generation comes from an explosive charge, we do not need larger machinery to form large parts, and we simply apply a larger charge. Current explosively hydro-formed parts range up to 6 ft.

The sheet metal work piece blank is lowered into a tank filled with water. The explosive charge is placed at some predetermined distance from the work piece (Figure 4.1).

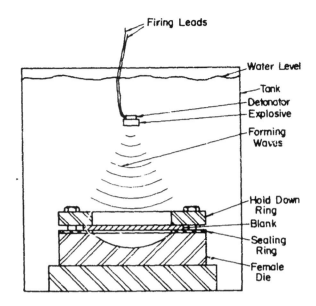

FIGURE 4.1 Scheme of explosive forming of the sheet metal.

On detonation of the explosive, a pressure pulse of very high intensity is produced. A gas bubble is also produced, which expands spherically and then collapses [4,5]. When the pressure pulse impinges against the work piece, the metal is deformed into the die with a high velocity. This velocity can be so high that the metal breaks away from the water. Cavitation arising in this case can greatly spoil the process of forming the work piece.

Ships and offshore structures. A study of the dynamic response of ships and offshore structures to the action of explosions and shock waves is required for safety and serviceability assessments, including habitability on ships. For merchant ships, accidental explosions have to be considered as possible events, as they could be severe enough to cause serious damage to structures and injuries to people. They are also becoming of great interest due to increasing public concern about ship safety and potential harm to the environment [6–8].

Basically, design methods adopted for naval and merchant ships are the same, but naval ships must fulfill some specific requirements [9–15]. In particular, they are required to retain a high standard of operational effectiveness when under attack.

Figure 4.2 shows the results of an underwater explosion of a torpedo. It was found that the results of the attacks are connected with the transient cavitation (the negative pressure) on the ship girder. Above-water weapon attacks include air blast loads. The negative pressure and the impulse caused by the expanding bubble are main reasons of the final anomalous sag distortion of the ship girder [10].

Underwater weapons usually also explode near to the hull (Figure 4.3a). As a result, an underwater shock wave can be formed in water. High-velocity interaction of it with the constructions is connected with many complex phenomena. In particular, the so-called "hull" cavitation can strongly change the shock pressure [16–19].

Extreme Effects and Waves

FIGURE 4.2 A few stages of the sinking of a ship after an explosion.

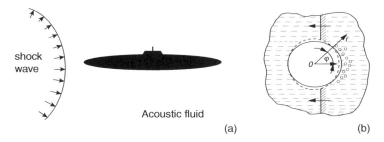

FIGURE 4.3 Sketch of shock wave induced by an underwater explosion impinging on a submarine (a). A qualitative picture of the cavitation interaction of underwater shock waves with the hull of a submarine (b).

Hull cavitation (near the hull) or the so-called bulk cavitation (far from the hull) can occur at points where the local pressure (the sum of shock wave pressure and the ambient static pressure at that depth) drops below the saturation vapor pressure. This element of the interaction is qualitatively shown in Figure 4.3b. The solid line (circle) corresponds to the undisturbed hull. The dashed line corresponds to the hull loaded by the underwater shock wave. The bubbles determine the cavitation zone [17,18].

The hull cavitation, owing to its specifics, is almost unknown in hydrodynamics, though some books and a number of articles are devoted to it [3,8–28]. In particular, a gas–water mixture may be generated due to the hull cavitation. The gas–water vapor appears because the water volume has locally emptied from the compression due to the hull element displacement [17–19]. The cavitation limits the reduction in the water pressure. As a result, cavitation can increase the pressure on the deforming surface and may determine the result of the ship hull–extreme underwater wave interaction.

4.2 UNDERWATER EXPLOSIONS AND EXTREME WAVES OF THE CAVITATION: EXPERIMENTS

Now we consider the influence of deformability of the water surface on the explosion pressure [19,20,23]. First, as a limiting case, we consider the free water surface.

The photographs presented in Figure 4.4 illustrate three instants following an underwater explosion near the water surface. The explosion product (a dark spherical spot) and the expanding spherical shock wave are shown in Figure 4.4a. The initial stage of the reflection of this wave from the free water surface is shown in Figure 4.4b. The dark zone located under the free surface and above the explosion product is the transient cavitation zone. This zone expands as shown in Figure 4.4c.

Let us consider the hull cavitation on the thin flexible solid surfaces. In Figure 4.5, these surfaces are modeled by a round plate. The plate was loaded by an underwater shock wave.

The circular closed pipe whose left-hand end is at the center of Figure 4.5 is surrounded by fluid and is orientated horizontally, extending to the right beyond the image. The experiments have been performed using a thin (0.02 in.) plate of cellulose acetate 6 in. in diameter backed by air, and photographing the water in front of the plate after reflection of a shock wave. The cavitation zone is shown by a multitude of bubbles. This zone arose following the impact of an underwater shock wave onto a flexible plate.

Thus, with sufficiently large amplitude, the pressure will become negative at points in the fluid. This corresponds to a rarefaction wave reflected from the accelerated surface of the plate. If the water cavitates, the subsequent motion of the plate will be modified [11,23].

The photos presented in Figures 4.6 and 4.7 illustrate the interaction of a tube wall with a shock wave generated by the electro-explosion of a metal wire in water filling the tube (cylindrical shell). The wire is spread along the tube axis. The pictures were

FIGURE 4.4 The interaction of underwater explosion with a free water surface. The labels a, b, c correspond different moments of time.

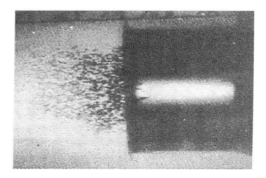

FIGURE 4.5 Cavitation beyond the end of a circular pipe closed by a plate after being subjected to a shock wave approaching from the left [23].

Extreme Effects and Waves

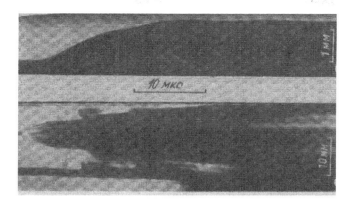

FIGURE 4.6 The radial displacement of an internal point of the wall of a cylindrical tube under the action of an internal underwater explosion (upper picture) and the expansion of the explosion product (bottom picture) [23].

FIGURE 4.7 A photograph of the cavitation interaction of a cylindrical shell–water–underwater explosion [23].

obtained by photographing from the end face of the cylindrical steel tube (radius 1.47 cm, shell thickness 2 mm, length 10 cm) immersed in water.

Figure 4.6 shows typical photographed scans of the tube section [19,21–24]. The upper photo shows a typical photographed scan of the deflection of an internal point of the cylindrical shell. The shell deforms, and its radius increases. As a result, the cavitation zones are formed (the lower photo and Figure 4.7). The inner surface of the tubes is seen as the dark lines clearly visible at the first instants after the explosion. The central dark zones are formed by the metal–water vapor (explosion product). The bright zones are water. The straight lines are the shock wave trajectories. The dark zones, located within the water and near the wall of the shell, are formed by the transient cavitation.

Figures 4.5–4.7 demonstrate some important peculiarities of the hull–wave interaction. The wave impacts the structure. It accelerates and the rarefaction (cavitation) zones are formed in the water [19,21–24].

The destructive effect of giant waves is well known, and hence, the prediction of where and when they will occur is of extreme importance to all who lives or works beside or upon the sea. The existence of these extreme waves makes it important to re-examine some fundamental ideas of merchant ship design, which were developed earlier [6–8]. In particular, experience from designing underwater naval ships (submarines), which are required to retain a high standard of operational effectiveness under explosion attack, may be used.

4.3 EXPERIMENTAL STUDIES OF FORMATION AND PROPAGATION OF THE CAVITATION WAVES

The process of interaction of extreme water waves with a ship hull is very complex, since it may be associated with large strains in the structural elements and the generation of cavitation in the water. This complex interaction process has not been studied fully, although now powerful numerical codes are available [9,10,13–16], which allow the response of a ship hull to the action of water waves to be calculated. However, there are large uncertainties in the determination of loads from extreme waves acting on the structures. Therefore, the results of the calculations need to be supported by experimental data. This is especially true for local elements of a hull such as panels and plates.

In this section, we consider some important aspects of the transient interaction of deformable surfaces with water waves. This problem is examined using an experimental approach, by creating a high pressure wave in water and impacting it on plates with various properties and thicknesses. The experiments are focused on these extreme and potentially damaging waves whose loadings cannot be predicted very accurately in simulations.

Many tests were conducted using an apparatus whose principal features are presented in Figure 4.8. The fundamental principle used in the apparatus is the conversion

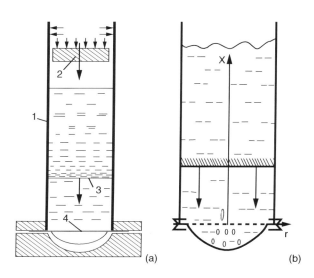

FIGURE 4.8 Sketches of the experimental apparatus (a) and a scheme of the plate–shock wave interaction (b) [23].

of the kinetic energy of a rapidly moving mass (a flat piston) into water pressure energy. The piston was accelerated by a compressed gas in the upper part of the cylinder.

The experimental apparatus consists of a vertical cylindrical metal tube (1) having an inner diameter of 112 mm. The tube wall thickness is 20 mm. The acoustical sound speed calculated for water in the tube was 1265 m/s. The plane piston (2) impacts the water surface and causes the compression wave (3). The tube bottom is closed by the circular plate that is clamped around its edge. This plate can be replaced during the experiments in order to examine an influence of its properties (thickness h and stiffness) on the resulting pressures. The compression wave deforms the plate as shown in sketch. The plate is shown in its undeformed state (4) (Figure 4.8a) and deformed state (Figure 4.8b). An air layer ahead of the piston was compressed during its acceleration. This diminished the wave amplitude in the water to about half compared to what it would have been in the absence of an air layer. The pressures were measured at different points in the water using piezoelectric pressure sensors and recorded on an oscilloscope screen. The piezoelectric pressure sensors and displays had previously been calibrated statically and dynamically, and both calibrations gave identical results.

4.3.1 Elastic Plate–Underwater Wave Interaction

A series of experiments was made when the mass of the piston was 13 kg, and the height of the water column was 500 mm. The impact velocity of the piston onto the water was kept constant at 1.8–2.0 m/s. The wave amplitude was approximately 1.25 MPa in the experiments with a wavelength of 2.5–3 m.

An oscillogram of the pressure measured on the surface of a steel plate with thickness $h = 20$ mm is shown in Figure 4.9. One can see that the pressure increases rapidly and then decays according to an exponential law. We will consider this pressure as the incident wave, which is the same for all tests with elastic plates.

Results for aluminum alloy plates with thicknesses $h = 8$, 4, and 2 mm are shown in Figures 4.10a–c, respectively [3,17,22–24]. The pressure oscillograms measured on the tube wall at points 255 mm and 55 mm from the plate surface correspond to curves 1 and 2, respectively. The left-hand graphs have a 2 ms timescale, and the right-hand graphs have a 0.5 ms timescale. It can be seen that a reduction in the plate thickness results in a radical change in the pressure curves with an increase in the interaction time.

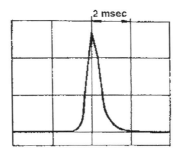

FIGURE 4.9 An oscillogram of pressure in the incident wave [23].

We will consider primarily the curves obtained by sensor *II* close to the plate. These curves consist of the first peak, a pressure drop section, and a second peak which is somewhat longer. The first peak results from the pressure occurring at the very beginning of the interaction when the plate just starts to move. The interaction is evident here in that the peak amplitude depends on the plate thickness and is less than the maximum pressure of the incident wave.

The pressure remains positive for the 8 mm thick plate (a). Cavitation occurs in the water for the 4 and 2 mm thick plates (curves 4.10b and c). The pressure then increases, which is explained by the collision between the water column and the plate when the cavitation zone near the plate collapses. The amplitude of the second peak is close ($h = 8$ mm (a)) to, or exceeds ($h = 4$ mm (b) and $h = 2$ mm (c)), the magnitude of the first peak and is less dependent on h. The length of the pressure decay section and the second peak is related to the deceleration of the plate by elastic forces and the velocity of its reverse motion, which are determined by the fundamental natural frequency of the plate.

Oscillograms *I* show the pressures measured further from the plate than oscillograms *II*. There the effects of the plate thickness are weaker. The first peak is determined by the frontal part of the incident wave. Therefore, this peak weakly depends on the plate thickness. The second peak is determined by the water–plate interaction. For $h = 4$ mm (b) and $h = 2$ mm (c), the cavitation is determined by the form and the amplitude of the second peak. The third peak is formed from the reverse motion of the plate and connects with the period of the natural oscillation of the plate.

Figure 4.10 shows that the cavitation measured near the plates strongly depends on mechanical and geometrical properties of the plates. In particular, if the plate is thick enough, cavitation effects are either absent or small. However, if the ratio of the radius R of the plate to its thickness h is large, the effect of cavitation on the resulting pressure may be very important. It is evident that the initial wave is damped by a thin plate. Therefore for thin plates, the first peak of the pressure may be much smaller than the amplitude of the initial wave (1.25 MPa). The subsequent two-dimensional waves generated by cavitation then move along the plate surface. These waves form the peaks measured in the plate center. Hence, the lifetime of the cavitation cavities strongly depends on the plate properties.

At the same time, cavitation may be generated far from the plate due to the resulting superposition of the incident (Figure 4.9) and the reflected wave. This so-called bulk cavitation is determined by curves 1 in Figure 4.10. During the initial period of the interaction, the water motion may be considered as one-dimensional. Then, because of the plate deflection, which strongly depends on the radial coordinate, the water motion near the plate transforms into two-dimensional motion, and radial waves of pressure and cavitation are formed. Therefore, oscillograms measured at different points of the plate could be expected to be different.

4.3.2 Elastoplastic Plate–Underwater Wave Interaction

The hull cavitation strongly depends on the amplitude of the incident wave and its duration. Therefore, we have studied this phenomenon examining the interaction of a circular plate with extreme waves which can rupture it.

Extreme Effects and Waves

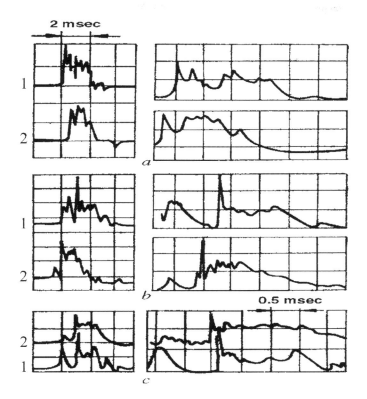

FIGURE 4.10 Effect of plate thickness on the water pressure [23]. The pressure oscillograms for plates with thicknesses $h = 8$, 4 and 2 mm are labeled as 1, 2 and 3, respectively. The pressure oscillograms measured on the tube wall at points 255 mm and 55 mm from the plate edge correspond to curves 1 and 2, respectively.

The same apparatus was used. The height of the water column was increased to 650 mm, the impact velocity of the piston was increased from 1.8–2 m/s to 30–33 m/s, and the piston mass was reduced from 13 to 2.3 kg. As a result, the peak pressure in the incident wave increased to 15–18 MPa.

The pressure oscillograms recorded on the tube surface at points distant by 255 mm (curve 1) and 55 mm (curve 2) from a metal plate with thickness $h = 20$ mm are shown in Figure 4.11. One division in curves 1 and 2 corresponds to 8 and 11 MPa, respectively. Curve 1 has two large peaks, explained by the passage of the incident and reflected waves; curve 2's first peak is large and is caused by an approximate doubling of the wave amplitude at the plate. There is also the second peak, which is not far from the first one. It is impossible to explain its appearance in terms of the wave reflected from the piston. The origin of the second peak is apparently related to the radial motion of the water at the plate surface, caused by a certain deformability of the plate and the tube wall, due to the very high pressure generated near the plate. There is a third insignificant peak on both the upper and lower curves.

Further experiments were carried out using plates where the loadings resulted in elastoplastic strains.

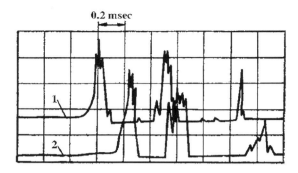

FIGURE 4.11 Pressure waves measured near (2) and far (1) from the thick plate [23].

Oscillograms of the pressure measured at the edges (curve 1) and at the center (curve 2) of plates are presented in Figure 4.12. The curves of Figure 4.12a correspond to $h = 4$ mm, and the curves of Figure 4.12b correspond to $h = 2$ mm. The pressure was measured at 15 mm from the plate surfaces. The rapid changes in the pressure curves are associated with the origination of hull cavitation in the water and the radial water motion caused by the plate deformation. Since the incident wave is very large in magnitude and short in duration, the plate accelerates very quickly at the beginning of the interaction. As a result, the pressure drops to zero and the plate separates from the water. The subsequent deceleration of the plate by the strain resistance forces, and the action of the earth's gravitation, results in the disappearance of cavitation, and the plate makes contact with the water by impact causing a new pressure peak. There is then the possibility of a second acceleration of the plate causing a separation of it from the water, and repetition of the process. Of course, this sequence is complicated by the radial motion of the pressure and the cavitation waves along the plate surface. The experiments showed that the peak pressure amplitudes depend on the thickness h of the plate, and the amplitude increases as h increases.

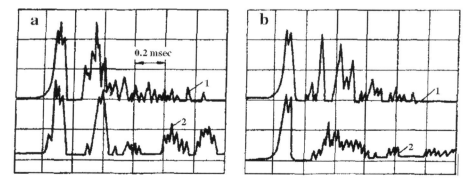

FIGURE 4.12 The pressure wave measured at the edges (curve 1) and at the center (curve 2) of the elastoplastic (aluminum alloy) plates [23]. The labels a and b correspond $h = 0.4$ mm and $h = 0.2$ mm.

4.4 EXTREME UNDERWATER WAVE AND PLATE INTERACTION

In this section, the influence of deformability of the plate and cavitation in the liquid on the process of the transient plate–liquid interaction is investigated. Cavitation is modeled as the instantaneous destruction of a liquid in a rarefaction wave. Wave processes in the experimental apparatus (Figure 4.8) are studied.

4.4.1 Effects of Deformability

It is assumed that the edge of the plate is rigidly clamped and that the front of the extreme wave is parallel to the plate surface (Figure 4.8). At the initial moment of the interaction, the pressure is constant on the plate surface. At the next instants, because of the plate deformation, the axisymmetric loading of the plate occurs. The influences of membrane forces and plastic deformations are ignored. In this case, the equation of plate motion has the following classical form (see Section 1.5):

$$w_{rrrr} + 2r^{-1}w_{rrr} - r^{-2}w_{rr} + r^{-3}w_r = -12(1-v^2)(\rho h w_{tt} + p)(Eh^3)^{-1}, \quad (4.1)$$

where w is the plate displacement, subscripts t and r indicate derivatives with respect to time and the radial coordinate, ρ is the plate material density, h is the plate thickness, E is Young's modulus, v is Poisson's ratio, and p is the water pressure.

In the case of a weak (acoustic) incident wave p_w and no cavitation [17,18],

$$p = 2p_w + \rho_0 c_0 w_t, \quad (4.2)$$

where ρ_0 and c_0 are the undisturbed density and sound velocity of the liquid, correspondingly, and w_t is the normal velocity of the plate. It is seen that the term $\rho_0 c_0 w_t$ takes into account the effect of the deformability of the plate.

It is assumed in the theoretical model that the edge of the circular plate is rigidly clamped at $r = R$. Therefore, the displacement w and slope w_r at the plate edge are both equal to zero so that

$$w = w_r = 0 \quad (r = R). \quad (4.3)$$

At the center of the plate, we use the symmetry condition and the limited bending moment

$$w_r = 0 \quad \text{and} \quad r(w_{rr} + vw_r/2) = 0 \quad (r = 0). \quad (4.4)$$

The initial conditions are zero. We give the incident wave pressure on the plate in the form:

$$p_w = A[1 - \exp(\beta c_0 t)]\exp(\alpha c_0 t). \quad (4.5)$$

The quantities A, α, and β characterize the amplitude, growth rate, and subsequent drop in pressure in the wave, respectively.

The solution of the presented system was obtained by the method of finite difference. The explicit approximation of the problem by finite differences was used. The plate radius R was divided into 20 segments h_r. The interaction time was divided into J small steps τ $(t = m\tau, m = 0, 1, 2, 3, ..., J)$. As a result, for the each pint of the plate i, we obtained the expression that is qualitatively written in the form [17,18]:

$$w^{m+2} = F\left(w^{m+1}, w^m, p^{m+1}\right), \tag{4.6}$$

where F is the known function that is different within and on the contour of the plate. Starting values for the calculation for instants $t = 0$ and $t = \tau$ are determined by the initial conditions.

The results of numerical study of the influence of the plate thickness on the interaction are shown in Figure 4.13 [17,18,23]. A steel circular plate ($R = 5.5$ cm) was considered. The following parameters describe the initial wave form: $A = 2.5$ MPa, $\alpha = -0.07$ cm^{-1}, $\beta = -5$ cm^{-1}, and $c_0 = 1500$ m/s (4.5).

In Figure 4.13, the curves labeled 0.2, 0.4, 0.8, and 1.2 correspond to the plate thicknesses of 0.2, 0.4, 0.8, and 1.2 cm. The crosses show the pressure p corresponding to an absolutely rigid surface.

As can be seen, the influence of the plate pliability manifests essentially immediately. The first pressure peak and the following form of the pressure curves strongly depend on the plate thickness. In particular, for $h \geq 0.08$ cm, the pressure does not drop below zero, while for $h \leq 0.4$ cm, a disruption (cavitation) of the fluid is possible if it is assumed that cavitation occurs at pressures in the range from -0.2 to -0.35 MPa. As the plate velocity approaches zero, the plate becomes effectively more and more rigid to the water wave, and the pressure starts to increase. The water is compressed to higher positive pressures during the reverse motion of the plate and the pressure starts to exceed the pressure that would occur for a rigid surface (crosses). The absolute value and the shape of the second series of pressure peaks in Figure 4.13 are dependent on the natural frequency of vibrations of the plate.

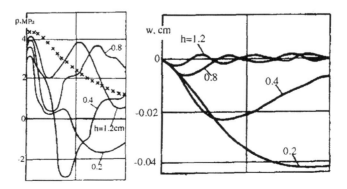

FIGURE 4.13 Pressure (left) and displacement (right) curves at the center of plates of thicknesses $h = 0.2, 0.4, 0.8,$ and 1.2 cm as a function of time t from 0 to 17 ms for the pressure, and from 0 to 33 ms for the displacement (no-cavitation model) [23].

Extreme Effects and Waves

An increase in the frequency (plate thickness) results in a diminution in the spacing between the pressure peaks. Thus, the effect of cavitation is expected to be important if the duration of the loading by the water pressure is less than the fundamental period of the plate vibration.

4.4.2 Effects of Cavitation on the Plate Surface

Let us consider the equation of plate motion (4.1), taking into account the possibility of cavitation, originating on the plate. In this case, the pressure p (4.2) is written in the form:

$$p = \delta(r,t)(2p_w + \rho_0 c_0 w_t). \qquad (4.7)$$

Equation (4.7) is rather similar to Eq. (4.2), but the former contains the delta function $\delta(r,t)$. This function takes into account the appearance and disappearance of cavitation on the plate surface. Equations (4.1) and (4.7) yield that

$$w_{rrrr} + 2r^{-1}w_{rrr} - r^{-2}w_{rr} + r^{-3}w_r = -12(1-v^2)\left[\rho h w_{tt} + \delta(r,t)(2p_w + \rho_0 c_0 w_t)\right](Eh^3)^{-1}. \qquad (4.8)$$

Here the case $\delta(r,t) = 1$ corresponds to *no* cavitation on the plate. Case $\delta(r,t) = 0$ corresponds to *yes* cavitation on some point (or points) of the plate. The function $\delta(r,t) = 1$ or 0 is determined during the numerical solution of Eq. (4.8).

It is assumed that at $t = 0$, the delta function $\delta(r,t) = 1$ as for the rigid plate. However, the plate is not rigid, but starts to deform and move under the influence of the pressure pulse. This changes the pressure pulse (see Eq. (4.2)). If the full pressure drops below some critical value p_k in certain point i, then we assume that this point of the wet plate surface separates from the water and cavitation is generated in this point according to the cavitation model. In this point, the delta function is changed to 0 in Eq. (4.8). If the following calculations show that in this point, the full pressure is above the critical value p_k, then in this point, the delta function is returned to 1 in Eq. (4.8).

Namely, at each time step $m + 2$, the values w^{m+2} are determined using earlier founded w^{m+1}, w^m, and δ^{m+1} for each point of the plate (4.6). Then, the pressure of the fluid is determined (for the step $m + 2$) according to the model of ideal liquid (4.2). It is assumed that cavitation exists at the points in which the pressure is lower than p_k. In this point, $\delta^{m+2}(r,t) = 0$. For another point, $\delta^{m+2}(r,t) = 1$. As a result, the function $\delta(r,t)$ is fully determined for the step $m + 2$, and we can continue calculation on the step $m + 3$.

The effects of cavitation on the pressure and displacement of plates were studied in [17–23]. Some results are presented in Figures 4.14–4.16 for the pressure and displacement curves calculated for a D16AT duralumin plate ($E = 7.15 \times 10^5$ kg/cm^2, $v = 0.3$, $\rho = 2.8$ kg/cm^3, $R = 5.5$ cm). The following parameters describe the initial wave form: $A = 16.9$ kg/cm^2, $a = -0.07$ cm^{-1}, $\beta = -0.3$ cm^{-1}, and $c_0 = 1200$ m/s. It was assumed that $p_k = -2$ kg/cm^2.

Pressure curves at the plate center are presented in Figure 4.14 (solid lines with cavitation, dashed lines with no cavitation). Curves of the pressure marked with the numbers 1, 2, 3 (no cavitation) and 4, 5, 6 (cavitation) are shown for plates with thicknesses $h = 0.8$, 0.4, and 0.2 cm, respectively. Curves 5 and 6, which are calculated according to the cavitation model, have sections that are coincident with the line $p = 0$. For $h = 0.8$ cm (lines 1 and 4), the pressure at the center reaches negative values but does not drop below the critical value; that is, cavitation does not occur. Cavitation appears for $h = 0.4$ cm (lines 2 and 5). The lifetime of the cavitation zone is 8.3 ms. For $h = 0.2$ cm (lines 3 and 6), the cavitation zone exists for approximately 13.3 ms.

The curves in Figure 4.15 characterize the changes in the pressure and displacement at the surface of the $h = 0.2$ cm plate. The dashed curves 1, 2, 3 (no cavitation) and solid curves 4, 5, 6 (with cavitation) correspond to the times 8.3, 16.6, 58.3 ms and 8.3, 16.6, 27 ms, respectively. The cavitation model showed that the displacement reached its maximum value earlier (curve 6) at $t = 27$ ms, compared to the non-cavitation model which gave the maximum value later (curve 3) at $t = 58.3$ ms. The maximum displacement found with cavitation (curve 6) was almost twice the displacement calculated according to the non-cavitation model (curve 3). It is evident from Figure 4.15 that the central part of the plate moves as a rigid body at the initial instant that the shock pressure is applied, when the displacement and pressure change rapidly only near the edge. Therefore, a low-pressure zone is generated near the edge (curves 1 and 4). At $t \approx 8.3$ ms, the cavitation forms a ring-like low-pressure zone (curve 4). At $t \approx 16.6$ ms, the cavitation zone covers the central part of the plate. The velocity of the plate motion later decreases as it reaches its maximum displacement (curve 6) and the cavitation zone diminishes. There is no cavitation at the beginning of the plate's return motion when $t \approx 27$ ms.

A curve of dependence of the maximum displacement of a plate from the alloy BrKMts ($h = 0.1$ cm, $E = 1.05 \times 10^6$ kg/cm^2, $v = 0.3$, $\rho = 8.5$ g/cm^3, $R = 5.5$ cm) on the amplitude A is shown by dots in Figure 4.16. As the wave amplitude diminishes, a

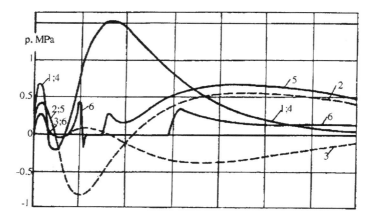

FIGURE 4.14 Curves characterizing the change in pressure on the wet surface of plates as a function of time t from 0 to 0.0583 seconds (solid lines with cavitation and dashed lines with no cavitation) [23].

Extreme Effects and Waves

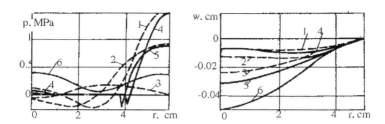

FIGURE 4.15 The changes in pressure (left) and displacement (right) on the surface of the $h = 0.2$ cm thick aluminum alloy plate as a function of radial coordinate r (solid lines with cavitation and dashed lines with no cavitation) [23].

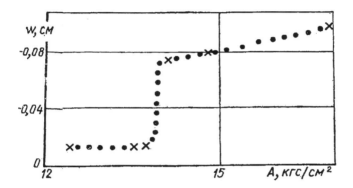

FIGURE 4.16 The displacement jump as cavitation effect.

slow drop (according to a linear law) is observed in the displacement amplitude down to $A \approx 14$ kg/cm^2, when a jump-like drop in the maximum plate deflection occurs (the calculation is made up to the time $t \approx 70$ cm/c_0).

Thus, the jump-like and significant diminution in the deflection amplitude for a moderate change in the amplitude A was found. This effect is associated with the phenomenon of cavitation in the water. For $A \geq 14$ kg/cm^2, cavitation occurs at points of the water abutting the plate, which causes a growth in the total pressure acting on the plate (as compared with the computation without cavitation): for $A < 14$ kg/cm^2, there is no cavitation on the plate surface, and the nature of the interaction is completely different.

Curves of the pressure change are constructed in Figure 4.17 for the case $A = 14.1$ kg/cm^2 (curve 1) and $A = 13.7$ kg/cm^2 (curve 2) with the possibility of the appearance of cavitation taken into account. For $A = 14.1$ kg/cm^2, the curve shape agrees with that found for $A = 16.9$ kg/cm^2, while for $A = 13.7$ kg/cm^2, the pressure curve is similar to that computed for $A = 16.9$ kg/cm^2 but without taking account of the cavitation. We connected this jump effect with generation and focusing of ring-like cavitation wave in the plate center (see, also, Figure 4.15 (left)). The calculation is made at the time $t \approx 70$ cm/c_0.

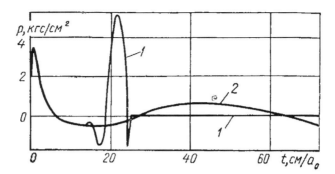

FIGURE 4.17 Extreme amplification of the pressure because of focusing the cavitation wave in the plate center.

On the whole, the calculations show that the effect of cavitation on the impact deformation of deformable elements may be very important. In particular, the cavitation tends to increase the wave pressure on the elements.

4.4.3 Effects of Cavitation in the Liquid Volume on the Plate–Liquid Interaction

Formula (4.7) was obtained taking account of the possibility of the origination of cavitation on the body surface (which corresponds to the case of separation of the surface from the water), but it does not take account of the influence on the interaction of cavitation within the water volume. This, in principle, is the constraint on Formula (4.7) since cavitation in the water volume can alter the pressure acting on the body substantially. The contribution of cavitation within the liquid volume will evidently be more significant if prolonged separation of the body surface from the fluid will not occur.

Qualitative analysis of the cavitation interaction process. A compression wave is reflected from a body upon which a wave is incident, and the body is set into motion. As the velocity of the body surface increases, the reflected compression wave goes over into a tension wave, which upon being added to the tail part of the incident wave can cause the appearance of tensile wave and discontinuity in the liquid (see Figure 4.18). Later, the tail part of the initial pressure wave incident on the surface of the discontinuity is reflected as a new tension wave. Thus, new and new discontinuities can form. On the other hand, tensile (rarefaction) waves are reflected from the free surface of discontinuity as compression waves. Complex interference between the incident and reflected tension and compression waves occurs, which is accompanied by the formation and joining together of discontinuities.

A qualitative picture of the one-dimensional cavitation interaction of the incident wave with a certain point of the plate is shown in the z, t plane in Figure 4.18. The solid straight lines characterize the waves moving toward the plate surface, while the dashes determine the reflected waves. The curve starting from the point 0 determines the plate deflection. The lines parallel to the t axis are the location and lifetime of the discontinuities in the fluid.

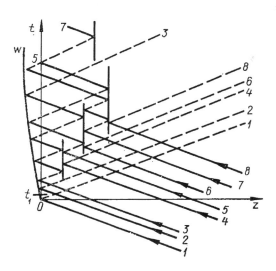

FIGURE 4.18 Wave scheme of the cavitation interaction of deformable body with underwater wave.

The plate is fixed for $t = 0$, and it later starts to be displaced becoming more and more "soft" for the incident wave. At a certain time t_1, the wave starts to be reflected as a tension wave. The intensity of the tension wave increases as the velocity of plate motion increases. As a result, the separation of the plate from the fluid can occur, or a discontinuity appears in the liquid.

It is seen that the cavitation interaction of deformable bodies with liquid may be very complex. Instead of the initial compression wave, the complex series of compression and rarefaction waves act on the body surface. As a result, the body surface begins to oscillate. This conclusion agrees with many observations (see, for example, Figures 4.19–4.22).

In order to better describe the results of the experiments, let us consider the interaction of a circular plate being the bottom of a cylindrical tank with an acoustic underwater wave (see an apparatus for hydrodynamic stamping of components from thin-sheet metals and alloys (Figure 4.8)). The liquid motion is considered potential, and the wall of the tank is not deformable. The pressure in the incident wave is given in the previous form (4.5), and the influence of the static pressure is neglected.

The problem consists in the joint solution of the equation of motion of the plate (4.1) and the wave equation for the liquid (water) in the tank:

$$\psi_{rr} + r^{-1}\psi_r + \psi_{xx} = c_0^{-2}\psi_{tt}, \qquad (4.9)$$

where ψ is the velocity potential of the waves reflected from the plate, and the x axis is passed through the center of the plate and is directed upward. The pressure is determined by the Cauchy–Lagrange integral (1.45). Cavitation appears in a liquid point if

$$p = p_w - \rho_0 \psi_t \leq p_k. \qquad (4.10)$$

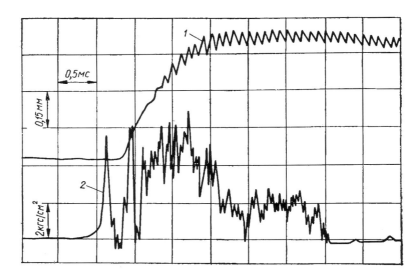

FIGURE 4.19 Displacement (curve 1) and pressure (curve 2) oscillograms near the center of a duralumin plate.

Within the cavitation zones, we assume that the pressure is zero:

$$p = p_w - \rho_0 \psi_t = 0. \tag{4.11}$$

Cavitation disappears in a liquid point if according to numerical calculations the pressure increases above p_k,

$$p = p_w - \rho_0 \psi_t > p_k. \tag{4.12}$$

Expressions (4.9)–(4.12) allow us to find boundaries of cavitation zones at any instant of time. We note that the instantaneous destruction described by (4.10)–(4.12) corresponds to certain approaches of Sections 4.4.2, 1.4.3, and 2.1.

The liquid pressure on the plate is the sum of the incident pressure (4.5) and the reflected waves:

$$p = p_w - \rho_0 \psi_t = A(t)\big[1 - \exp\beta(c_0 t + x)\big]\exp(\alpha c_0 t) - \rho_0 \psi_t. \tag{4.13}$$

The boundary conditions for the plate remains as before (4.3) and (4.4). The kinematic condition on the contact of the water with the plate takes the form:

$$w_t = \psi_x + u_w \quad (x = 0). \tag{4.14}$$

In (4.14), the value u_w is the velocity of the incident wave

$$u_w = -\rho_0^{-1} \int \partial p_w / \partial x \, dt. \tag{4.15}$$

Extreme Effects and Waves

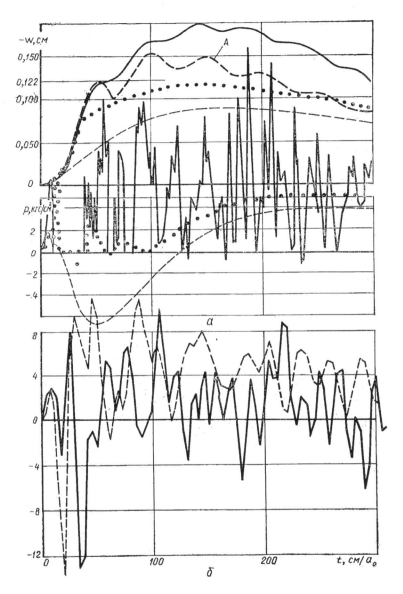

FIGURE 4.20 Results of numerical modeling of the oscillograms.

Generally speaking, (4.14) is valid only for small displacement of the contact surface. However, our researches [17–19] showed that the linear form of the condition is valid enough for the displacement much more than the plate (shell) thickness.

If $\alpha = 0$, $A(t) = A = $ const in p_w (4.5) (see, also, Eq. (4.13)), then (4.15) is rewritten in the form:

$$u_w = -c_0^{-1}\rho_0^{-1}p_w = -c_0^{-1}\rho_0^{-1}A\left[1 - \exp\beta(c_0 t + x)\right] \quad (4.16)$$

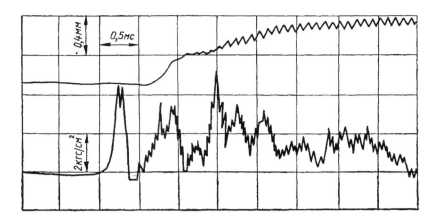

FIGURE 4.21 Displacement (curve 1) and pressure (curve 2) oscillograms near the center of the BrKMts alloy plate.

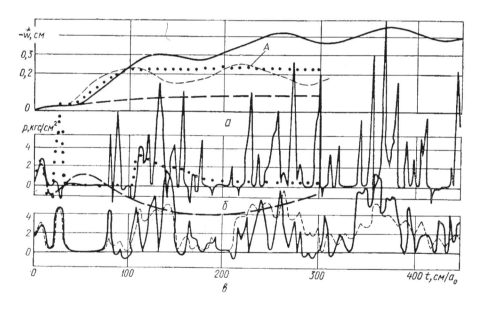

FIGURE 4.22 Results of numerical modeling of the oscillograms of Figure 4.21.

The liquid radial velocity is

$$\psi_r = 0 \qquad (4.17)$$

on the cylindrical tank surface ($r = R$) and on the line $r = 0$. There are no perturbations from the plate at the infinite distant point $\psi = 0$ ($x \to \infty$). The initial conditions are zero.

Extreme Effects and Waves

Method of solution. The solution was sought by the finite difference method. The region occupied by the plate and the liquid was divided into a rectangular grid. The central differences or the one-sided differences (on the boundaries) were used.

The effect of cavitation was taken into account in accordance with the recommendations of [17–19]. At each time step $m + 2$, first the formulated aforementioned boundary problem was solved using w, u, and ψ found at the nodes of the grid on steps $m + 1$ and m. Then, the pressure of the fluid was determined (for the step $m + 2$). The regions in which the pressure was lower than the tensile strength p_k were assumed to be cavitation zones. Thus, while determining the cavitation zones, the liquid was viewed as ideally elastic. It is emphasized that the values ψ were used that had been found at the preceding steps by taking the possibility of cavitation into account. If there was cavitation according to the calculation for the step $m + 2$, then inside the cavitation zones, the values ψ were redefined by Eq. (4.11). Analogously, the calculation of the liquid was made at each time step of solving the problem.

Some results of the calculation and comparison with the experimental data. Now we estimate the accuracy of the approaches developed above to cavitation effects. An attempt is hence made to take account of the influence of the air layer originating ahead of the piston on the pressure in the water. The increase in the air pressure in the fluid results in an exponential increase in p_w to the time t_1 when the piston impacts on the fluid surface. It is assumed that the pressure prior to $t + x/c_0 = t_1$ increases according to the law:

$$p_w = A\left[e^{\beta(c_0 t + x)} - 1\right]\left[\left(\exp(\beta c_0 t_1) - 1\right)\right]^{-1}, \quad \beta > 0. \tag{4.18}$$

For $t_1 \leq t + x/c_0$,

$$p_w = A \exp \alpha\left[c_0(t - t_1) + x\right], \quad \alpha < 0. \tag{4.19}$$

Computations were performed for the quantities $A = 12.5$ kg/cm^2, $\alpha = -0.015$ cm^{-1}, $\beta = 0.07$ cm^{-1}, $t_1 = 10$ cm/c_0, and $p_k = -2$ kg/cm^2. In this case, the shape and length of the wave corresponded sufficiently well (Figure 4.9) to those observed. The length of the numerical cell along the cylindrical axis was 0.6 cm, the length of the numerical cell along radius was $R/10$, and the time step was 0.15 cm/c_0, where $c_0 = 1200$ m/s.

The plate thickness and material during the experiment were selected so as to assure sufficient displacement and the appearance of cavitation effects in the zone of elastic operation of the chosen material.

Duralumin plate. Displacement (curve 1) and pressure (curve 2) oscillograms near the center of a duralumin plate with $h = 0.2$ cm are presented in Figure 4.19. The pressure was measured by a sensor at the end of a thin rod fastened to the cylinder wall at 12 mm from the plate. The maximum values of the displacement and pressure were approximately 0.52 mm and 7 kg/cm^2.

The displacement curve has a significant lag relative to the pressure curve (on the order of 0.35–0.40 ms). Despite the action of a considerable pulse (the first peak on the pressure curve), the plate remains absolutely fixed. The plate starts to move opposite rather in the direction of action of the pressure. Such a nature of the interaction is impossible. The strain gauge therefore does not record the initial stage in plate motion.

Indeed, according to the calculations (see Figure 4.15), the center of the plate moves at the first instants as the undeformable body. Then, the wave of the bending comes to the center. As a result, the plate begins to move back for very short time. Only after this, the plate begins to move again in the loading direction and the strain gauge begins to fix this motion. On the whole, the plate motion is oscillatory. Acceleration and deceleration sections can be distinguished on the deflection curve. The peak pressure change in the liquid is explained by cavitation effects.

Results of the displacement and pressure computations at the central point of the plate are presented in Figure 4.20. Let we stop on curves of 4.20a. Curves found taking into account the cavitation in water volume are superposed by solid lines. However, the curve A is found according to the 2D theory of the ideal elastic model of the liquid (water).

Curves found taking into account the cavitation only on the plate surface are superposed by dots. The dashed curves were calculated according to the model of Section 4.4.1 without taking into account the cavitation. Qualitative agreement between the solid curves and the experimental results is observed. The lines superposed by the dots and dashes correspond worse to the experimental data. If cavitation is not taken into account, then the rarefaction in the water reaches -7 kg/cm², which contradicts the experiment.

The quantitative discrepancy between the maximum values of the solid and experimental curves reaches 300% for the displacement and 200% for the pressure. This is primarily explained by the inaccuracy in the experiment. In the initial stage of the interaction when the central part of the plate moves as a rigid body, the wire strain gauges are unsuitable, in principle, for the determination of the deflection at the center. The pressure sensor (of approximately 11 mm diameter) and the elastic rod on which it is disposed affect the measured pressure, and this distortion is not taken into account in the theory. Theoretical and experimental curves determined the pressure at different, albeit nearby, points of the water. On the other hand, the used numerical cells do not assure high accuracy of the calculations. It is assumed in the computations that the height of the water column exceeds 180 cm, while the height in the experiments was 50 cm. Therefore, the influence of the water upper boundary was not taken into account. Finally, the theory of cavitational interaction used here does not itself take account of the whole complexity of the process occurring here in the water.

An analysis of the theoretical and experimental curves permits the conclusion that the strain gauge starts to measure the strain at $t = 55 - 60$ cm/c_0 when $w \approx -0.122$ cm (Figure 4.20a). Namely, then does the displacement diminish temporarily and a trough, seen well on the displacement curves, appears. The occurrence of trough is not directly dependent on the arrival of a bending wave from the edge to the center, which is reached at $t \approx 36$ cm/c_0 for the incident wave taken. It apparently appears because of deceleration of the center part, when a gas cavity originates above it, and a -2 kg/cm² negative pressure is reached with a subsequent acceleration during collision between the plate and the water, when the cavitation cavity vanishes and the pressure wave is fixed at the center.

Let us consider the displacement curve starting with the line $w \approx -0.122$ cm and the pressure curve computed using the cavitation model. These curves are in good

Extreme Effects and Waves

agreement with the experimental curves. In particular, the agreement is observed between the maximum displacement of 0.0504 cm, measured from $w \approx -0.122$ cm, and that found experimentally.

Two curves of the pressure calculated for water without cavitation are presented in Figure 4.20б. We used different numerical schemes for these calculations. In contrast with Figure 4.20a, it is seen that the results do not agree with experiments (Figure 4.19).

The BrKMts alloy plate. Oscillograms for plates from the BrKMts alloy ($h = 0.1$ cm) are presented in Figure 4.21. The curves differ from those obtained for the duralumin plate. The lifetime of the cavitational cavities grew; the plate deceleration and subsequent acceleration sections due to pressure occurring during collapse of the cavitational cavities became more noticeable. The deflection curve lags by approximately 0.80–0.85 ms as compared to the pressure curve. The displacement curve starts with a trough. The measured maximum displacement and pressure values are approximately 0.7 mm and 5.5 kg/cm².

Displacement (a) and pressure (б, в) curves, computed for a plate from the alloy BrKMts ($h = 0.1$ cm), are constructed in Figure 4.22. The notation in curves (a) and (б) corresponds to that in Figure 4.20. They were computed for $p_k = -2$ kg/cm². As p_k increases, the deflection calculated with cavitation taken into account in the volume of the water does not change, but the maximum values of the pressure diminished. This is seen from the pressure curves (Figure 4.22в) constructed for $p_k = -1.3$ kg/cm² (solid curve) and $p_k = 0$ (dashed curve).

Comparison the curves in Figure 4.22, computed with cavitation in the water taken into account, and the oscillograms in Figure 4.21 show that the pressure curves are in good qualitative agreement. This also refers to the location of the peaks and cavitation troughs on the pressure curves (the first, moderate peak on the theoretical curves is not determined in the experiment). The maximum values of the pressure computed for $p_k = 0$ and $p_k = -1.3$ kg/cm differ by 20% and 35%, respectively, from those measured. As p_k diminishes, the divergence of the maximum values of the pressure increases. The theoretical displacement exceeds the measured value almost fourfold. Such a significant difference between the displacements when agreement between the pressure curves is good is explained by experimental error, as in the case of the duralumin plate with $h = 0.2$ cm. The deflection starts to be measured from the time $t \approx 130 - 140$ cm/c_0. The deflection of the center point of the plate hence equals 0.3 cm. Taking this into account, the discrepancy between the maximum deflections found theoretically and experimentally diminishes to 100%. Starting from $t \approx 130$ cm/c_0, the section of the displacement curve (see Figure 4.22) is in good qualitative agreement with the experiment. There is a trough, and the repeated deceleration of the plate is clearly defined.

The difference observed between the theoretical and experimental curves can be explained by the same reason as for the case of the duralumin plate. However, here the error in the assumption that the deflections are small, used in deriving (4.1), can be perceptible since the deflection exceeds the plate thickness by more than threefold.

The displacement and pressure curves computed with cavitation on the plate surface taken into account (superposed by dots in Figure 4.22) are in worse agreement with experimental data than those constructed with cavitation taken into account in the volume of the water. Still more significant is the discrepancy between experiment and the results of a computation without taking account of the influence of cavitation (Figures 4.22a and 6, dashed lines).

Conclusion. There is extreme importance of the cavitation effect for problems considered here.

4.4.4 Effects of Plasticity

Now, on the basis of a previously formulated models (Sections 1.2 and 1.5), we can investigate the process of hydrodynamic pressing of plates in a press gun (see Figure 4.8 as the scheme). We will form the large plastic deflection of plates to more correctly compare the results of calculations and experiments.

Below the equations of motion of the plate are presented for the case of large displacements. The plastic properties of the material are described by the theory of flow. The motion of the liquid is examined on the basis of the wave equation. The possibility of cavitation is taken into account.

Statement of the problem. We write the equations of motion of the plate in the form:

$$\frac{h^2}{12}\nabla^2\nabla^2 w + F_1(u,w) + F_1^p = \frac{1-v^2}{Eh}\left(p_z - \rho h \frac{\partial^2 w}{\partial t^2}\right),$$

$$\nabla^2 u - \frac{u}{r^2} + F_2(u,w) + F_2^p = \frac{1-v^2}{Eh}\left(\rho h \frac{\partial^2 u}{\partial t^2} - p_r\right).$$

(4.20)

Equations (4.20) corresponding to (1.87) are written for w and u and for axisymmetric deformation of circular plate. In (4.20), w and u are the normal and radial displacements, respectively; p_z and p_r are the normal and radial loads, respectively; h is the thickness; p is the density; E is the modulus of elasticity; v is Poisson's ratio, r is the radius.

$$F_1(u,w) = -\nabla^2 w \left[\frac{\partial u}{\partial r} + \frac{1}{2}\left(\frac{\partial w}{\partial r}\right)^2 + \frac{1}{2}\left(\frac{\partial u}{\partial r}\right)^2\right]$$

$$- \frac{\partial w}{\partial r}\left(\frac{\partial^2 u}{\partial r^2} + \frac{\partial w}{\partial r} \times \frac{\partial^2 w}{\partial r^2} + \frac{\partial u}{\partial r} \times \frac{\partial^2 u}{\partial r^2} + \frac{v}{r} \times \frac{\partial u}{\partial r}\right) - v\frac{u}{r} \times \frac{\partial^2 w}{\partial r^2};$$

$$F_2(u,w) = \frac{\partial w}{\partial r} \times \frac{\partial^2 w}{\partial r^2} + \frac{\partial u}{\partial r} \times \frac{\partial^2 u}{\partial r^2} + \frac{1-v}{2r}\left[\left(\frac{\partial w}{\partial r}\right)^2 + \left(\frac{\partial u}{\partial r}\right)^2\right];$$

$$F_1^p = \frac{1}{r} \times \frac{\partial}{\partial r}\left(\varepsilon_2^\beta - \varepsilon_1^\beta\right) + \frac{1}{r} \times \frac{\partial}{\partial r}\left(r\frac{\partial}{\partial r}\varepsilon_1^\beta\right) + \frac{1}{r} \times \frac{\partial}{\partial r}\left(r\varepsilon_1^\alpha \frac{\partial w}{\partial r}\right),$$

$$F_2^p = \frac{1}{r}\left(\varepsilon_2^\alpha - \varepsilon_1^\alpha - r\frac{\partial}{\partial r}\varepsilon_1^\alpha\right),$$

(4.21)

where

$$\varepsilon_1^\alpha = \sum_{n=1}^{N}\left(\Delta_n^\alpha \varepsilon_r + \nu\Delta_n^\alpha \varepsilon_\varphi\right) \quad \varepsilon_2^\alpha = \sum_{n=1}^{N}\left(\Delta_n^\alpha \varepsilon_\varphi + \nu\Delta_n^\alpha \varepsilon_r\right), \quad (4.22)$$

$$\Delta_n^\alpha \varepsilon_l = \frac{1}{h}\int_{-h/2}^{h/2} \Delta_n \varepsilon_l^p \, dz, \quad \Delta_n^\beta \varepsilon_l = \frac{1}{h}\int_{-h/2}^{h/2} \Delta_n \varepsilon_l^p \, z \, dz. \quad (4.23)$$

In (4.23), $l = r$ or φ. The expressions for ε_1^β and ε_2^β are written analogously. Equations (4.20)–(4.23) follow from the expressions presented in Sections 1.2 and 1.5 and the assumption that the time of loading of the plate is divided into a finite number k of steps. For an arbitrary step N, we have the correlation between stresses $\sigma_l\left(l = r \text{ or } \varphi\right)$ and deformations $\varepsilon_l\left(l = r \text{ or } \varphi\right): \varepsilon_l = \varepsilon_l^e + \sum_{n=1}^{N}\Delta_n\varepsilon_l^p \, (n = 1,2,3...,N,...k)$. The last expression yields

$$\varepsilon_l^e = \varepsilon_l - \sum_{n=1}^{N}\Delta_n\varepsilon_l^p \quad (4.24)$$

Now we can write Hook's law (1.26) in the form:

$$\sigma_r = \frac{E}{1-\nu^2}\left[\varepsilon_r + \nu\varepsilon_\varphi - \sum_{n=1}^{N}\left(\Delta_n\varepsilon_r^p + \nu\Delta_n\varepsilon_\varphi^p\right)\right],$$

$$\sigma_\varphi = \frac{E}{1-\nu^2}\left[\varepsilon_\varphi + \nu\varepsilon_r - \sum_{n=1}^{N}\left(\Delta_n\varepsilon_\varphi^p + \nu\Delta_n\varepsilon_r^p\right)\right], \quad (4.25)$$

The deformations are expressed through displacements

$$\varepsilon_r = u_r + \tfrac{1}{2}w_r^2 + \tfrac{1}{2}u_r^2 - zw_{rr}, \quad \varepsilon_\varphi = u/r - zw_r/r, \quad (4.26)$$

where z is the coordinate normal to the middle surface of the plate.

The above equations, supplemented by the relations of the theory of plastic flow and the experimental data, allow us to calculate the plastic deformation of a circular plate by an impact of liquid (see, also, Sections 1.2 and 1.5).

We consider the potential motion of the liquid. Let the compressive wave produced by the piston impact is acoustic. When the wave is reflected by the plate, cavitation may occur. Outside the cavitation zone, Eq. (4.9) is correct. The appearance and disappearance of cavitation are determined by Expressions (4.10) and (4.12). Inside of cavitation zones is used (4.11).

Let us formulate the boundary conditions of the problem. Here, we should take into account the possibility of large displacements of the plate. In this case, the kinetic condition on the plate has the form [17,18]:

$$\frac{\partial w}{\partial t} - \frac{\partial u}{\partial t}\frac{\partial w}{\partial r}\left(1+\frac{\partial u}{\partial r}\right)^{-1} = u_w + \frac{\partial \psi}{\partial x} - \frac{\partial w}{\partial r}\frac{\partial \psi}{\partial r}\left(1+\frac{\partial u}{\partial r}\right)^{-1}, \qquad (4.27)$$

where u_w is determined as (4.15). Approximately, (4.27) yields $\dfrac{\partial w}{\partial t} = u_w + \dfrac{\partial \psi}{\partial x}$ (4.14). The dynamic conditions determine the load acting on the plate:

$$p_r = \frac{\partial w}{\partial r}\left(p_w - \rho_0 \frac{\partial \psi}{\partial t}\right), \quad p_z = \left(1+\frac{\partial u}{\partial r}\right)\left(p_w - \rho_0 \frac{\partial \psi}{\partial t}\right). \qquad (4.28)$$

The liquid radial velocity on the cylindrical tank surface ($r = R$) and on the line $r = 0$ is zero (4.17). Approximately, (4.28) yields $p_r = 0$ and $p_z = p_w - \rho_0 \dfrac{\partial \psi}{\partial t}$ (4.13). The initial conditions are zero. On the boundaries of cavitation zones, the velocity potential changes continuously.

We will examine two cases.

The first is the case when the column of liquid is sufficiently high. A wave impacts the plate. The pressure in the wave increases according to (418):

$$p_w = A\left[\exp\beta(c_0 t + x) - 1\right]\left[\exp(\beta c_0 t_1) - 1\right]^{-1}, \quad \beta > 0, \qquad (4.29)$$

if $t + xc_0^{-1} \geq t_1$. For $t + xc_0^{-1} < t_1$, we have (4.19) and

$$p_w = A\exp\alpha(c_0 t - c_0 t_1 + x), \quad \alpha > 0. \qquad (4.30)$$

Here the constants A, α, β, and t_1 can be found experimentally. Value t_1 determines the section of pressure increasing in the wave. Ahead of the front of the wave p_w, the potential $\psi = 0$. As the initial instant of time, we take the instant of loading of the plate.

In the second case, the piston strikes the surface of the liquid with speed $v_0\,(x = L)$:

$$Mv_t = p, \qquad (4.31)$$

where v and M are the speed and mass of the piston, respectively. In the process of impact, the speed of the piston is equal to the speed of the adjacent liquid particles $v = \psi_x$. With a view to this and the condition $\psi = 0\,(t = 0)$, we represent Relation (4.31) in the following form:

$$M\psi_x = -\rho_0 \psi + Mv_0 \quad (x = L). \qquad (4.32)$$

The solution of the formulated problem is complicated largely because the position of the surface for which Conditions (4.27) and (4.28) are written is unknown at every instant of time. It can be determined by introducing the deformed coordinates β_2 and β_3 [18,19],

$$r = \beta_2, \quad x = \beta_3 + w(\beta_2, t)\left(1 + \beta_3 s^{-1}\right)^{-1}, \qquad (4.33)$$

where s is the unit of length measurement. Following [18,19], it is possible to rewrite Eq. (4.9) in the new system of coordinates.

The problem consists in the simultaneous solution of the equations of motion of the plate (4.20)–(4.26) and Eq. (4.9) written in coordinates β_2 and β_3 using the above-mentioned boundary and initial conditions. It is emphasized that according to [18,19], the geometric nonlinear terms in Eqs. (18.20), (18.27), and (18.28) have to be taken into account only for very large displacements.

Method of solution. The solution was sought by the finite difference method. The region occupied by the plate and the liquid was divided into a rectangular grid. The central differences or the one-sided differences (on the boundaries) were used. Numerical calculation algorithms are briefly described in Section 4.4.2.

In determining the plastic strain component, schematized tensile stress–strain diagrams with linear strain hardening were used (see Figure 1.3). At each step N, we carried out the iteration refinement of the cumulative plastic deformations by an algorithm described in [18,19,25] (see, also, Section 1.2). The step for correction of the plastic deformations was one order larger than the step of the difference grid in time.

Results of the calculations. Below, we investigate the interaction process and the effects of nonlinearity and cavitation on the nature of deformation of the plate. The plate was assumed to be elastic or elastoplastic. Some additional results of the investigation of the dynamics of cavitation zones, and the effects of the nonlinearity and the thickness of the plate on the displacements and the pressure in the liquid are presented in [18,19,25].

Elastic plate. Steel plates with $R = 10$ cm and variable thickness were examined. The calculations were made at the time step $\tau = 0.1 - 0.4$ cm$/c_0$, $h_x = 0.8$ cm, $h_r = R/10$, $p_k = -1$ kg/cm^2, and $c_0 = 1460$ m/s (h_x and h_r are the grid steps). The column of liquid above the plate was taken as infinite.

Figure 4.23 presents the results obtained for $h = 0.1$ cm. The pressure in the wave (4.29) and (4.30) was determined from the following parameters: $A = 900$ kg/cm^2, $\beta = 0.07$ cm^{-1}, $\alpha = -0.4$ cm^{-1}, and $t_1 = 10$ cm$/c_0$. The positions of the cavitation zones in the water volume bounded by the wall, the bottom, and the axis of the cylinder are shown. Cavitation began at the surface of the plate at $t = 10$ cm$/c_0$, and then, the cavitation zone grew rapidly. According to the calculation, its thickness attained 8 cm with $t = 12$ cm$/c_0$, then the zone moved upward, while up to $t = 20$ cm$/c_0$,

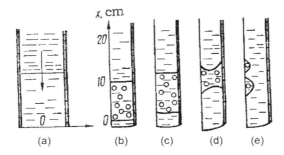

FIGURE 4.23 Pictures of cavitation interaction of an underwater wave with the elastic plate for the instant of time $t < 0$ (a), $t = 12$ (b), $t = 16$ (c), $t = 24$ (d), and $t = 32$ cm$/c_0$ (e).

its volume remained practically unchanged. The size of the zone decreased when $\underline{t} = 24$ cm/c_0, and then, it vanished.

Calculations revealed that the dimensions of the cavitation zone and the time of its existence depend on the length of the incident wave. When the wavelength increases, the thickness of the cavitation zone and the time of its existence increase.

The effect of cavitation on nonlinear deformation of the plate was investigated. The results confirm the conclusion about the increased influence of cavitation on deformation with decreasing ratio of the time of action of the incident wave to the period of the natural vibrations of the obstacle. This influence may be substantial both with large deflection and with deflection not exceeding the thickness of the plate. In the case of sufficiently short waves, the influence of cavitation is always substantial. Calculations showed that the geometric nonlinear terms in Eqs. (4.20) have to be taken into account only for very large displacements.

Elastoplastic plate. First, we present a comparison between results of calculations and experimental data. The latter are obtained with the help of the apparatus shown in Figure 4.8. The weight of the piston was 13 kg, and the height of the liquid column was 50 cm. The impact velocity on the water was 2 m/s. In Figure 4.24 [17–19], the results for a bronze alloy plate ($h = 1$ mm) are presented. The dashed curves are calculated according to the elastic model of the plate material, and the solid curves are calculated by means of the elastic–plastic model. The thick solid line for the pressure p is calculated at the plate center. The curve A represents the displacement w measured near the plate center. The thin solid curve A is the pressure measured in the water at a distance of about 1.5 cm from the plate center.

One can see that there is a good qualitative agreement between the experimental data and the calculations done according to the elastic–plastic model. The lifetime of the cavitation zones is determined by the intervals between the peaks in the pressure curves. It is seen that the locations of the peaks and the cavitation zones on the pressure curves practically coincide. The elastic model may be used only to predict the first moments of the interaction. The maximum displacement of the elastic plate, however, is close to the residual displacement of the elastic–plastic plate.

We recall that the deformation sensors used in the experiments practically do not fix the initial movement of the plate. Therefore, for small deformations, we have a strong discrepancy in the experimental and theoretical values of deflections, as

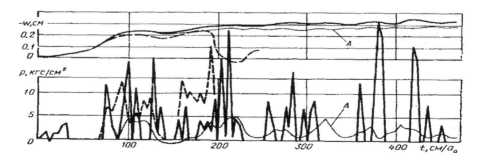

FIGURE 4.24 A comparison of measured and calculated data for a thin metal plate.

shown in Figs. 18. 20, 18. 22, and 18. 24. With increasing deflection of the plate, this inaccuracy associated with the initial motion of the center of the plate, when it moves practically without deforming, decreases.

Let us consider the cases of large plastic displacements of the plate ($R = 5.5$ cm, alloy D16AT) and make a comparison of them with the calculations. The below the values of residual displacements found experimentally [20] (numerator) and theoretically (denominator) are shown for case $M = 2.3$ kg, $\nu_0 = 32$ m/s, $L = 50$ cm, amplitude of the compression wave in the liquid 503 kg/cm^2, and the water column length 50 cm:

h, cm	0,2	0,4	0,6
w, cm	$\dfrac{1{,}65}{1{,}38}$	$\dfrac{0{,}93}{0{,}96}$	$\dfrac{0{,}65}{0{,}79}$

It can be seen that the results of experiments and calculations correlate well enough. The effects of cavitation are demonstrated in Figures 4.25 and 4.26 [25].

The curves of displacement of the plate in Figure 4.25 are not to scale. Smaller weight and higher speed of the piston increased the effect of cavitation. Figure 4.26 shows the curves of deflection calculated for $M = 0.3$ kg, $h = 0.1$ cm, $h_x = 0.2$ cm, $L = 18$ cm, and $\tau = 0.1$ cm$/c_0$, taking and not taking cavitation into account. In the cases $\nu_0 = 32$, 50, and 70 m/s, the amplitudes of the waves in the water were 482, 754, and 1050 kg/cm^2, respectively; since the weight of the piston is constant, the wavelength did not change. Because of the instability of the calculations, curve 3′ could be plotted only up to $t = 32$ m/s. The curves in Figure 4.26, obtained with and without taking cavitation into account, differ substantially from each other. Taking cavitation into account improved the stability of the numerical algorithm and made it possible to examine the case $\nu_0 = 70$ m/s, which could not be calculated on the basis of the model of perfectly elastic liquid. This last circumstance is apparently

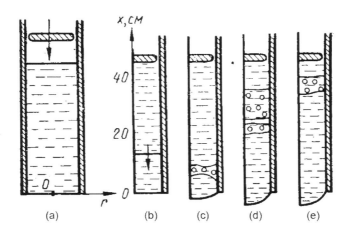

FIGURE 4.25 Hydroelastoplastic processes in the press gun at the instants of time $t < 0$ (a), $t = 32$ (b), $t = 64$ (c), $t = 88$ (d), and $t = 112$ cm$/c_0$ (e) ($h = 0.2$ cm).

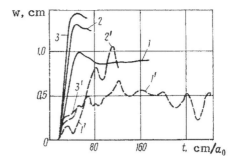

FIGURE 4.26 Effect of cavitation and speed of the piston on residual deflection taking (solid lines) and not taking (dashed lines) cavitation into account: (1 and 1') $\nu_0 = 32$ m/s; (2 and 2') $\nu_0 = 50$ m/s; (3 and 3') $\nu_0 = 70$ m/s.

connected with the large negative pressures and strong variability of the functions arising in the calculations; this led to reduced accuracy of the difference operators, a cumulation of errors, and instability of the calculations.

Comment to this section. It was shown that the hull cavitation can have a large effect on the transient structure–liquid interaction. Therefore, any comparison of experiments and calculations interests us. Figure 4.27 shows the comparison.

Figure 4.27 shows the final plastic displacement of the middle section of metal plates (width = 78 mm, length = 78 mm, thickness = 0.6 mm) loaded by underwater shock wave having different energy contents E. The numerical calculations were made using the 2D cavitation model proposed by Galiev [17,25] (curve 1), Shauer's model [16] (curve 2), and the model of an ideally elastic liquid (curve 3). The experimental data are represented by the dots. It can be seen that cavitation can increase the final displacement of the plate by a factor of 2–3. Curve 1 describes the experimental data very well. Shauer's model (curve 2) is valid only for small plastic deformation.

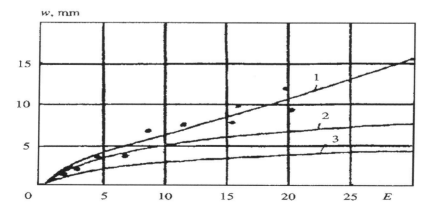

FIGURE 4.27 Comparison of the maximum displacements of the plate center. Black points (dots) are the experimental points; the lines are calculated by means of Shauer's model [16], Galiev's model [17], and the elastic liquid model (3) [23,26].

4.5 MODELING OF EXTREME WAVE CAVITATION AND COOL BOILING IN TANKS

Speaking about gas-hydroelastic system, we mean a structure that interacts substantially with both gas located in liquid and the liquid. Simple example of such constructions is a container containing bubbly liquids. In this section, we present the results of a numerical study of the unsteady interaction of a bubble liquid with an elastic bottom of a cylindrical tank.

Generally speaking in this case, the formulation of the problem given in Section 4.4.3 partly remains valid. However, the equations describing the waves in the cavitating fluid must be replaced by the equations for a bubble liquid. In view of what has been said, we do not give here a complete mathematical formulation of the problem. It suffices to introduce equations describing waves in a bubbly mixture. Such equations were considered in Section 1.4 and Chapter 3. We assume that the flow of the mixture is potential and is described by the following equation:

$$\psi_{rr} + r^{-1}\psi_r + \psi_{xx} = c_0^{-2}\psi_{tt} + nV_t, \qquad (4.34)$$

where $V = 4\pi R^3/3$ is the volume of a bubble and n is the number of the same bubbles in unit volume of the mixture. If $n = 0$, then (4.34) corresponds to the model of ideal liquid. Oscillations of radius of bubbles are described by the Rayleigh equation (1.53):

$$R\ddot{R} + \tfrac{3}{2}\dot{R}^2 = (p_g - p)/\rho_l, \quad p_g = p_0(R_0/R)^{3\gamma}, \qquad (4.35)$$

where p_g is the pressure of the gas in the bubble and γ denotes the adiabatic exponent for gas. In (4.35), $p = -\rho_0\psi_t$.

It should be noted that Eqs. (4.34) and (4.35) were written with the following assumptions:

1. The bubbles do not interact;
2. The number of bubbles does not change; that is, they do not merge or subdivide;
3. There is no relative motion of the bubbles in the liquid.

The results of calculations are given when the non-wetted surface of the plate or free surface of a liquid (4.19) is subjected to a pressure pulse:

$$p_w = A\exp(\alpha t). \qquad (4.36)$$

The amplitude of the load was usually 1 MPa, and the duration of the load was of the order of 5×10^{-5} seconds. In the case of small vibrations, Eq. (4.35) takes the form:

$$\ddot{V} + \omega_0^2 V = \sigma\rho_0\dot{\psi}, \qquad (4.37)$$

where $\omega_0^2 = 3\gamma p_0 \rho_l^{-1} R_0^{-2}$ (3.47), and $\sigma = 4\pi R_0 \rho_l^{-1}$.

We will use three methods of solving the Rayleigh equation for gas-hydroelastic problems:

1. Method of linearization;
2. The analytical method. The analytical model of solution of Eq. (4.35) with $p = $ const is

$$\dot{R} = \left\{ R^{-3} \left[\tfrac{2}{3} \rho^{-1} \left(p_0 (1-\gamma)^{-1} R^{3(1-\gamma)} R_0^{3\gamma} - pR^3 \right) + C \right] \right\}^{0.5}. \quad (4.38)$$

The first integral (4.38) can also be used in the case of variable pressure p. To do this, we subdivide the time interval of motion of the bubble into a finite number of time steps.

Here, we assume that the pressure changes suddenly from step to step while remaining constant in each time layer. Data from calculations made for the previous step is used as the initial conditions for the next integration step. This allows us to determine C with allowance for the next integration step. This allows us to determine C with allowance for a change in p:

$$C = R^3 \dot{R}^2 - \tfrac{2}{3} \rho^{-1} \left[p_0 (1-\gamma)^{-1} R^{3(1-\gamma)} R_0^{3\gamma} - pR^3 \right]. \quad (4.39)$$

Equation (4.38), with condition (4.39), approximately determines the solution of the Rayleigh equation for the case $p = p(t)$.

3. The Runge–Kutta method.

4.5.1 IMPACT LOADING OF A TANK

We studied an influence of gas concentration and bubble dynamics on deformation of the tank. Considering problems were solved by the finite difference method. Using of it was described earlier.

Influence of gas content. Calculations were made assuming that the volume content of gas in the mixture does not change with time and is equal to $G = 1.2 \times 10^{-4}$ or $G = 10^{-3}$. For the indicated values G, sound velocities in the medium were equal to 918 and 320 m/s, correspondingly. First, the tank dynamics was calculated using the ideal liquid model, and the presence of gas was taken into account by the corresponding setting of the speed of sound.

Figure 4.28 shows the variation in the displacement of the central point of the plate as a function of time and gas content, calculated for the case of impulse loading of the steel bottom with a thickness of 0.1 cm and $R = 5.5$ cm [19]. The amplitude of the load (4.36) was 2 MPa. The plate in this and other variants of the calculation was divided into 15 parts, the numerical step along the mixture column was 0.05 cm, and the time step was 0.3×10^{-6} seconds. Curve 1 was constructed for a pure liquid, and curves 2 and 4 for $G = 1.2 \times 10^{-4}$ and $G = 10^{-3}$, respectively. Comparison of these curves shows that the plate deflection is essentially dependent on the gas content. Moreover, deflections also increase with increasing G. An increase in the plate radius by a factor of 6 did not change these results.

Extreme Effects and Waves

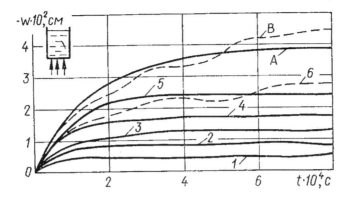

FIGURE 4.28 Dependence of plate displacements from time and gas content: (1) pure liquid; (2, 4) gas content $G = 1.2 \times 10^{-4}$ and $G = 10^{-3}$, respectively, without allowance for bubble dynamics; (3, 5) same, with allowance for bubble dynamics; (6) gas content $G = 10^{-3}$, numerical–analytical method.

Thus, the presence of a gas in a liquid even in a very small amount ($G = 1.2 \times 10^{-4}$ and $G = 10^{-3}$) leads to a very significant (in two and three times) growth of deflections in the considered case.

Influence of dynamics and concentration of bubbles. Let the concentration of bubbles $n = 10^3$ in $1\,\text{cm}^3$. The initial radii of the bubbles that correspond to $G = 1.2 \times 10^{-4}$ and $G = 10^{-3}$ are $R_0 = 10^{-3}$ и $R_0 = 0.66 \times 10^{-2}\,\text{cm}$. They are found by the formula $R_0 = [3G(4\pi n)^{-1}]^{1/3}$. The curves of tank bottom displacements calculated according to the indicated values are shown in Figure 4.28 by numbers 3 and 5. Comparing these curves with curves 2 and 4, calculated for a constant gas content G, we notice a significant influence of bubble dynamics on the tank behavior. For $G = 10^{-3}$, the model of undeformable bubbles leads to an increase in the displacements by a factor of 3.5. If the oscillations of bubbles are taken into account, an increase in the displacements is by a factor 4.9 as compared with the case of pure liquid (curve 1). It is emphasized that during above calculations, the linearized Rayleigh equation was used.

We note that the effect of the nonlinearity of the oscillations of the bubbles on the deformation of the bottom was also studied. Allowance for the nonlinearity of these oscillations led in the cases considered above to an increase in the displacement by 8%–10% (compare the interrupted curve 6 and curve 5). With an increase in the initial size of the bubbles, the effect of their nonlinearity increased. Figure 4.28 shows the curves calculated for $G = 0.5 \times 10^{-2}$ according to the linearized A and nonlinear B equations of bubble dynamics. The maximum divergence of the curves A and B reaches 15%.

Thus, the calculations showed that the influence of the bubble nonlinearity on the deformation of the structure is not very significant. The deformation of the structure, in addition to the gas content and bubble oscillations, is influenced by the dimensions of the bubbles. Large but rarely placed bubbles have little effect on the unsteady behavior of bubble-hydroelastic systems. When the bubbles and the distance between them are reduced, the dynamics of the bubbles begins to influence the displacement more and more strongly. However, if the bubbles become very small, then the stiffness of them becomes significant and the influence of the oscillations of the bubbles can be neglected.

4.5.2 Impact Loading of Liquid in a Tank

We studied the case of impact loading of tank bottom. Let us consider the impact loading of water surface containing in the tank. The method of research is not changed. The pressure acting on the liquid surface is illustrated in (4.36). Our aim is to study waves of cavitation and cool boiling in the liquid.

Figure 4.29 shows curves depicting the change in displacement of the central point of the plate [19]. We varied the height of the column of liquid H in the calculations. Curve 1 was constructed for a pure liquid with $H = 15$ cm, while curve 2 was constructed with the same value of H but with allowance for cavitation. It is evident that curves 1 and 2 are quite different. To understand these results, we conducted tests with $H = 50$ cm. It is intuitively clear that an increase in the height of the liquid column should not appreciably affect the deformation of the tank bottom for the given load. In the last case, curve 3 was constructed for a pure liquid, while curve 4 was constructed for a liquid in which cavitation may occur. It should be noted that curves 2 and 4 nearly coincide for different values of H.

The same cannot be said about curves 1 and 3. The displacement of curve 2 relative to curve 4 is connected with the fact that the compression wave in the liquid reaches the plate later when $H = 50$ cm than when $H = 15$ cm.

Thus, it can be concluded that the data calculated from the cavitation model agrees well with intuitive predictions. We repeated the calculations using a bubbly liquid model. The results obtained are plotted in Figure 4.29 as curves A and B. It is seen that they almost coincided with curves 2 and 4, respectively.

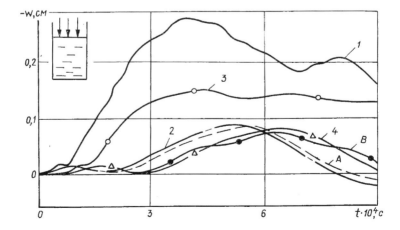

FIGURE 4.29 Change in displacements of the plate center over time: (1, 3) model of ideal liquid; (2, 4) model of liquid with cavitation; (A) bubbly liquid ($H = 15$ cm); (B) bubbly liquid ($H = 50$ cm).

Extreme Effects and Waves 125

It is emphasized that in Figure 4.29, curves 1 and 3 are located above curves 2, 4, A and B, which are calculated for not perfectly elastic liquids. This does not correspond to the results given in the previous sections of this chapter, according to which the account of cavitation phenomena increases the deflection. However, the last results were obtained for the case of the absence of influence on the plate of other deformable liquid boundaries. Thus, changing the boundaries of the liquid volume can greatly affect the dynamic destruction of liquids in tanks and its deformation.

The low degree of accuracy of the results calculated according to the model of an ideal elastic fluid is connected with the effect of boundaries on the repeated reflections of compression waves in the liquid (liquid oscillations in the tank). The greater such re-reflections, naturally the less accurate the calculations. In connection with this, it is interesting to compare compression waves in different models of the liquid. Figure 4.30 shows the propagation of these waves in the liquid column.

Let us analyze this figure. We proceed from the fact that during the time of the action of the pulse, a compression wave is excited in the liquid, the length of which is less than 15 cm. That is, in all variants of calculation, the wavelength is less than H. Therefore, when the free surface is unloaded, a rarefaction wave is formed. It is similar to that excited in solids after short-time loads (see Chapter 22). After the beginning of the interaction of the compression wave with the plate, a second rarefaction wave propagates from the bottom.

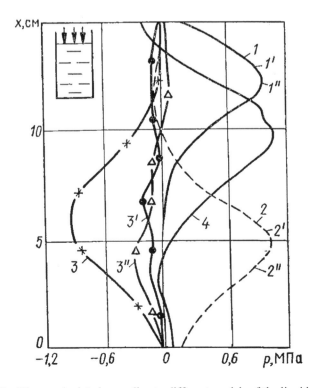

FIGURE 4.30 Waves calculated according to different models of the liquid.

This is illustrated by the diagrams of the pressure variation across the liquid column calculated for the instants of time 1.5×10^{-4}, 3×10^{-4}, and 4.5×10^{-4} seconds. Curves 1–3 correspond to the three points of time, and in the ideal fluid model, curves 1′–3′ correspond to the model of the liquid to be destroyed.

According to the latest model, the wave processes in the liquid are damped at the moment 4.5×10^{-4}. It does not agree with the model of an ideal fluid. In accordance with the last model, a rarefaction wave propagates from the plate and, after reaching the free surface, becomes a compression wave. It is shown in Figure 4.30 for the moment 7.5×10^{-4} (curve 4).

We repeated the calculations using a bubbly liquid model. The results obtained are plotted in Figure 4.29 as curves A and B. In Figure 4.30, the results are plotted as 1″ ($t = 1.5 \times 10^{-4}$s), 2″ ($t = 3 \times 10^{-4}$s), and 3″ ($t = 4.5 \times 10^{-4}$ seconds) [19]. It is seen that curve 3″ almost coincided with curve 3′.

Thus, the cavitation and bubble models are in good agreement with each other.

It is evident that curves 2′ and 3′, constructed with the use of the model considering the possibility of cavitation, differ from curves 2 and 3, constructed from the pure liquid model.

The difference can be attributed to the fact that the cavitation model limits the increase in negative pressures in the liquid. The inaccuracy of the ideal model in the solution of hydroelastic problems is reinforced in our example due to the repeated reflections of the compression waves from the free surface where cavitation effects are strong.

Thus, calculations show that in the design of structures interacting with a liquid, it is necessary to consider the development of cavitation and cool boiling in the liquid. Ignoring such phenomena will lead to inaccurate calculations and an inaccurate physical representation of the problem.

In particular, we found that even a small amount of gas in the liquid has a significant effect on the deformation of structures holding a bubble-containing liquid.

Influence of the bubble screen. The effect of bubble screen on the displacement of the tank bottom was investigated. A bubble screen with a 10 cm thickness was placed 20 cm from the bottom. The amplitude of the impulse load applied to the free surface of the liquid does not change (2 MPa). Calculations showed that the bubble screen reduced the maximum displacement by 20% compared to the case of pure liquid.

Numerical experiments showed the possibility of controlling the behavior of hydroelastic systems using bubble screens. In the cases considered, the wavelength after the passage of the screen increased, but the amplitude and velocity of propagation of the wave decreased. The screen changed the wave properties of the system.

5 Shells and Cavitation (Cool Boiling) Waves

5.1 INTERACTION OF A CYLINDRICAL SHELL WITH UNDERWATER SHOCK WAVE IN LIQUID

Let us consider the action of a wave on the lateral surface of a circular cylindrical shell immersed in water. The wave front moves perpendicular to the axis of the shell (Figure 5.1).

Basic equations. The equations of motion of an infinite cylindrical shell (a ring) follow from Section 1.5 if we put in the last $u = 0$, $k_x = 0$, and $k_y = 0$ and introduce a circle coordinate φ. Discarding the nonlinear terms in the equations and taking into account the equality to zero of the axial derivative, we obtain

$$v_\varphi + \tfrac{1}{12} h^2 R^{-2} \left(w_{\varphi\varphi\varphi\varphi} + 2 w_{\varphi\varphi} + w \right) + w = -\left(1 - v^2\right) R^2 \left(\rho h w_{tt} + p \right) (Eh)^{-1}, \quad (5.1)$$

where R is the radius of the shell, and w and v are the radial and circle displacements of the shell surface, correspondingly. The liquid pressure is

$$p = p_w - \rho_0 \psi_t. \quad (5.2)$$

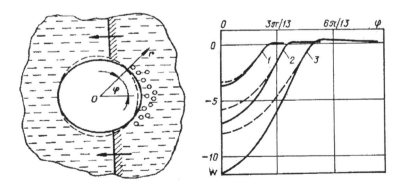

FIGURE 5.1 A qualitative picture of the cavitation interaction of underwater shock waves with the hull of a submarine. The solid line (circle) corresponds to the undisturbed hull. The dashed line corresponds to the hull loaded by the underwater shock wave. The bubbles determine the cavitation zone (left). The radial displacement of the shell calculated for three instants of time (right) [23].

We find the potential ψ from the following equation:

$$\psi_{rr} + r^{-1}\psi_r + r^{-2}\psi_{\varphi\varphi} = c_0^{-2}\psi_{tt}. \tag{5.3}$$

The appearance and disappearance of cavitation are described in Section 44.3 (see Eqs. (4.10)–(4.12)). In particular, the velocity potential in cavitation zones is described by the following equation:

$$p_w - \rho_0 \psi_t = 0. \tag{5.4}$$

We formulate the boundary conditions of the problem. On the surface of the shell, the equality of the normal velocities of the points of the body and the adjacent particles of the liquid yields

$$w_t = \psi_r \quad (r = R). \tag{5.5}$$

The deformed state of the shell and the motion of the liquid must be symmetrical with respect to the line passing through the points $\varphi = 0$ and $\varphi = \pi$. At infinity, the liquid is at rest. The initial conditions are zero.

A shock wave falls on the shell:

$$p_n = A\exp\left[\alpha(c_0 t - R + r\cos\varphi)\right]/R. \tag{5.6}$$

The velocity of the liquid in the incident wave is determined as

$$u_n = -p_n c_0^{-1} \rho_0^{-1} \cos\varphi. \tag{5.7}$$

The problem is solved by the finite-difference method using the explicit difference scheme. Its short description is given in Chapter 4 and publications [17,18].

Results of calculations. The influence of cavitation on the pressure field in the liquid and shell displacements was investigated. The results are presented in [17,18]. Here, we only discuss the effect of cavitation on the change in pressure on the shell surface and its deflections.

Calculations were made for a cylinder made of borosilicate glass: ($E = 6650$ kg/cm², $v = 0.2$, $\rho = 2.23$ g/cm³, $R = 1$ m, and $h = 0.01$ m). The sound speed c_0 was equal 1500 m/s, the pressure $A = 100$ kg/cm², $\alpha = -3$ cm⁻¹, and $\beta = -5$ cm⁻¹.

It was found that the presence of the deformable shell qualitatively changes the pressure in the liquid. The deformability of the shell leads to a strong drop in pressure on the shell and to overtaking by the pressure wave of the front of the incident wave. The pressure in the rarefaction zones according to the model of an ideal fluid falls below −110 kg/cm². Accounting for cavitation leads to smoothing of the pressure curves calculated taking into account the interaction and the disappearance of strong liquid motion along the cylindrical surface.

Shells and Cavitation Waves

The displacement *w* calculated for different instants of time is given in Figure 5.1. The dashed lines are constructed without taking into account cavitation, while solid lines taking into account cavitation for three different time instants. In each instant, the solid lines lie below the dashed ones. The effect of cavitation on *w* increases with time. The excess of *w* calculated with allowance for cavitation over the value found using the ideal liquid reaches 40%. The latest results are in accordance with those obtained for the plates. Cavitation limits the values of negative pressures in the liquid. Therefore, *w* found taking into account the finite strength of the liquid in the rarefaction zones exceeds *w* found for the ideal liquid.

The appearance of extremely low pressures shows the inapplicability of the model of the ideal liquid to a given problem.

5.2 EXTREME WAVES IN CYLINDRICAL ELASTIC CONTAINER

We examine the reaction of two coaxial cylindrical shells, separated with a liquid layer, to pulsed loading applied on part of the surface of the external shell (Figure 5.2). The problem is examined in the two-dimensional formulation. Movement of the shells is determined by the equations written in relation to the displacements, whereas the behavior of the liquid is described by hydrodynamic equations rewritten for potential motion. The problem is solved by the finite-difference method.

The effects of nonlinearity of liquid on the properties of considered system are studied.

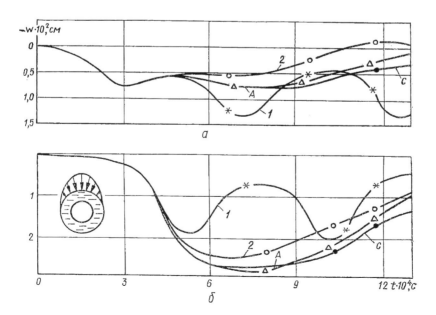

FIGURE 5.2 Curves of changes in deflections of cylindrical shells, calculated with the use of various models of liquids (*a* – external, *б* – internal shells).

5.2.1 EFFECTS OF CAVITATION AND COOL BOILING ON THE INTERACTION OF SHELLS

Governed equations [19]. We use the equations presented in Section (5.1). However, instead of single shell, two shells are considered:

$$\frac{\partial v_\alpha}{\partial \varphi} + \tfrac{1}{12} h_\alpha^2 R_\alpha^{-2}\left(\frac{\partial^4 w_\alpha}{\partial \varphi^4} + 2\frac{\partial^2 w_\alpha}{\partial \varphi^2} + 1\right) + w_\alpha = -\left(1 - v_\alpha^2\right) R_\alpha^2 h_\alpha^{-1} E_\alpha^{-1} \left(p_\alpha + \rho_\alpha h_\alpha \frac{\partial^2 w_\alpha}{\partial t^2}\right). \tag{5.8}$$

Here, P_α is a sum of the hydrodynamic reaction and the incident wave pressure. The shells are numbered in the order of the increase in their radii ($\alpha = 1$ for the internal shell, $\alpha = 2$ for the external shell); w_α and v_α, are, respectively, the radial and annular displacements of the shells; E_α is Young's modulus; v_α is Poisson's ratio; ρ_α is the density of the material of the shells; and R_α and h_α, are, respectively, the radius and thickness of the shells. The liquid contains bubbles. Wave motion of this mixture is potential and described by the following equation:

$$\psi_{rr} + r^{-1}\psi_r + \psi_{\varphi\varphi} = c_0^{-2}\psi_{tt} + nV_t, \tag{5.9}$$

where $V = 4\pi R^3/3$ is the volume of a bubble and n is the number of the same bubbles in unit volume of mixture (see Section 1.4 and Chapter 3). We emphasize that R is the running radius of the bubble. Oscillations of radius of bubbles are described by the Rayleigh equation (1.53):

$$R\ddot{R} + \tfrac{3}{2}\dot{R}^2 = (p_g - p)/\rho_l, \quad p_g = p_0(R_0/R)^{3\gamma}. \tag{5.10}$$

On the surface of the shells, conditions of impenetrability are recorded that have Form (5.5).

A part of the external surface of the structure is subjected to the effect of a pressure pulse whose intensity along the angular coordinate varies in accordance with the sinusoidal law (Figure 5.2) [19]. The pressure is determined by expressions $p_{1w} = Ata^{-1}\cos\varphi$, where $0 \le t \le a$, and $p_{1w} = A(2a-t)a^{-1}\cos\varphi$, where $a < t \le 2a$. Thus, the change in pressure with respect to time corresponds to an isosceles triangle. Here, the value a determines the time of loading. In the last expressions, $0 \le \varphi \le \pi/2$ and $3\pi/2 < \varphi \le 2\pi$, correspondingly. In the equation for the second shell, $p_{2w} = 0$.

It is easy to see that the problem formulated for the two coaxial cylindrical shells is fully consistent with the case of the tank considered in Section 4.5. Therefore, we will not stop on details of boundary and initial conditions and numerical method of solution. We note just that the problem was solved by the finite-difference method.

Calculations were made for steel shells at $R_1 = 50$ cm, $h_1 = 0.5$ cm and $R_2 = 100$ cm, $h_2 = 0.3$ cm.

The loading time was 2.6×10^{-4} s. The main calculations were made when the liquid region was divided by radii into 13 parts, and by the angle into 30 parts, and when the time step was $\tau = 10^{-6}$ s. In this case, high accuracy of calculations was provided.

Model of liquid with cavitation. First, we appreciate the use of the model of the instantly destructible (cavitation) liquid. Corresponding equations of Section 5.1 were used. We give some results of the calculations. The curves of the changes in deflections of frontal points of shells are shown in Figure 5.2 [19]. The amplitude of the impulse load is $A = 1$ MPa. It can be seen that curve 2 obtained using the cavitation liquid model differs substantially from curve 1 obtained using the ideal liquid model. We note that an increase in the pulse amplitude to 5 MPa increased the mismatch of the maximum values of w calculated using the two above-mentioned liquid models. So, for $t = 7 \times 10^{-4}$ s in the case of $A = 1$ MPa for the outer shell, the mismatch was 25%, and for the inner shell, the mismatch was 30%. In the case of $A = 5$ MPa for the outer shell, the mismatch was 32%, and for the inner shell, the mismatch was 45%.

The wave character of the variation of curve 1 in Figure 5.2 is connected with the use of the ideal fluid model. This model leads to an accumulation of errors at each wave reflection from the boundaries (shells).

Thus, the calculation for two coaxial cylindrical shells confirmed the earlier conclusion (Section 4.5) that in cases of deformable containers, the ideal liquid model should be approached with even greater caution than in cases of undeformable boundaries or infinite liquid.

Influence of the cool boiling. We assume that the liquid between the cylindrical shells in the initial state is saturated with bubbles of gas of radius $R = 2.25 \times 10^{-5}$ cm in a concentration $n = 10^3$ of 1 cm. Such gas volume content in a liquid has practically no effect on the sound speed, which is 1490 m/s. The specified amount of gas in the liquid in the initial state can be considered perfectly clean. On the passage of the compression wave, this gas content does not exert any influence. But when a pressure wave is reflected from deformable obstacles, a rarefaction wave appears in which the gas bubbles begin to grow rapidly, changing the mechanical characteristics of the liquid medium and thus affecting the deformation of the structure.

The results of calculations using the bubbly liquid model are shown in Figure 5.2 by curve A. In the calculations, the linearized Rayleigh equation was used. To illustrate the accuracy of the calculations in Figure 5.2, curve C is plotted for the shell using the nonlinear Rayleigh equation.

From the analysis of the curves, it follows that taking into account the presence of bubbles in the liquid (in the indicated amount they are always present there) leads to a significant change in the nature of shell deformation. We emphasize that results of the calculations using the bubble and cavitation liquid models practically coincided, although their results were very different from the results using the ideal fluid model.

The curves of the deformation of the frontal point of the inner shell are shown in Figure 5.3.

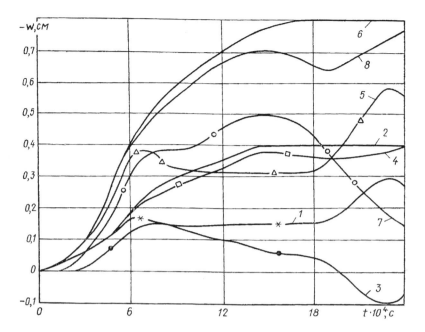

FIGURE 5.3 Effects of the amplitude of the impulse load and the used liquid model on the deflection of the inner shell.

Curves 1–4 correspond to $A = 1\,\text{MPa}$, while curves 5–8 correspond to $A = 5\,\text{MPa}$. Curves 1 and 5 are calculated using the ideal fluid model, curves 2 and 6 are calculated using the cavitation liquid model, and curves 3, 7, and 4, 8 are calculated using the bubbly liquid model. Moreover, curves 3 and 7 are calculated on the assumption that the bubble can be reduced to volume much smaller then its original size. Curves 4 and 8 are calculated on the assumption that the bubble can be reduced only to its original size. Note that in the case $R < R_0$, the computation requires very small steps in time, and the accuracy of the calculations decreases.

The curves of the deformation of the frontal points of the outer (*a*) and inner (*б*) shells, as well as the pressure change curves at the point of the liquid lying opposite the frontal point of the inner shell, calculated for the case $A = 100\,\text{MPa}$ are given in Figure 5.4. Curves 1, 2, and 3 were obtained using the ideal, cavitation, and bubbly liquid models, respectively [19].

Curves 1 were calculated using the time step $\tau = 10^{-6}$. Curves 2 were calculated using $\tau = 10^{-6}/2$.

The strong nonlinearity of bubble oscillations in the case of curve 3 required a reduction in time step. For the latter case, it was equal $\tau = 10^{-6}/2.7$. We emphasize that a further decrease τ to $\tau = 10^{-6}/3$ and $\tau = 10^{-6}/4$ worsened the accuracy of the calculations (curves 3′ and 3″ in Figure 5.4, respectively).

We note that the bubble calculations for $A = 100\,\text{MPa}$ were made by the Runge–Kutta method. Other algorithms considered in [19] for such A turned out to be

Shells and Cavitation Waves

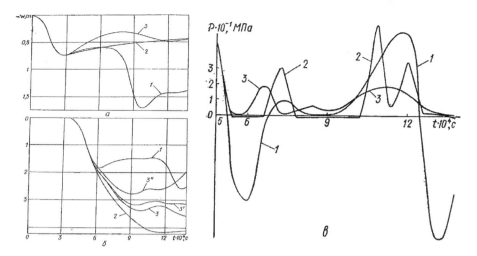

FIGURE 5.4 Displacement of the internal shell and variation in pressure.

unsuitable. Note that the difference between the minimum and the maximum values of the radius of the bubble strongly depended on A. At A of the order of 1 MPa, this difference was 2–3 times; at A of the order of 10 MPa, it reached 4 times; and at A of the order of 100 MPa, it reached 6–7 times. We also note that the displacements in Figure 5.4 far exceed the thickness of the shells and, therefore, should be calculated from nonlinear theories of shells. This was not done here since the inaccuracy in the calculation of the shells does not affect the basic conclusion about the applicability of various liquid models.

5.2.2 Features of Bubble Dynamics and Their Effect on Shells

Calculations showed the complexity of the influence of bubbles on the behavior of the system under consideration. As an example of this complexity, we give here some results of calculations for uniformly distributed bubbles and bubble screens [27].

Evenly distributed bubbles. We emphasize that we are interested here in the influence of cavitation and bubbles on the construction. We will continue to consider the action on the shells of a pulse that is triangular in time. The amplitude of the pulse A varied from 1 to 20 MPa. The gas content was 1% of the volume of the liquid. The presence of gas led to the fact that in the case of an amplitude of 1 MPa, the pressure wave did not reach the inner shell and it remained undeformed. With an increase in A up to 5 MPa and higher, the inner shell begins to deform, and its displacements begin to exceed the displacements obtained according to the ideal liquid model.

The effect of the amplitude on the speed of propagation of waves in a bubble liquid should be emphasized. At amplitudes of 5, 10, and 20 MPa, the velocity of the emerging wave in water was 500, 700, and 1000 m/s, respectively (for pure water, it was 1500 m/s). Bubbles also strongly influenced the shape of the pressure waves arising between the shells.

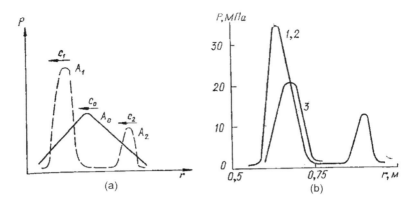

FIGURE 5.5 Effect of bubble dynamics on the shape and velocity of waves in a liquid. (a) shows qualitatively waves formed in an ideal liquid (solid line) and bubble liquid (dashed line). (b) Curves 1, 2, and 3 are calculated for different versions of the Rayleigh equation [27].

Figure 19.5a shows waves qualitatively formed in an ideal liquid (solid line) and bubble liquid (dashed line). It can be seen that a two-velocity wave appears in the bubble liquid (the wave is "torn" into two unequal parts having different amplitudes and moving at different speeds) [27]. In particular, at $A = 10$ MPa, the velocities of these waves were $c_1 = 700$ m/s and $c_2 = 330$ m/s, and their amplitudes were 16 and 3 MPa. If $A = 20$ MPa, we found $c_1 = 1000$ m/s and $c_2 = 670$ m/s, and 32 and 12 MPa, respectively. It is interesting that the amplitude of the first, more "fast" wave in the medium is more than 1.5 times higher than the maximum value of the external action, while in the case of an ideal liquid, the pulse passes into the liquid without changing the amplitude.

Calculations have shown that a decisive role in the formation of a two-speed wave is played by a term in the Rayleigh equation (5.10) that takes into account the acceleration of the walls of the bubble. Thus, the appearance of the peaks is associated with the vibration motion of the bubbles. Curves 1, 2, and 3 are calculated for different versions of the Rayleigh equation. In particular, curve 1 corresponds to

$$R\ddot{R} + \tfrac{3}{2}\dot{R}^2 = (p_g - p)/\rho_l, \tag{5.11}$$

curve 2 corresponds to $R\ddot{R} = (p_g - p)/\rho_l$, and curve 3 corresponds to $0.1 R\ddot{R} = (p_g - p)/\rho_l$.

Wave processes in the presence of bubble screens. We have already noted the influence of bubble screens on the behavior of a structure (see Section 4.5.2). Here, this effect is additionally investigated. The shell and loading parameters remain the same. Amplitude A varied. It was believed that the gas is located in the annular regions at different distances from the inner shell. In those places where the gas is absent, the liquid was described by the ideal fluid model.

Calculations were made for screens with a thickness of 8 cm with an initial gas concentration of 1%, which are located in three places: near the inner shell (1), in the middle of the liquid layer (2), and near the outer shell (3).

Shells and Cavitation Waves

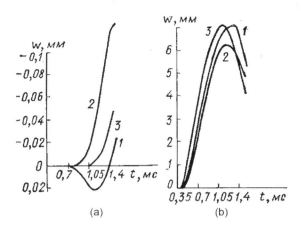

FIGURE 5.6 Curves of changes in displacement of the frontal point of the inner shell for a weak (a, $A = 1$MPa) and strong (b, $A = 20$MPa) pulse.

Figure 5.6 shows the change in the deflections of the frontal point of the inner shell in time for the three above-mentioned screen locations. Two loading cases were considered: weak ($A = 1$ MPa) and strong ($A = 20$ MPa).

It turned out that for large A, the arrangement of the screens does not significantly affect the deformation of the inner shell. For small A, the maximum deflection of the inner shell is when the screen is located in the middle of the liquid layer. If the screen location is near the inner shell, then not only the size of the deflection, but also its sign is changed. The shell first protrudes outward and then begins to move toward the center of curvature.

A wave of small amplitude after the passage of the gas layer almost completely disappeared, while a strong wave could amplify almost twice in the layer [27].

5.3 EXTREME WAVE PHENOMENA IN THE HYDRO-GAS-ELASTIC SYSTEM

Here, we continue our consideration of cylindrical shells. In contrast to Section 5.2, liquid motion is described by the equations of the hydrodynamics [19,27–29]. The effects of the reduction (increase) in the loaded section of the external shell and also those of the nonlinearity of the hydrodynamic equations on the deformed state of the examined system are considered.

The liquid is assumed to be ideal. Its motion is described by the following equations written in the polar coordinate system r, θ:

$$\frac{\partial v_r}{\partial t} = -\frac{1}{\rho}\frac{\partial p}{\partial r}, \qquad (5.12)$$

$$\frac{\partial v_\theta}{\partial t} = -\frac{1}{\rho r}\frac{\partial p}{\partial \theta}, \quad \frac{\partial \rho}{\partial t} + \frac{\rho}{r}\left[\frac{\partial (rv_r)}{\partial r} + \frac{\partial v_\theta}{\partial \theta}\right] = 0, \qquad (5.13)$$

where v_r and v_θ are the speeds of the particles of the liquid in the radial and annular directions; p and ρ are the pressure and density of the liquid. The equation of state is used in Tait's form (1.36). This equation can be used to examine the propagation of waves with relatively high amplitudes (to 3000 MPa) in the liquid.

A part of the external surface of the structure is subjected to a pressure pulse whose intensity along the angular coordinate varies in accordance with the sinusoidal law (Figure 5.7),

$$p_w(t,\theta) = p_{nw}(t)\sin \pi\theta L^{-1}, \quad 0 \le \theta \le L, \tag{5.14}$$

where L determines the angle of the loaded section.

On the shells, the equalities of the normal velocities of the points of the bodies and the adjacent particles of the liquid (impermeability conditions) take place:

$$\dot{w}_1 = v_r \quad (r = R_1), \tag{5.15}$$

$$\dot{w}_2 = v_r \quad (r = R_2). \tag{5.16}$$

Cavitation arose if the pressure at the point of the liquid dropped below the critical value. In cavitation zones, the pressure was equal to zero. Cavitation at the point of the liquid disappeared if there, according to the calculation, the pressure exceeded the critical value. In general, the described algorithm corresponds to the one used in the case of potential fluid motion (Section 4.4.3).

Parameters of the load and the shells are described in Section 5.2.1. The results of some numerical investigations are presented below.

Results of calculations. Figure 5.8 shows the curves of the radial displacements of the front points of the cylinders calculated for various distributions of the extreme load on the surface of the structure. Curves 1–4 correspond to $L = \pi$, $5\pi/6$, $2\pi/3$, and $7\pi/12$. Solid curves correspond to the external shell, while the broken curves correspond to the internal shell. Point T_1 on the abscissa in Figure 5.8 corresponds to the end of the loading. Consequently, the deformation time of the structure is considerably longer than the loading time. The load amplitude $A = 1000\,\text{MPa}$.

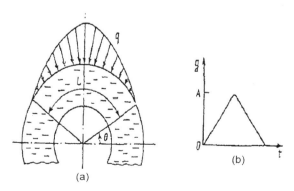

FIGURE 5.7 Cross section of the system (a) and the type of loading (b).

Shells and Cavitation Waves

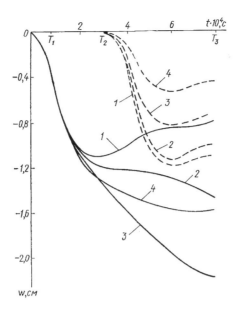

FIGURE 5.8 Effect of variation in the loading area on the displacement of the front points of the external (solid lines) and internal (broken lines) shells.

Deformation of the outer shell with respect to the angle is represented in Figure 5.9 [27–29]. Solid and broken lines denote the displacements for the moments shown by points T_2 and T_3 in Figure 5.8. Curves 1 (and 1'), 2 (2'), 3 (3') and 4 (4') correspond to $L = \pi$, $5\pi/6$, $2\pi/3$, and $7\pi/12$ in (5.14). Namely, curves 1–4 correspond to the instant T_2, while curves 1'–4' the instant T_3.

It can be seen that the variation in the loading area has a strong effect on the nature of deformation of the external shell, with the smallest deflections detected at large and small loading areas. This is associated with the fact that the stress state at the large loading area is close to the momentless (membrane) state, whereas in the case of the small loading area, the applied pulse is insufficient to cause extensive deformation of the structure. Maximum displacements occur when the pulse of the forces is sufficiently intensive and the stress state of the shell greatly differs from the stress state of the membrane. The dimensions of the loading area have almost no effect on the deformation of the internal shell. The stress state of the internal shell is close to that of the membrane shell for all loading cases: the loading of the external shell is local, while the loading of the internal shell is not local since the wave excited in the liquid diverges. Consequently, the deflection of the internal shell does not depend on the method of load application but increases when the load area is increased.

The displacements of the shells did not exceed half the thickness. This justifies the use of the equations of the linear shell theory.

According to the calculations, deformation of the walls in the liquid leads to the formation of the zones of negative pressures (Figure 5.10) [29]. The rarefaction forms in the vicinity of the point $\theta = 0$ of the external shell. Then, with decreasing strain rate, the rarefaction forms in the zone adjacent to this shell. In the internal shell, the negative

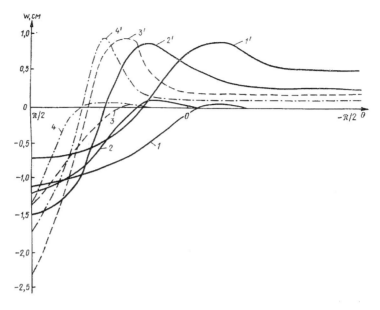

FIGURE 5.9 Variation, in respect of the angle, in the displacements of the external shell in relation to the loading area: $L = \pi$, $5\pi/6$, $2\pi/3$, and $7\pi/12$ (solid and broken lines denote the displacements for the moments denoted by points T_2 and T_3 in Figure 5.8).

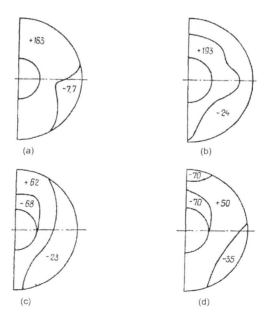

FIGURE 5.10 Distribution of the zone of cavitation at various moments: (a) $t = 300$ μs, (b) $t = 450$ μs, (c) $t = 600$ μs, and (d) $t = 800$ μs. (The figures give the highest absolute values of the positive and negative pressures in the loading on the area $L = \pi$).

Shells and Cavitation Waves

pressures formed only in the vicinity of the front point. It should be mentioned that regardless of the large dimension of the rarefaction zone, the absolute values of the negative pressures in the vicinity of the front point are considerably lower than those of the pressures in the compression zones. However, these values are sufficient for cavitation to take place. The appearance of cavitation was taken into account using the method described above. The effect of cavitation on the given loads and parameters of the shells resulted in a slight (5%–15%) reduction in the displacement of the external shell and an increase in the displacement of the internal shell. The results similar to those shown in Figure 5.10 were also obtained for other cases of the loading areas of the external shell.

5.4 EFFECTS OF BOILING OF LIQUIDS WITHIN RAREFACTION WAVES ON THE TRANSIENT DEFORMATION OF HYDROELASTIC SYSTEMS

The above-presented calculations of transient deformations of the systems are based on the so-called normal conditions for a liquid. However, the structures fairly often contain either very hot liquids under high static pressures or liquefied gases (nitrogen, oxygen, helium, etc.) [30,31].

Using our previously developed methods, we will investigate here the transient deformation of two coaxial cylindrical shells, separated by a liquid at a high temperature, under the action of sources of pressure waves in the liquid. Moreover, examples of calculation of impulse deformation of shells containing cryogenic liquids are given. We will consider the effect of the boiling-up of liquefied helium, nitrogen, and oxygen in low-pressure waves on the behavior of cylindrical shells bounding these media. The effect of the temperature of the shells on their deformation is not considered.

The problems stated above are reduced to the simultaneous solution of linear equations from the moment theory of shells and conservation equations from liquid mechanics. For describing the thermodynamic behavior of liquids heated to a high temperature that were initially under high hydrostatic pressure in a state close to boiling, we use the equations, which hold in a wide range of liquid parameters for a condensed, two-phase, or gaseous state of the medium; the temperature is assumed to be constant.

The calculations were made for steel shells. A detailed analysis of the effect of the grid dimensions, the convergence criterion, and the stability on the calculation accuracy is given in [30,31].

Behavior of shells containing a heated liquid and sources of pressure waves. Water is the most widely used coolant in practice. Therefore, water in the state close to boiling was considered as the heated liquid. Water can remain in a liquid state up to very high temperatures (>600°K) if it is kept under high pressure.

The radius and the thickness of the inner shell are assumed to be equal to 0.5 and 0.01 m, and those of the outer shell, 1 and 0.01 m, respectively. Assume that at $t = 0$, the water temperature is everywhere, with the exception of certain volumes (perturbation sources), equal to 500°K and that the static pressure is equal to 5 MPa. The initial temperature and the initial pressure of the medium in these volumes (sources) are equal to 600° and 20 MPa, respectively. The perturbation sources are located in the middle between the shells.

We consider the cases of one and two sources. We assume that there is not enough time for the thermal conductivity processes to manifest themselves during the straining of the structure; that is, the temperature of the medium in the sources and the surrounding liquid remains at its initial level, regardless of the drastic changes in the density and pressure. We assume in calculations that, in the case of a single source, it occupies two cells adjacent to the symmetry line $\theta = \pi/2$. In the case of two sources, they occupy one cell each, which are located symmetrically relative to the $\theta = \pi/2$ line, and the angle between them is equal to $\pi/3.8$.

The aim of the calculations was to illustrate the possibility of numerical analysis of the thermal shock process within a hydroelastic structure on the basis of Kuznetsov's equations (1.49)–(1.52).

Using the initial values of p and T, we find the value of ρ_0 from Eq. (5.16), which fully defines the initial conditions necessary for solving the hydrodynamics equations.

Figure 5.11 shows the deflections of the head points in time for the outer and the inner shells [27,31]. The initial strained state of the shells is caused by the static pressure of the high-temperature liquid. It is evident that there is no difference in the character of straining for these shells in the above-mentioned cases. The curves have a two-wave character. It should be mentioned that, with respect to the angle, the deformations also form a two-wave configuration.

A numerical investigation of the propagation pattern of perturbations throughout the liquid has shown that, in the case of a single source, the boiling-up occurs in the neighborhood of the head points of the shells due to wall deformation. The boiling-up zones propagate together with the deformation waves along the shells. In the presence of two sources, other boiling-up zones besides those mentioned above develop in the liquid volume between the sources. This is connected with the fact that two hydraulic shocks moving away from the sources collide, after which low-pressure zones develop at the point of collision.

Effect of the boiling-up of cryogenic liquids. Cryodynamics, the discipline concerned with dynamic phenomena in cryogenic liquids, promotes new lines of development in modern technology and, in particular, facilitates the calculation of shells containing nitrogen, oxygen, helium, etc. In order to describe the thermodynamic

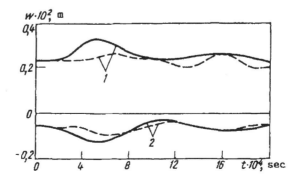

FIGURE 5.11 Displacements of the head points of the outer (1) and the inner (2) shells, calculated for one (solid curves) and two (dashed curves) perturbation sources.

Shells and Cavitation Waves

behavior of these media, we use the averaged equations of state of Vukalovlch, Altunin, and Spiridonov [32] in the following form:

$$z = 1 + \sum_{i=1}^{I}\sum_{j=1}^{J} b_{ij}\omega^{i}\tau^{-j}, \qquad (5.17)$$

where b_{ij} are the experimentally determined adjustment constants, which are not related to each other and are independent of the thermodynamic parameters [32], $z = p/\rho RT$ is the compressibility factor, $\tau = T/T_{cr}$ is the reduced temperature, and $\omega = \rho/\rho_{cr}$ is the reduced density.

The following values of the critical parameters and the gas constant were used in calculations:

a. For nitrogen, $T_{cr} = 126.2°K$, $p_{cr} = 3.4\,MPa$; $\rho_{cr} = 313.1\,kg/m^3$; $R = 296.8 \times 10^{-6}\,MPa/(°K\,kg/m^3)$;
b. For oxygen, $T_{cr} = 154.581°K$, $p_{cr} = 5.043\,MPa$; $\rho_{cr} = 436.2\,kg/m^3$; $R = 259.835 \times 10^{-6}\,MPa/(°K\,kg/m^3)$;
c. For helium, $T_{cr} = 5.19°K$, $p_{cr} = 0.22746\,MPa$; $\rho_{cr} = 69.64\,kg/m^3$; $R = 2.077252 \times 10^{-6}\,MPa/(°K\,kg/m^3)$.

The radius and the thickness of the inner shell were assumed to be equal to 0.75 and 0.02 m, and those of the outer shell, 1 and 0.01 m, respectively [27,31]. The initial pressure for each liquid exceeded the saturation pressure by a factor of 1/10, while the initial temperature was equal to $4/5 \times T_{cr}$. The saturation pressure values were determined for each medium on the basis of the initial temperature values in accordance with data from a manual [32]. The outer shell was subjected to the lateral impulse load with a duration of 300 μs, whose variation in time corresponded to an isosceles triangle, while the distribution along the circular coordinate varied according to the $\sin\varphi$ law at the semicircle points $0 \leq \theta \leq \pi$, and it was equal to zero at other points.

Figure 5.12 shows the curves of head point deflections for the shells, calculated for a load with an amplitude of 30 MPa. It is evident that the characteristics of the outer shell's deformation are heavily dependent on the liquid filling the cavity. On the whole, it can be said that the larger the deflection magnitude, the lower the liquid density. At the same time, the displacement of the inner shell is less heavily dependent on the liquid characteristics than that of the outer shell (Figure 5.12). This is connected with the specifics of wave generation by the outer shell in various liquids.

Figure 5.13 shows the pressure curves calculated for the vicinity of the head point of the inner shell for nitrogen, oxygen, helium, boiling water, and water under normal conditions. (The scale of curve 3 along the axis of ordinates is magnified by a factor of 4.) Regardless of the considerable differences between the liquid characteristics, the amplitudes and shapes of the waves excited in the liquids differ slightly from each other. Helium is somewhat of an exception; the wave amplitude for helium amounts to approximately one-fourth of that excited in other liquids. However, the wavelength of this wave is almost twice as large as the wavelength in oxygen and nitrogen.

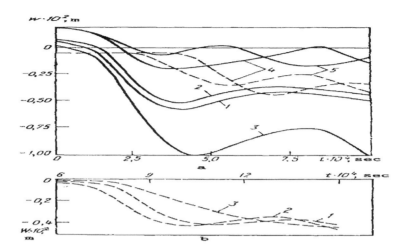

FIGURE 5.12 Effect of the boiling-up of cryogenic liquids on the deflection of the head points of the outer (solid curves) and the inner (dashed curves) shells: (1) nitrogen, (2) oxygen, (3) helium, (4) water at $T = 517°K$, and (5) water at $T = 0°$ (cavitation is neglected).

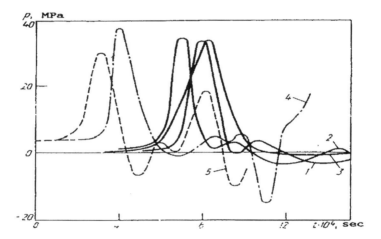

FIGURE 5.13 Time variation in pressure in the liquid near the head point of the inner shell: (1) nitrogen, (2) oxygen, (3) helium, (4) boiling water, and (5) water under normal conditions (cavitation is neglected).

It is evident that if Eq. (5.12) is used, the pressure in cryogenic liquids does not drop to considerable negative values. This cannot be said about water. In all these cases, the maximum amplitudes of the inner shell's deflection were approximately equal.

Remark. We examined the wave processes in shock-excited shells containing highly heated or cryogenic liquids. Recall that we model the problems using model equations describing the properties of liquids. When modeling the behavior of shells, the influence of high or low temperatures on the properties of their material is not taken into account.

Shells and Cavitation Waves

5.5 A METHOD OF SOLVING TRANSIENT THREE-DIMENSIONAL PROBLEMS OF HYDROELASTICITY FOR CAVITATING AND BOILING LIQUIDS

Numeric methods currently known for the solution of the problems of dynamic cavitation hydroelasticity can be conditionally divided into two groups.

1. The first involving direct integration of a system of differential equations in partial derivatives, for example, by the finite-difference method for a discrete grid. In this case, cavitation can be considered by different means: on the basis of a model of instantaneous fluid failure, on the basis of a model of a bubble fluid using a nonlinear or linearized Rayleigh equation, etc.
2. The second involving the use of expansions of all the unknown and known variables in a series of terms of orthogonal functions in terms of one of the spatial coordinates with subsequent use of the numeric method. This method has been termed numerical–analytical method.

The numerical methods of the first group make it possible to account readily for cavitation effects, but they are uneconomical, since large outlays of online computer memory and computer time are required. Numerical–analytical methods that make it possible to reduce the dimensions of the problem to unity are more preferable from this standpoint; with their use, however, it is more difficult to account for cavitation phenomena.

A method for solving 3D physically nonlinear problems of hydroelasticity in the cylindrical coordinate system, which is referred to the second group, is described below.

5.5.1 Governing Equations

Cavitation failure of the fluid can be treated as an instantaneous event. The equation of state of the liquid is shown graphically as the solid line in Figure 5.14. The broken line corresponds to Tait's equation, and the inclined rectilinear segment to its linearized variant. The tangent of the angle between the inclined section and the horizontal axis is equal to the second degree of the sound speed in the fluid. Let us assume that the fluid pressure cannot drop below a certain critical value P_c; this is just what makes the problem nonlinear. When crossing the critical point B, the liquid is failed by means of cavitational boiling and the pressure inside fluid becomes equal to some constant value P_s. Generally, P_s is the pressure of saturated steam. When the pressure increases, the liquid restores its continuity again. Thus, the variables varied linearly everywhere, with the exception of the vicinity of vertical segment BC, where the equation of state is essentially nonlinear.

In this connection, it is necessary to develop an algorithm that makes it possible to continue the problem's solution on passing through the indicated interval BC (from the inclined segment to the horizontal one, and conversely).

Let us examine a structure consisting of two coaxial elastic cylindrical shells of infinite length, between which a liquid is located. The fluid can be considered nonviscous, and its motion potential [33].

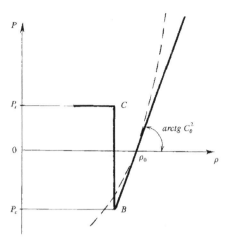

FIGURE 5.14 Curve of fluid pressure P versus density ρ.

The displacements of the shells can be considered small. Let us use the cylindrical coordinate system.

At the initial time $t = 0$, a pulsed load of the form

$$Q(z,\varphi,t) = q(t)\left(1 - z^2/z_0^2\right)\left(1 - \varphi^2/\varphi_0^2\right), \quad |z| \leq z_0, \quad |\varphi| \leq \varphi_0,$$

$$Q(z,\varphi,t) = 0, \quad |z| > z_0 \quad \text{and} \quad (\text{or})\,|\varphi| > \varphi_0,$$

(5.18)

where

$$q(t) = At/T, \quad 0 \leq t < T; \quad q(t) = A(2 - t/T), \quad T \leq t < 2T; \quad q(t) = 0, \quad t \geq 2T$$

begins to act on a portion of the external lateral surface of the outer shell.

Let us present the mathematical formulation of the problem. The equations of motion of the shells are written as follows [33]:

$$R_i^2 u_{i,zz} + (1 - v_i) u_{i,\varphi\varphi}/2 + (1 + v_i) R_i v_{i,z\varphi}/2 + v_i R_i w_{i,z} = \left(1 - v_i^2\right) \rho_i E_i^{-1} R_i^2 u_{i,tt},$$

$$v_{i,\varphi\varphi} + (1 - v_i) R_i^2 v_{i,zz}/2 + (1 + v_i) R_i v_{i,z\varphi}/2 + w_{i,\varphi} = \left(1 - v_i^2\right) \rho_i E_i^{-1} R_i^2 v_{i,tt},$$

$$v_i R_i u_{i,z} + v_{i,\varphi} + h_i^2 R_i^{-2} \left(R_i^4 w_{i,zzzz} + 2R_i^2 w_{i,zz\varphi\varphi} + w_{i,\varphi\varphi\varphi\varphi} + 2w_{i,\varphi\varphi} + w_i \right)/12 + w_i$$

$$= -\left(1 - v_i^2\right) h_i^{-1} E_i^{-1} R_i^2 \left(P_i + \rho_i h_i w_{i,tt} \right),$$

(5.19)

where $i = 1$ for the inner shell and $i = 2$ for the outer shell; subscripts t, φ, and z, respectively, indicate the time and space derivatives; the positive direction of displacement w coincides with the positive direction of the radial coordinate r.

The external pressure P_i on the structure is determined in the following manner for the outer shell:

Shells and Cavitation Waves

$$P_2 = Q(z,\varphi,t) + \rho_0 \lim_{r \to R_2} \psi_t.$$

For the inner shell

$$P_1 = -\rho_0 \lim_{r \to R_1} \psi_t.$$

To find the value of ψ, we follow [33]:

a. The wave equation

$$\psi_{rr} + r^{-1}\psi_r + \psi_{zz} + r^{-2}\psi_{\varphi\varphi} = c_0^{-2}\psi_{tt}, \quad (5.20)$$

where the speed of sound is valid in the zones where P depends on ρ in proportion (the inclined segment in Figure 5.14). The condition

$$P = -\rho_0\psi_t \geq P_c \quad (5.21)$$

must be satisfied in this case.

b. In areas where $P < P_c$, the gas bubbles (cavitation) appear. It is known that the speed of sound in a bubbly fluid is lower than the speed of sound in pure gas. For example, for the gas concentration as small as 4%, c_0 is only 50 m/s (see Chapter 3). We assume approximately that $c_0 = 0$ in the cavitation area. As a result from Eq. (5.20)

$$-\rho_0\psi_t = P_s = \text{const}. \quad (5.22)$$

So the behavior of the cavitation fluid is described by the wave equation (5.20), where c_0 is the unknown discontinuous function. Relations (19. 20)–(19. 22) can be united:

$$\psi_t = \begin{cases} \Omega & \text{if } \Omega \leq -P_c/\rho_0 \\ -P_s/\rho_0 & \text{if } \Omega > -P_c/\rho_0 \end{cases}, \quad (5.23)$$

where

$$\Omega = (\psi_t)_{t=t_f} + c_0^2 \int_{t_f}^{t} \Delta\psi \, dt,$$

t_f is the fixed time moment, $0 \leq t_f < t$.

5.5.2 Numerical Method

The problem was numerically integrated in the following way: wave equation (5.20) was satisfied initially for each time interval. If the value of ψ obtained in this case did not satisfy Condition (5.21), the potential was redetermined proceeding from Eq. (5.22). This made it possible to account for both the development and disappearance of cavitation at each point in the space of the calculated region.

The system of Equations (5.18)–(5.23) was closed by boundary and initial conditions for the wetted surfaces of the shells. The equality of normal velocities of the points of the body and adjacent fluid particles is written as follows:

$$w_{i,t} = \psi_{i,r} \quad (r = R_i, \; i = 1,2).$$

The deformed state of the shells and the motion of the fluid under the type (5.18) load should be symmetric with respect to the planes $\varphi = 0$, $\varphi = \pi$, and $z = 0$. The initial conditions were assumed to be zero.

The given system of partial differential equations was solved by the method of trigonometric Fourier series expansions in terms of the angle φ. The components of the displacement vector of the shells and the velocity potential of the fluid under the load symmetric with respect to the plane $\varphi = 0$ should be sought in the form:

$$f = \sum_{k=0}^{K} f_k \cos k\varphi, \quad v_i = \sum_{k=1}^{K} (v_i)_k \sin k\varphi, \tag{5.24}$$

where $f = \{w_i, u_i, \psi\}$, $i = 1, 2$.

The pressure pulse acting on the outer cylinder can also be arranged in a Fourier series:

$$Q = \sum_{k=0}^{K} Q_k \cos k\varphi. \tag{5.25}$$

In the case of Eq. (5.18), this expansion assumes the form:

$Q_0 = 2\varphi_0 \pi^{-1} \left(1 - z^2/z_0^2\right) q(t)/3, \quad |z| \leq z_0;$

$Q_k = 4\pi^{-1} k^{-2} \varphi_0^{-1} \left(1 - z^2/z_0^2\right) q(t) \left(k^{-1} \varphi_0^{-1} \sin k\varphi_0 - \cos k\varphi_0\right), \quad |z| \leq z_0, \quad k = 1, 2, \ldots, K;$

$Q_0 = Q_k = 0, \quad |z| > z_0.$

Substituting Expansions (5.24) and (5.25) into Eqs. (5.19)–(5.22), the 3D problem can be reduced to the one of searches for the coefficients $(w_i)_k, (v_i)_k, (u_i)_k$, and ψ_k, which depend on r, z, and t, from $K + 1$ systems of equations and corresponding boundary and initial conditions, as well as the conditions favorable to cavitation.

These two-dimensional dynamic boundary problems were numerically solved by the finite-difference method in accordance with an explicit scheme. A rectangular grid can be introduced to apply the method to the range of variations assumed by the variables r, z, and t. The values of the variables at the nodes of the grid are as follows:

$$f_{l,j}^m = f(z_l, r_j, t_m), \quad f = (w_i)_k, (u_i)_k, (v_i)_k, \psi_k,$$

$l = 1, 2, \ldots L$; $j = 1, 2, \ldots J$. The equations for the coefficients of the kth harmonic and the method of their finite-difference approximation were previously described

Shells and Cavitation Waves

in detail in [27] and are omitted here. Note that [27] examined the transient cavitation problem. Since cavitation was accounted for by the use of a linearized Rayleigh equation for small vibrations of gas bubbles in the fluid, the initial system of equations was dissociated into $K+1$ independent two-dimensional systems.

In this paper, the system becomes nonlinear as a result of accounting for cavitation ruptures of the fluid; it is no longer broken down into independent equations, and all $K+1$ harmonics must be considered jointly for each time increment.

Let us dwell in greater detail on the proposed method of accounting for cavitation effects in numeric calculations. For a fixed node l, j of the calculated region, the function

$$\Phi(\varphi) = -\rho(2\tau)^{-1} \sum_{k=0}^{K} \left(3\psi_{l,j}^{m+1} - 4\psi_{l,j}^{m} + \psi_{l,j}^{m-1}\right)_k \cos k\varphi,$$

which approximates the fluid pressure at time t_{m+1} at a point of the medium with the coordinates z_l, r_j, φ, (where τ is the time increment) can be plotted for the $(m+1)$th time layer.

For each step of the problem's solution, a new function

$$F(\varphi) = \Phi(\varphi) \quad \text{if} \quad \Phi(\varphi) \geq P_c \quad \text{and} \quad F(\varphi) = P_s \quad \text{if} \quad \Phi(\varphi) < P_c$$

can be constructed and arranged in a Fourier series,

$$F(\varphi) = \sum_{k=0}^{K} F_k \cos k\varphi,$$

where

$$F_0 = \frac{1}{\pi} \int_0^\pi F(\varphi) \, d\varphi, \quad F_k = \frac{2}{\pi} \int_0^\pi F(\varphi) \cos k\varphi \, d\varphi. \tag{5.26}$$

Let us divide the closed interval $[0, \pi]$ uniformly into N segments with a length of $\delta = \pi/N$. Then, the integrals of Eq. (5.26) can be calculated approximately, for example, from the rectangle formula.

Now the values of $\psi_{l,j}$, which were calculated earlier in accordance with the explicit scheme from the wave equation, can be defined more precisely as a result of the development or existence of cavitation:

$$\left(\psi_{l,j}^{m+1}\right)_k = \left(4\psi_{l,j}^{m} - \psi_{l,j}^{m-1}\right)_k / 3 - 2F_k \tau / 3\rho_0.$$

5.5.3 Results and Discussion

Bleich–Sandler plate problem [34] was considered to verify the accuracy of using mathematical model of the cavitating liquid. The response of a horizontal layer of mass on the surface of a half-space of fluid (see Figure 5.15) was determined. A plane pressure wave with a sudden rise and an exponential decay moves toward the surface, reaching the mass at the time $t = 0$. The system is subject to gravity and atmospheric pressure, all particles being at rest prior to arrival of the shock.

The parameters used in the problem are $c_0 = 1420$ m/s; $\rho_0 = 1000$ kg/m^3; $P_c = P_s = 0$; $A = 0.7$ MPa; the decay length of the pressure wave $D = 1.45$ m; surface mass per unit area $M = 145$ kg/m^3 in accordance with [34].

The motion of the mass is of interest. The history of its velocity is shown in Figure 5.15. Curve 1 corresponds to Bleich–Sandler one; curve 2 corresponds to Galiev's method with the velocity potential; curve 3 corresponds to Newton's method [35] with the displacement potential; curve 4 corresponds to the ideally elastic fluid.

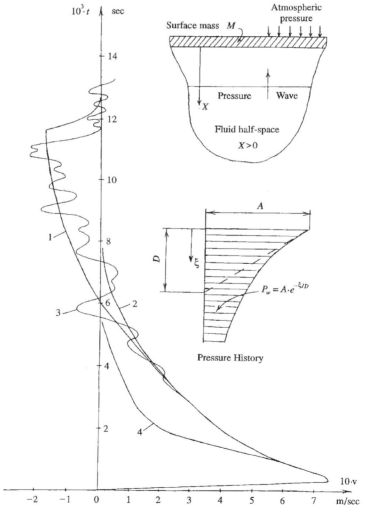

FIGURE 5.15 Bleich–Sandler plate problem showing the velocity of mass (1, 2, and 3 correspond to the case $P_c = \infty$).

Shells and Cavitation Waves

The finite-difference grid steps were also chosen in accordance with the recommendations of [34]. In the case of the ideally elastic fluid, the above methods gave the same results (as shown in Figure 5.15, curve 4). In the case of cavitation, the model with the velocity potential agreed with results of [34], which was worse than the model with the displacement potential [35]. Nevertheless, Galiev's method takes into account the main effect of cavitation phenomena. This model had a quite good numerical stability (as shown in Figure 5.15). This fact is very important in case of 2D and especially 3D problems. As it will be shown further, the suggesting model (based on the velocity potential) has no difficulties when expanding for 3D problems.

Thus, we additionally estimated a precision of Galiev's method, which is the basis of the new algorithms of this chapter. The accuracy of the algorithm developed by the authors was verified using trial calculations.

A two-dimensional method, which makes it possible to calculate coaxial cylindrical shells of infinite length with a cavitating fluid for a non-axisymmetric pulse effect (the case when $z_0 = \infty$ in Eq. (5.18)), was described above in detail. This algorithm does not assume expansions of the unknowns in trigonometric series: the system of differential equations was numerically integrated directly over a two-dimensional regular curvilinear grid with r_j = const and φ_n = const. This algorithm will hereinafter be called approach 1, and the 3D method proposed in the present study approach 2. The reliability of the results obtained using approach 1 has been confirmed earlier, and hereinafter, these results will be used as the data for comparisons.

The parameters of the finite-difference grids employed are as follows: $h_r = (R_2 - R_1)/12$; $h_z = 2h_r$; $h_\varphi = \delta = \pi/30$; $\tau = h_r/2c_0$, where h_r, h_z, and h_φ are the increments with respect to the corresponding spatial coordinates. A curvilinear grid with the increments h_r and h_φ was used in the case of trial calculations made in accordance with approach 1, and a rectilinear one-dimensional grid with the increment h_r in the case of trial calculations made in accordance with approach 2; in this case, the first eight harmonics were computed since the accounting of 12 harmonics altered the results by less than 3%. The ratio of the computer times required for the calculations by method 1 or 2 was approximately proportional to the ratio of the number of the grid nodes along the angle coordinate to the number of the considered harmonics. In the case of $z_0 < \infty$, the length L_c of the computed region was limited by assuming $L_c = 8(R_2 - R_1)$. The boundary condition of symmetry was satisfied in section $z = 0$. The time during which the behavior of the structure was investigated was rather short (not more than 2 ms); after this period, the excitation from the load application point $(z < z_0)$ could not reach the plane $z = L_c$; therefore, the selection of the boundary condition in this section had no effect on the wave pattern. For the simplicity of calculations, the condition of symmetry was used when $z = L_c$.

Let us examine a system consisting of two steel cylinders ($E_1 = E_2 = 20,0000$ MPa, $v_1 = v_2 = 0.3$ and $\rho_1 = \rho_2 = 7800$ kg/m), having the dimensions: $R_1 = 0.5$ m, $R_2 = 1$ m, $h_1 = 0.003$ m, and $h_2 = 0.005$ m, between which water was located under normal conditions: $c_0 = 1500$ m/s; $\rho_0 = 1000$ kg/m^3; $P_c = -0.2$ MPa; and $P_s = 0$. The load characteristics for the trial calculations were as follows: $A = 20$ MPa, $T = 0.13$ ms, $\varphi_0 = \pi/2$, $z_0 = \infty$.

The curves showing the variation in the radial deflections of frontal (with the coordinate $\varphi = 0$, $w < 0$) and rear ($\phi = \pi$, $w > 0$) points of the shells and the fluid

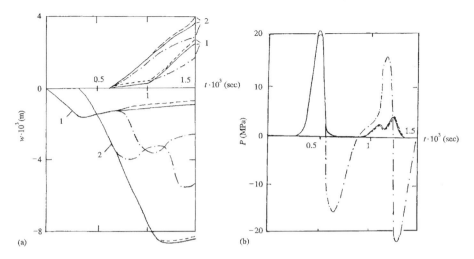

FIGURE 5.16 Variation in deflections of outer (1) and inner (2) shells (a) and fluid pressure (b), calculated in accordance with 2D approach 1 (solid lines) and 3D approach 2 (broken lines). Dash-dotted lines denote the calculation using the model of ideally elastic fluid $(L_c = -\infty)$.

pressure in the vicinity of the zone $\phi = 0$, $r = 0.6$ m with time are presented in Figure 5.16. It follows from Figure 5.16 that approximately by the time moment $t = 0.6$ ms, the excitations reach the rear points of both shells, and when considering the pipes' deformed state, it is necessary to take into account the progressive motion of the pipes as a unit in the direction of the acting load. It is apparent that the results obtained from different approaches are in good agreement. Thus, the proposed numerical–analytical method can be used with a rather high degree of accuracy for the calculations involving the dynamic behavior of structures containing a cavitation fluid. When the model of an ideally elastic fluid is used, nonphysical spikes of high negative pressure in water, which lead to distorted patterns of deformation of both shells, are seen distinctly.

As an example of the calculation made in accordance with the developed algorithm, let us examine the local loading of a structure by a pulse with the following characteristics: $T = 0.13$ ms, $\varphi_0 = \pi/6$, and $z_0 = R_2$. The amplitude of the load varied. The calculations were made using the three-dimensional methods (approach 2). We investigated primarily the deformation of the inner shell.

Curves of the quantity w_{max}/A versus load amplitude A, where w_{max} is the maximum deflection of the frontal point of the internal shell, which can be acquired during the first 1.5 ms after the start of loading (solid line), are shown in Figure 5.17. Also presented here is the curve of the time t^* at which this maximum is attached (broken line) versus the quantity A. An interesting fact is that depending on the same parameter, the curves have approximately the same character, although the physical nature of the quantities which they describe is completely different. A zone of rigorous nonlinearity for small loads compared with $|P_c|$ is observed. The nonlinearity of

Shells and Cavitation Waves

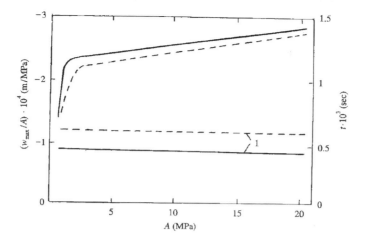

FIGURE 5.17 Curves of w_{max}/A and t^* versus load amplitude (curve 1 corresponds to ideally elastic fluid).

the curves diminishes sharply with increasing load. In the case of an ideally elastic fluid, both curves degenerate into horizontal segments (curve 1 in Figure 5.17).

Isolines of the pressures that develop in the fluid, which are divided by the pulse amplitude (nondimensional pressure), are shown in Figure 5.18 for three amplitude variants: strong $(A = 20\,\text{MPa})$, moderate $(A = 1\,\text{MPa})$, and weak $(A = 0.3\,\text{MPa})$ pulses. Each pattern is plotted at the time t^* corresponding to it. When $A = 20\,\text{MPa}$, $t^* = 1.2\,\text{ms}$ and $w_{max} = 0.00024\,\text{m}$; when $A = 1\,\text{MPa}$, $t^* = 1.1\,\text{ms}$ and $w_{max} = 0.00024\,\text{m}$; and when $A = 0.3\,\text{MPa}$, $t^* = 0.71\,\text{ms}$ and $w_{max} = 0.000036\,\text{m}$. One can see a nonlinear dependence of w_{max} from A. It is a result of the cavitation effect. (We did not take into

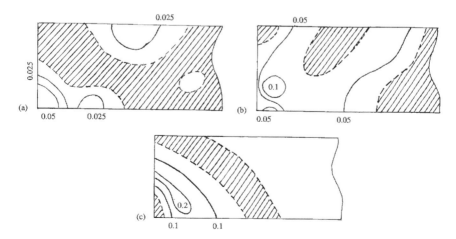

FIGURE 5.18 Isolines of fluid nondimensional pressure for strong (a), moderate (b), and weak (c) pulses. (Zones of cavitation are hatched.)

account the geometrical nonlinear properties of the shell for the case $A = 20\,\text{MPa}$ since we wanted to demonstrate only the cavitation effect.) The wave patterns are plotted in the section $\varphi = 0$, the lower boundary corresponds to the coordinate $r = R_1$, and the upper one to $r = R_2$; the left vertical line corresponds to $z = 0$, the z axis is directed to the right, and the scales along the longitudinal axis and radius are similar. Note that in all three cases, the region of high pressures is localized in the lower left corner in the vicinity of the point with the coordinates $r = R_1$, $z = 0$, and $\varphi = 0$, although the load is initially applied to the outer surface $r = R_2$.

In the other zones of the structure, the patterns of pressures and cavitation failures of the fluid bear little resemblance to one another. Away from the plane $\varphi = 0$, the pressures are damping rapidly. Thus, for an instant, the maximal fluid pressures occurring in the plane $\varphi = 0$ during the whole calculated period $(t \leq 1.5\,\text{ms})$ never exceed 6% of the corresponding impulse amplitude.

Figure 5.19 demonstrates the variations in radial deflections of the inner shell along z axis at different time moments in the planes $\varphi = 0\,(w < 0)$ and $\varphi = \pi\,(w < 0)$. The curves have been drawn for strong and weak pulses. It is apparent that under weak impulse, an evident wave propagation of w along the longitudinal axis is observed, while in case of strong pulse, this phenomenon disappears almost completely. It happens, as it seems to us, due to the fact that when the load amplitude increases, the ratio of the irreversible energy losses in the cavitational zones to the general impulse energy also increases. For a weak pulse, the formation of insignificant corrugation in the vicinity of section $\varphi = \pi/2$, which dampens rapidly with time, was observed in the initial moments of deformation of the inner shell. This effect vanishes for high amplitudes.

Thus, a numerical–analytical method was developed for the solution of a single class of nonlinear nonstationary three-dimensional problems of hydroelasticity, which employs the expansion of functions in trigonometric Fourier series, the

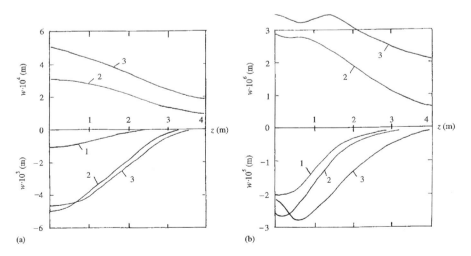

FIGURE 5.19 Curves of w versus z in different time moments: (1) – 0.5 ms, (2) – 1 ms, and (3) – 1.5 ms for strong (a) and weak (b) pulses.

accuracy of which is proven by numerical experimentation. The suitability of the algorithm is demonstrated for the solution of a rather wide range of problems of dynamic hydroelasticity with the allowance for cavitation. The method elaborated can also be employed for the solution of dynamic problems that are similar to those described and take into account failures in solid bodies or their plastic properties.

Conclusion. We carried out in this chapter additional tests of the cavitation models that showed their acceptable accuracy even when solving complex 3D hydroelasticity problems.

6 Interaction of Extreme Underwater Waves with Structures

Important elements of structural resistance to the action of an underwater shock wave and an underwater explosion are considered. Effects of fracture and cavitations are studied. A simple model of the underwater wave is introduced which was formed as a result of the explosion. The influence of plastic deformations, large displacements, and reinforcing elements on the deformation of the structure is studied.

6.1 FRACTURE AND CAVITATION WAVES IN THIN-PLATE/UNDERWATER EXPLOSION SYSTEM

We will examine a two-dimensional problem concerning the deformation and fracture of an infinite plate lying on a compressible fluid and subjected to hydrodynamic shock waves created by the explosion of a cylindrical pentolite charge.

Basic relations. The dynamic behavior of the elastoplastic plate will be described by the equations of a geometrically nonlinear theory of thin shells of the Timoshenko type (see Section 1.5). We use the equations of motion of the plate written in forces and moments (Chapter 1 and [36]):

$$\frac{1}{r}\frac{\partial}{\partial r}(rN_1) - \frac{1}{r}N_2 = \rho h \frac{\partial^2 u}{\partial t^2};$$

$$\frac{1}{r}\frac{\partial}{\partial r}(rQ) + \frac{1}{r}\frac{\partial}{\partial r}\left(rN_1 \frac{\partial w}{\partial r}\right) + p = \rho h \frac{\partial^2 w}{\partial r^2}; \quad (6.1)$$

$$\frac{1}{r}\frac{\partial}{\partial r}(rM_1) - \frac{1}{r}M_2 - Q = \rho \frac{h^3}{12}\frac{\partial^2 \psi}{\partial r^2}.$$

where w and u are the displacements of the middle surface in the normal and tangential directions, respectively; ψ is the angle of rotation of the normal to the middle surface; r is a coordinate reckoned along the radius of the plate; h is the thickness of the plate; ρ is the density of the material; and p is the hydrodynamic load.

The normal forces, moments, and shearing force were determined by integrating the stresses over the thickness:

$$\{N_1; M_1; N_2; M_2; Q\} = \int_{-h/2}^{h/2} \{\sigma_1; \sigma_1 \zeta; \sigma_2; \sigma_2 \zeta; \sigma_{13}\} d\zeta, \quad (6.2)$$

where ζ is the distance measured along the normal to the middle surface.

The plastic properties of the plate were studied on the basis of the algorithms described in Chapter 1 and Section 4.4.4. The relationship between the stresses and the elastoplastic strains was established by means of the formulas (1.34) and (1.83) (see, also, Section 1.5):

$$\sigma_1 = \frac{E}{1-v^2}\left[\varepsilon_1 + v\varepsilon_2 - \sum_{n=1}^{l}\left(\Delta_n\varepsilon_1^p + v\Delta_n\varepsilon_2^p\right)\right];$$

$$\sigma_2 = \frac{E}{1-v^2}\left[\varepsilon_2 + v\varepsilon_1 - \sum_{n=1}^{l}\left(\Delta_n\varepsilon_2^p + v\Delta_n\varepsilon_1^p\right)\right]; \quad (6.3)$$

$$\sigma_{13} = \frac{E}{2(1+v)}\left[\varepsilon_{13} - \sum_{n=1}^{l}6\left(\frac{1}{4} - \frac{\zeta^2}{h^2}\right)\Delta_n\varepsilon_{13}^p\right].$$

The strain tensor components are connected with the displacements by the following geometric relations:

$$\varepsilon_1 = \varepsilon_1^0 + \xi\varepsilon_1^1; \quad \varepsilon_2 = \varepsilon_2^0 + \xi\varepsilon_2^1; \quad \varepsilon_{13} = 6\left(\frac{1}{4} - \frac{\xi^2}{h^2}\right)\varepsilon_{13}^0;$$

$$\varepsilon_1^0 = \frac{\partial u}{\partial r} + \frac{1}{2}\left(\frac{\partial w}{\partial r}\right)^2; \quad \varepsilon_2^0 = \frac{u}{r}; \quad \varepsilon_1^1 = \frac{\partial \psi}{\partial r}; \quad \varepsilon_2^1 = \frac{\psi}{r}; \quad \varepsilon_{13}^0 = \frac{\partial w}{\partial r} + \psi. \quad (6.4)$$

The liquid and gas are assumed to be inviscid and ideal.

The hydrogasdynamics equations will be written in the cylindrical coordinate system r, z for the case of axial symmetric motion:

$$\frac{\partial v_r}{\partial t} = -\frac{1}{\rho}\frac{\partial p}{\partial r}, \frac{\partial v_z}{\partial t} = -\frac{1}{\rho}\frac{\partial p}{\partial z}, \frac{\partial \rho}{\partial t} + \rho\left[\frac{\partial(rv_r)}{r\partial r} + \frac{\partial v_z}{\partial z}\right] = 0. \quad (6.5)$$

The Lagrangian approximation was used in (6.5) [37]. We adopt Tait's form (1.36) as the equation of state of the liquid. The axis of symmetry passes through the center of the plate, in the direction normal to its surface. At the axis of symmetry, there is a volume of the gas of cylindrical form. Its motion is governed by (6.5). The closing equations are represented by the adiabatic equations (1.55), where $\gamma = 3$. The initial speed, pressure, and density of the gas are assumed to be specified. At $t = 0$, the gas volume starts to expand, and the resultant underwater wave and the hydraulic flow act on the sheet causing its deformation.

We assume that the explosive transformation of the explosive material and the expansion of the detonation products occur in accordance with the scheme of instantaneous wave detonation. During detonation, the pressure p_0 and the density of the detonation products ρ_0 – equal to the initial density of the explosive – are instantaneously established throughout the volume of the charge.

Interaction of Extreme Underwater Waves

The system of equations of hydroelastoplasticity is augmented by boundary and initial conditions. We use the condition of impermeability on the contact surface between the plate and the liquid, while we apply pressure equal to 1 atmosphere on the external surface of the plate. We adopt zero initial conditions (except for pressure in its source). The scheme used for numerical solution of the problem is based on the Wilkins method [38] and difference relations presented in [37] for determining the displacements in thin plates. The simple model of cavitation phenomena in the liquid was used. If the pressure drops in a grid cell less than the critical value, it is assumed to be equal to some constant.

Plate failure was accounted for by means of the following criterion: if the intensity of the stresses in a plate cell exceeds the ultimate tensile strength of the material, this cell is considered to have failed.

Results of calculations. Calculations were performed for a steel plate with a thickness $h = 0.02$ m and for water. The thickness and radius of the pentolite charge were assumed to be equal to 0.08 and 0.2 m, respectively. The density of the charge was taken to be 1650 kg/m^3, the detonation rate 7655 m/s, and the initial pressure 12 GPa. The symmetry axis of the gaseous region coincides with the symmetry axis of the isolated volumes of the plate and water.

The dynamics of the given hydroelastoplastic system can be studied on the basis of the above equations if the z axis is directed into the liquid along the symmetry axis. All of the other principles underlying the calculation are analogous to those described earlier in this chapter.

Let us examine the results of the calculation, in which we varied the distance between the gas volume and the plate: $d = 12h$, $16h$, and $20h$. First, let us study the dynamics of the hydrodynamic shock waves. Figure 6.1 shows diagrams depicting the change in the pressure of the underwater wave over time at the point $r = 0$, $z = 0$.

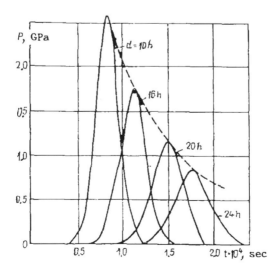

FIGURE 6.1 Change in pressure in the liquid at the point $r = 0$, $z = 0$.

These results were calculated for different distances of the gas volume from the plate. It can be seen that with increasing distance from the source of the perturbations, peak pressures of the shock wave decay exponentially (dashed lane). Another feature of the problem is related to the change in pressure on the plate surface.

Figure 6.2 shows the pattern of peak pressures on the plate surface with $d = 16h$. It is evident that with an increase in r from zero, the pressure acting on the plate decreases exponentially.

When the underwater wave had traveled 2.5–3 m on the plate surface, we assumed in the calculations that all of the energy of the liquid had been transferred to the plate. The calculation was subsequently performed without allowance for the presence of the liquid, under the influence of inertial forces. During the deformation of the plate, a large cavitation region develops in the liquid. This region occupies the entire deformed surface of the plate and grows into the liquid. Failure to allow for cavitation led to rapidly increasing pressure pulsations and loss of stability of the computing algorithm.

Figure 6.3 shows the deformation and fracture regions of the plate (the fractured cells in the plate are hatched). Fracture of the plate began at the point $r = 0$. The fracture zone then propagated from the symmetry axis. For example, with $d = 12h$, a plate with the radius $r = 0.48$ m fractured completely at the moment $t = 18 \times 10^{-4}$ second (Figure 6.3a). As the distance between the source of the underwater waves and the plate increased, the fracture region gradually disappeared and, with $d = 20h$, was not seen at all up to the moment of time $t = 44 \times 10^{-4}$ second [36].

According to the calculations, we can conclude the following: the cavitation zone propagates over the plate with a deformation wave while simultaneously developing deeper in the liquid; fracture of the plate begins at the point $r = 0$ and subsequently extends from the symmetry axis; it is necessary to allow for cavitation when calculating large strains of plates exposed to an underwater wave.

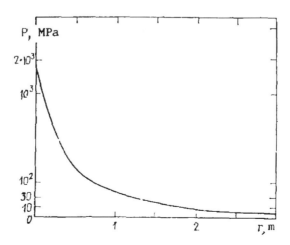

FIGURE 6.2 Theoretical values of peak pressure on the plate surface with $d = 16h$.

Interaction of Extreme Underwater Waves

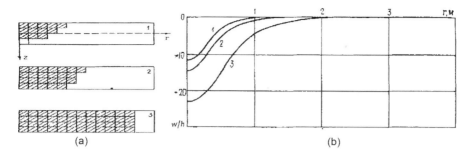

FIGURE 6.3 Regions of the fracture (a) and displacement (b) of the plate calculated for $d = 12h$: 1) $t = 6 \times 10^{-4}$ second; 2) $t = 8 \times 10^{-4}$ second; 3) $t = 18 \times 10^{-4}$ second.

6.2 FRACTURE AND CAVITATION WAVES IN PLATE/UNDERWATER EXPLOSION SYSTEM

Here, we numerically study the dynamic failure of a solid layer lying on a liquid under the influence of a rapidly expanding gas bubble in the liquid. We will also examine the effect of the distance between the gas bubble and the layer on the failure of the latter. We consider the possibility of very large displacements of the three interacting media (solid, liquid, gas) relative to each other [27,38].

Basic relations. Let a deformable plate of the thickness h lies on the surface of a liquid. A volume of gas of cylindrical form is located in the liquid near the plate. We will assume that the gas volume begins to expand at a high rate at $t = 0$.

In solving the problem, it is necessary to consider the interaction of the gasdynamic system and the plate, as well as the possibility of very large displacements of the different media relative to each other. This can be done on the basis of the numerical method of integrating the equations of continuum mechanics described in Ref. [38].

The nonsteady interaction of these media is described by using conservation equations written in the Lagrangian form in the cylindrical coordinate system z, r, θ:

$$\frac{\dot{V}}{V} = \frac{\partial \dot{z}}{\partial z} + \frac{\partial \dot{r}}{\partial r} + \frac{\dot{r}}{r};$$

$$\frac{\partial \sigma_{zz}}{\partial z} + \frac{\partial \tau_{zr}}{\partial r} + \frac{\tau_{zr}}{r} = \rho \ddot{z}; \qquad (6.6)$$

$$\frac{\partial \tau_{zr}}{\partial z} + \frac{\partial \sigma_{rr}}{\partial r} + \frac{\sigma_{rr} - \sigma_{\theta\theta}}{r} = \rho \ddot{r}.$$

Here, (6.6) are the equations of continuity and motion, respectively; $\sigma_{\alpha\alpha} = s_{\alpha\alpha} - (p+q)$ ($\alpha = z, r, \theta$); where V is the relative volume; τ_{zr} shear stress; σ_{zz}, σ_{rr}, and $\sigma_{\theta\theta}$ total stresses; s_{zz}, s_{rr}, and $s_{\theta\theta}$ components of the stress deviator; p pressure; q artificial viscosity; and ρ density of the medium. The dots above the parameters denote a partial derivative with respect to time for a fixed particle.

Since the plate material has shear strength, calculation of yielding in the plate involves determination of the stress tensor. This tensor is represented as the sum of the spherical and deviatoric terms. The spherical component of the stress tensor – pressure – was determined from the equation of state:

$$p = a(\eta - 1) + b(\eta - 1)^2 + c(\eta - 1)^3, \qquad (6.7)$$

where $\eta = \rho/\rho_0 = 1/V$, ρ_0 is the initial density of the plate material, and a, b, and c are material constants.

The quantities σ_{zz}, σ_{rr}, $\sigma_{\theta\theta}$, and τ_{zr} were calculated on the basis of Hooke's law written in the following form [38]:

$$\dot{s}_{\alpha\alpha} = 2\mu\left(\dot{\varepsilon}_{\alpha\alpha} - \frac{1}{3}\cdot\frac{\dot{V}}{V}\right), \dot{\tau}_{zr} = \mu\dot{\varepsilon}_{zr}, \alpha = z, r, \theta, \qquad (6.8)$$

where μ is the shear modulus. The geometric relations linking the strain rate with the rate of displacement have the following form:

$$\dot{\varepsilon}_{zz} = \partial\dot{z}/\partial z, \dot{\varepsilon}_{rr} = \partial\dot{r}/\partial r, \dot{\varepsilon}_{zr} = \partial\dot{r}/\partial z + \partial\dot{z}/\partial r, \dot{\varepsilon}_{\theta\theta} = \dot{r}/r. \qquad (6.9)$$

The transition from elastic to plastic strain is made by means of the von Mises yield the equation:

$$s_{zz}^2 + s_{rr}^2 + s_{\theta\theta}^2 + 2\tau_{zr}^2 = 2\sigma_0^2/3, \qquad (6.10)$$

where σ_0 is the yield point in simple tension.

The gas and liquid are assumed to be ideal, while their deviatoric terms are equal to zero. Thus, the equations of state $p = p(\rho)$ are sufficient for these media in order to close the considered system of the equations. In the case of the gas, we used the adiabatic approximation (1.55):

$$p = p_0\left(\rho/\rho_0\right)^\gamma, \qquad (6.11)$$

where $\gamma = 3$. For the liquid, we took the equation of state in Tait's form (1.36).

System (6.6)–(6.11) is closed by the initial and boundary conditions. The initial conditions are zero (except for the pressure in the gas). We assumed the existence of sliding contact between the plate and the gas–liquid system. We also assumed that the external surface of the layer is a free surface.

It should be noted that, in accordance with the Wilkins approach [38], Eqs. (6.6) and (6.7) are written on the Lagrangian grid. However, the tensors of strain rate (6.9) and stress intensity (6.8) are assumed to be independent of the Eulerian variables. Such a form of representation is very convenient in the numerical solution of dynamic problems. The relationship between the Lagrangian and Eulerian coordinates can be followed during the solution, and small differences between these coordinates within a small time interval can be ignored.

We will not stop on the algorithm for solution of the problem, since the finite-difference scheme we used was described in detail in Ref. [38].

Cavitation and failure. We should note the possibility of local disturbance of the continuity of the liquid (cavitation). A simple model (see Chapter 4) is used to allow for cavitation effects. The model is based on the assumption that pressure in the theoretical cell is zero if it is less than the critical value.

A large number of mathematical models may be used for the dynamic failure of solids (Chapter 2). However, these models usually contain empirical parameters that are unknown for many materials; it is difficult to compare models with regard to their accuracy. In such cases, it is useful to examine the simplest failure models. These models are valid for a broad range of materials and are accurate enough for approximate calculations.

Here, we analyzed plate failure by using a simple mechanical criterion of failure [37]. The model on which the construction of this criterion was based assumes that if the maximum stress in a microscopic volume of the medium exceeds the material critical constant σ_{cr} – representing the tensile strength of the material – then a microcrack directed perpendicular to this normal stress instantaneously forms in the material. Since the maximum normal stress is one of the principal stresses and there are only three principal stresses, the given model allows no more than three cracks to develop in the microvolume.

During the numerical calculation, a microvolume of the medium is associated with the Lagrangian finite-difference cell, and the initiation of cracks is followed in each such cell. In the presence of a crack, the stress normal to its surface is equated to zero. We assume that the stresses change due to filling of part of the microvolume with cavities, and the volume of these cavities is regarded as the volume of the given crack. This results in a change in the effective strain of the microvolume and a corresponding change in the quantities p, σ_{zz}, σ_{rr}, and $\sigma_{\theta\theta}$ so that they correspond to the stresses in the failed material. This sequence of events is based on a linear relationship between the stresses and strains in the region bordering the fractured material. During the calculations, we follow the change in the volumes of all nucleated cracks. If any of these volumes becomes negative, this is interpreted as welding of the crack, and the volume is equated to zero. We will henceforth assume that if a compressive stress develops perpendicular to the crack, the material will behave in a manner similar to the unfractured material. When a tensile stress develops, the crack reopens, and the procedure of observing the change in volume is repeated.

Results of calculations. All of the calculations were performed for a steel plate of 0.02 m thickness with the following parameters: density $\rho_0 = 7850\,\text{kg/m}^3$, shear modulus $\mu = 80\,\text{GPa}$, elastic modulus $E = 200\,\text{GPa}$, yield point $\sigma_0 = 0.6\,\text{GPa}$, tensile strength $\sigma_{cr} = 1.5\,\text{GPa}$, and $a = -b = c = 240\,\text{GPa}$ [37]. A gas bubble was formed in the explosion of pentolite (a high explosive made up of trotyl and PETN in a 50:50 ratio) with a density $\rho_0 = 1650\,\text{kg/m}^3$ at a detonation velocity $D = 7655\,\text{m/s}$ and an initial pressure $p_0 = \rho_0 D^2 / 8$. The thickness and the radius of the charge were assumed to be equal to the thickness of the plate.

We established the goal of using numerical calculations to study the effect of the method of modeling the interaction of a liquid arid gas with a plate and the distance from the gas bubble to the plate on the stress–strain state (SSS) and strength of the plate.

First, we examined an underwater bubble which expanded on the surface of the plate. The effect of the interaction of the liquid and gas with the plate on the character of the SSS of the latter was evaluated by comparing the results of calculations performed with the use of one of the following assumptions:

1. The cells of the solid, liquid, and gas cannot slip relative to each other.
2. The liquid does not slip relative to the contact surface (phase boundary) during deformation of the plate, while the gas cells can be displaced relative to the cells of the solid.
3. The cells of the solid, liquid, and gas can slip relative to each other.

It was found that the fracture zone of the plate had roughly the same size in all three cases. Here, the method used to describe the interaction had a significant effect on the pressure in the liquid. For example, the pressure in the neighborhood of the point of contact of the liquid, gas, and solid after 4 μs was 3.25, 3.38, and 3.04 GPa in the above cases, respectively.

While studying the effect of the distance of the gas bubble from the plate on the character of failure of the plate, first we examined the case of explosion on the plate. We compared calculations performed for expansion of the gas in a vacuum and in the liquid. It turned out that the presence of the liquid increases the amplitude of the wave σ_z propagating along the z axis by more than 10%–15%. Failure began from the free surface. It subsequently propagated from the external surface and the symmetry axis. The fracture zones calculated for the case of explosion in a vacuum turned out to be larger than those for the case of explosion in the liquid. This has to do with the effect on the fracture zone of unloading waves emanating from the gas–liquid, solid–liquid, gas–vacuum, or solid–vacuum contact surfaces. The amplitude of these rarefaction waves will naturally be higher in the latter cases (with a vacuum), and this evidently explains the more destructive effect realized for expansion of the gas in a vacuum.

Figure 6.4 shows diagrams of the waves in the plate ($\Sigma = -\sigma_z$) and in the gas and liquid ($\Sigma = p$) for different moments of time ant $r = 0$, with distances between the charges and the plate $d = 0$, 0.02, and 0.03 m. While constructing the curves, we used different scales for the coordinate: one unit of length in the gas and liquid regions is twice as great as a unit of length in the solid region.

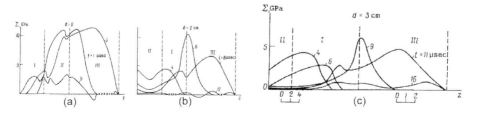

FIGURE 6.4 (a–c) Effect of the distance between the gas and the cylinder on the formation of waves Σ and fracture zones in an axial section: I, II, and III are liquid, gas, and solid regions, respectively (the asterisks denote fracture zones).

Interaction of Extreme Underwater Waves 163

It can be seen that the size of the fracture zones and the character of wave formation in the above-examined media depend heavily on the position of the gas bubble relative to the plate.

The calculations were performed with and without allowance for cavitation in the liquid. The failure processes in the solid and liquid phases were not observed to have had any effect on each other.

6.3 GENERATION OF CAVITATION WAVES AFTER TANK BOTTOM BUCKLING

We use a numerical approach to analyze the effect of these factors on the nonsteady deformation of a spherical panel – the bottom of a conical tank filled with water. A gas bubble undergoes rapid expansion near the panel. The motion of the panel is described by equations of the Timoshenko type, whereas the motion of the liquid is described by the equations of hydrodynamics (see, also, Sections 1.5 and 20.1).

Formulation of the problem. We will assume that the shell is shallow. Its motion is determined by nonlinear equations of the Timoshenko type (Section 1.5 and Ref. [27,39]):

$$\frac{1}{s}\frac{\partial}{\partial s}(sN_1) - \frac{1}{s}N_2 = \rho h \frac{\partial^2 u}{\partial t^2};$$

$$\frac{1}{s}\frac{\partial}{\partial s}(sQ) + \frac{1}{s}\frac{\partial}{\partial s}\left[\left(\frac{\partial w}{\partial s} + \frac{s}{R}\right)sN_1\right] + p = \rho h \frac{\partial^2 w}{\partial t^2}; \qquad (6.12)$$

$$\frac{1}{s}\frac{\partial}{\partial s}(sM_1) - \frac{1}{s}M_2 - Q = \rho \frac{h^3}{12}\frac{\partial^2 \psi}{\partial t^2}.$$

Here,

$$\{N_1; N_2; Q\} = \int_{-h/2}^{h/2} \{\sigma_1; \sigma_2; \sigma_{13}\} dz \text{ are the forces;}$$

$$\{M_1; M_2\} = \int_{-h/2}^{h/2} \{\sigma_1; \sigma_2\} z\, dz \text{ are the moments;}$$

$$\sigma_1 = E(\varepsilon_1 + v\varepsilon_2)/(1-v^2),\ \sigma_2 = E(\varepsilon_2 + v\varepsilon_1)/(1-v^2),\ \sigma_{13} = E\varepsilon_{13}/2(1+v);$$

$$\varepsilon_1 = \varepsilon_1^0 + z\varepsilon_1^1,\ \varepsilon_2 = \varepsilon_2^0 + z\varepsilon_2^1,\ \varepsilon_{13} = 6(1/4 - z^2/h^2)\varepsilon_{13}^0,$$

$$\varepsilon_1^0 = \partial u/\partial s - w/R + (\partial w/\partial s)^2/2;\ \varepsilon_2^0 = u/s - w/h;$$

$$\varepsilon_1^1 = \partial \psi/\partial s;\ \varepsilon_2^1 = \psi/s;\ \varepsilon_{13}^0 = \partial w/\partial s + \psi. \qquad (6.13)$$

We use also notations: w and u are the normal and tangential components of the displacement of the middle surface of the shell, ψ is the angle of rotation of the normal to the middle surface of the shell, E is the elastic modulus, v is the Poisson's ratio, s is a coordinate reckoned along the radius of the bearing contour of the shell, z is the distance measured along the normal to the middle surface of the shell, h is the thickness, f is the shell rise, and R is the radius of the shell.

The liquid in the tank and the gas in the expanding bubble are assumed to be inviscid media. In this case, the conservation equation (5.13) can be used to describe their flow [39]:

$$\frac{\partial v_r}{\partial t} = -\frac{1}{\rho}\frac{\partial p}{\partial r}, \frac{\partial v_\theta}{\partial t} = -\frac{1}{\rho r}\frac{\partial p}{\partial \theta}, \frac{\partial \rho}{\partial t} + \rho\left[\frac{\partial(r v_r)}{r \partial r} + \frac{\partial v_\theta}{\partial \theta}\right] = 0. \qquad (6.14)$$

Tait's equations (1.36) can be used to close Eq. (6.14) in the case of water, whereas the equation of the adiabatic curve (6.11) was used in the case of the gas. Cavitation effects were considered in accordance with Section 6.2.

The conditions of impermeability

$$v_n = 0 \text{ and } V_n = v_n \qquad (6.15)$$

are satisfied on the surface of contact of the undeformable cone with the liquid and the shell and the liquid, respectively. Here, $V_n = (\overline{V} \cdot \overline{n})$, where $\overline{V}$ is the velocity of the shell element, and $v_n = (\overline{v} \cdot \overline{n})$ is the normal component of the velocity of the liquid $\overline{v}$ relative to the boundary surface. We assign conditions $u = w = \psi = 0$ corresponding to fastening on the contour of the shell. At $t = 0$, the density of the gases is equal to the density of the water. The remaining parameters of the problem remain undisturbed at $t = 0$.

Algorithm for numerical integration. To solve the above-formulated boundary-value problem, we will use the finite-difference method with an implicit scheme of integration over time. We will use the Wilkins scheme to solve equations of flow of the liquid and gas (6.14) (see, also, Figure 6.9). The equations of motion of the shell will be integrated using an explicit difference scheme [27,39].

Results of calculations [27,39]. The following were taken as the geometric dimensions and mechanical characteristics of the shell material (aluminum): $R_0 = 0.4$ m (R_0 is the radius of the shell contour), $h = 0.01$ m, $f = 0.05$ m, $E = 75,600$ MPa, and $v = 0.3$; $\rho_0 = 2640$ kg/m^3. The gas bubble is formed on the symmetry axis of the tank at a distance of $0.16 R$ from the pole point of the shell. The volume of the space containing the gas was calculated for each time step in accordance with the Wilkins algorithm. We then again determined the pressure in this volume in accordance with the law $p = A \exp(\alpha c_2 t)$, where $A = 50$ MPa, $c_2 = 1500$ m/s, and $\alpha = -3$.

Let us examine the effect of hydroelastic processes on the dynamics of buckling (slamming) of the shell. Under the influence of the underwater wave propagating from the gas bubble and the hydraulic flow, the shell undergoes a deflection which reaches $-2f$. This corresponds to its nearly complete buckling. Here, we assumed that Eqs. (6.12) were valid up to deflections on the order of $10h$. The changes in the deflection of the pole of the shell and the pressure in the adjacent volume of liquid over time are shown in Figure 6.5.

Interaction of Extreme Underwater Waves 165

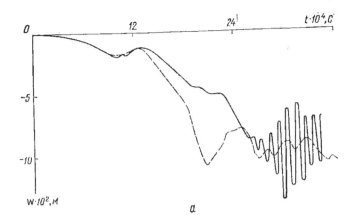

FIGURE 6.5 Change in the deflection of the pole point of the shell (the solid lines show results without allowance for cavitation, whereas the dashed lines show results with allowance for the same).

The results shown in Figure 6.5 indicate that the shell is completely inverted at $t \approx 26.5 \times 10^{-4}$ second. It then oscillates relative to this position. It should be noted that the shell was designed without allowance for ductility. Thus, retention of the inverted form of the shell can be explained only by pressure and the damping effect of the liquid.

The calculation also shows the substantial impact of cavitational effects on the dynamics of the shell. If we use the model of an ideal elastic fluid, we find that the shell undergoes high-frequency vibrations relative to the inverted position. These vibrations are accompanied by powerful pulsations of the pressure in the liquid. It follows from calculations performed with the above-described liquid model that the pressure in the rarefaction zones reaches −500 MPa, whereas that in the compression zones 1300 MPa. Thus, the model of an ideally elastic fluid leads to completely erroneous results while calculating the flow of the liquid. Appreciably more accurate data on deflections and the pressure in the liquid can be obtained by taking cavitation into account. This conclusion is consistent with the well-known results obtained for plates (see Chapters 4 and 5).

Figure 6.6 shows the process of inversion of the shell and the development of the cavitation zone in the adjacent liquid. Deformation of the shell begins in the central part and propagates outward to the contour.

It should be noted that besides the zones of very low negative pressure, calculations with the ideally elastic fluid model showed that other rarefaction regions also exist. The negative pressure in these regions reaches −1 MPa. However, calculations performed with allowance for cavitation did not reveal any other cavitation zones besides those noted in Figure 6.6 [39].

Thus, we can conclude the following: the liquid prevents development of the process of nonlinear vibration of the shell after buckling; after the loss of stability, the dynamics of the shell depends appreciably on cavitation processes in the liquid;

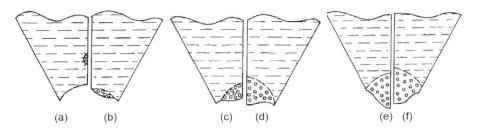

FIGURE 6.6 Dynamics of development of cavitation zones in the liquid and the elastic deformation of the shell over time: (a) $t = 0$, (b) $t = 12 \times 10^{-4}$ second, (c) $t = 15 \times 10^{-4}$ second, (d) $t = 18 \times 10^{-4}$ second, (e) $t = 20 \times 10^{-4}$ second, and (f) $t = 26 \times 10^{-4}$ second.

the volume occupied by the cavitation cavity may be an order of magnitude greater than the volume bounded by the deformed and undeformed surfaces of the spherical panel; and allowance for cavitation phenomena in the liquid is a necessary condition for the proper dynamic design of hydroelastic systems.

6.4 TRANSIENT INTERACTION OF A STIFFENED SPHERICAL DOME WITH UNDERWATER SHOCK WAVES

The characteristic features of the nonstationary interaction between stiffened structures and a fluid are investigated. Nonlinear equations of the theory of shells and hydrodynamics, which make it possible to account for the discreteness of stiffener arrangement and the possibility of cavitation developing in the fluid, are used. A method of solving the interaction problems is proposed for the case of a shell reinforced with a crossing system of stiffeners [40–44].

In our study, we describe models, methods, and the results of investigation of the nonstationary behavior of stiffened spherical shells that cover the ends of pipes, and their interaction with a fluid. Numeric calculations are performed.

6.4.1 THE PROBLEM AND METHOD OF SOLUTION

Let us examine an absolutely rigid semi-infinite cylinder, which is submerged in water and bounded by a stiffened elastic spherical dome. Let us investigate the nonstationary wave processes that take place in the hydroelastic system: liquid+dome+rigid cylinder when a plane shock wave of pressure strikes the dome (Figure 6.7a). The flow of the nonviscous fluid is described using equations of hydrodynamics with allowance for the possibility that cavitation will develop. The motion of the dome was investigated within the framework of the geometrically nonlinear Timoshenko-type theories. Two forms of dome reinforcement are analyzed: with annular stiffeners and a crossing system of stiffeners with allowance for the width and discreteness of their arrangement.

In the case of a shell reinforced by a crossing system of stiffeners, it is required to use three-dimensional equations of hydrodynamics to determine the hydrodynamic load; this is associated with excessive expenditure of machine time for the

Interaction of Extreme Underwater Waves

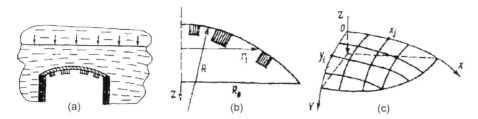

FIGURE 6.7 Object of investigation (a) and schematic diagram showing reinforcement of dome by annular stiffeners (b) and by crossing system of stiffeners (c).

numeric calculation. The three-dimensional problem is reduced to a two-dimensional one on the basis of an algorithm, which includes the following steps:

1. The search for a smooth shell (equivalent shell), which is close to the stiffened shell in terms of character of deformation [44].
2. Determination of the hydrodynamic load on the dome by the combined solution of two-dimensional equations of hydrodynamics and one-dimensional equations of the motion of the equivalent shell.
3. Determination of the SSS of the stiffened dome, proceeding from the hydrodynamic load determined in the previous step.

Motion equations of the shell. Let us examine a shallow dome reinforced from the inside. The lining and stiffeners are fabricated from the same material. The median surface of the lining can be adopted as the coordinate surface. The motion equations of the shell, which is reinforced by a crossing system of stiffeners, can be written in a Cartesian coordinate system [40–43]:

$$\frac{\partial N_x}{\partial x} + \frac{\partial N_{xy}}{\partial y} = \rho\left[(h+F)\ddot{u} + S\ddot{\psi}_x\right];$$

$$\frac{\partial N_y}{\partial y} + \frac{\partial N_{xy}}{\partial x} = \rho\left[(h+F)\ddot{v} + S\ddot{\psi}_y\right];$$

$$\frac{\partial Q_x}{\partial x} + \frac{\partial Q_y}{\partial y} + \frac{N_x + N_y}{R} + \frac{\partial}{\partial x}\left(N_x \frac{\partial \omega}{\partial x}\right)$$

$$+ \frac{\partial}{\partial y}\left(N_y \frac{\partial \omega}{\partial y}\right) + P = \rho(h+F)\ddot{\omega}; \qquad (6.16)$$

$$\frac{\partial M_x}{\partial x} + \frac{\partial M_{xy}}{\partial y} - Q_x = \rho\left[\left(\frac{h^3}{12}+J\right)\ddot{\psi}_x + S\ddot{u}\right];$$

$$\frac{\partial M_y}{\partial y} + \frac{\partial M_{xy}}{\partial x} - Q_y = \rho\left[\left(\frac{h^3}{12}+J\right)\ddot{\psi}_y + S\ddot{v}\right],$$

where universally adopted notations of the theory of shells are employed, and F, S, and I are geometric characteristics of the stiffener's section; the direction of the stiffeners coincides with the coordinate axes X and Y (Figure 6.7c). The load P is $P = p + q$, where p is the pressure of liquid–shell interaction, and q is the pressure of the plane shock wave (6.19). The forces, moments, and intersecting forces are represented as

$$N_x = \begin{cases} N_x^0 & \text{(at points where there is no stiffener);} \\ N_x^0 + N_x^p & \text{(at points where there is a stiffener).} \end{cases} \quad (6.17)$$

Here, the quantities that are determined from the theory of smooth shells [41] are denoted by the underscript "0", and the quantities that apply to the stiffeners are denoted by the underscript "p". The stiffened structure can be treated as a shell with a stepwise changed thickness (Figure 6.8). The height of the stiffeners and their arrangement can be given in accordance with the approach previously described in detail by Il'in and Karpov [43].

Shell models (A, B, C). Let us briefly describe the three methods used to model the influence exerted by stiffeners on the shell's behavior. Two of these are based on consideration of the discreteness of stiffener arrangement and the third on the use of the method of structural orthotropy.

A. A stiffened structure can be treated as a shell with a thickness that varies in a stepwise manner [43]. The height of the stiffeners and their arrangement can be assigned using single columnar functions:

$$H(x,y) = \sum_{j=1}^{M} h^j(x,y)\bar{\delta}(x-x_j) \\ + \sum_{i=1}^{N} h^i(x,y)\bar{\delta}(y-y_i) - \sum_{j=1}^{M}\sum_{i=1}^{N} h^{ij}(x,y)\bar{\delta}(x-x_j)\times\bar{\delta}(y-y_i). \quad (6.18)$$

Here, quantities with the subscript $i(j)$ apply to stiffeners positioned parallel to the $X(Y)$ axis [43].

B. Let us assume that the lining and stiffener are in contact along a line. In this case, the effect of the stiffeners on the shear and torsion of the median

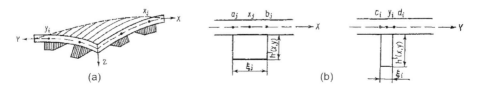

FIGURE 6.8 General appearance of shell (a) and cross section of stiffener (b).

surface can be neglected. The discreteness of stiffener arrangement can be accounted for δ-functions. This model is a special case of the previous one, and its relationships can be obtained by limiting conversion from single columnar functions to δ-functions. Model B is described in greater detail in Ref. [42,43].

C. According to this model, the stiffened shell can be considered orthotropic. The stiffness of the ribs is distributed or, in other words, "spread out" over the length of the section between ribs. Model C has been widely used by Vol'mir [41].

Hydrodynamic equations. Two-dimensional equations of conservation for the fluid were written in the forms (6.14) and (6.15). Computed region of fluid is shown in Figure 6.9.

Fluid failure in rarefaction zones is studied using the simplest model, whose essence consists in the fact that in the case when the pressure drops below the critical value in a microvolume of fluid ($P_{cr} = -0.2$ MPa), it is assumed equal to P_{cr}.

The shell's perimeter is considered as fixed. The initial time ($t = 0$) is the moment of contact of the front of the underwater wave and the surface of the shell. When $t = 0$, the parameters of the dome's SSS are known.

6.4.2 Numeric Method of Problem Solution

The problem can be solved using the finite-difference method with an explicit scheme of integration over time. Wilkins' numeric scheme was used while calculating (6.14) and (6.15) [38]. Equations. (6.17) were integrated on the basis of the numeric scheme [39,44].

The computed region of the fluid is shown in Figure 6.9. Its dimensions were selected so that the region boundaries did not affect the parameters of the processes under study in the time interval under consideration. To shorten the machine time

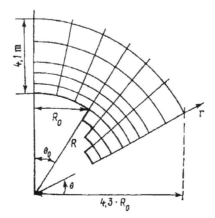

FIGURE 6.9 Computed region of fluid ($R = 2$ m; $R_0 = 0.5$ m).

required for the calculation, we used a nonuniform grid (see the grid region in Figure 6.9).

The median surface was divided into circular elements when calculating smooth shells or shells reinforced with the annular stiffeners, and into tetragonal cells in the case of the crossing system of stiffeners. The grid step was reduced in zones near stiffeners; in this case,

$$\min \Delta x \leq \min_{1 \leq j \leq M} \xi_j/2, \quad \min \Delta y \leq \min_{1 \leq i \leq N} \xi_i/2,$$

where the quantities with the subscript $i(j)$ apply to stiffeners situated parallel to the $X(Y)$ axis, $\xi_i(\xi_j)$ is the width of the ith (jth) stiffeners, and $N(M)$ is the number of the ith (jth) stiffeners.

In order to stabilize the calculation performed in accordance with the different equations, the time interval must conform to the condition $\tau \leq \min(\tau_1, \tau_2)$, where τ_1 and τ_2 are the permissible steps for the difference equations of the shell and liquid, respectively. The values τ_1 and τ_2 were found from the Curant condition and were defined more precisely through numeric experiments.

6.4.3 RESULTS OF CALCULATIONS

The calculations were performed for the following geometric and mechanical dome characteristics: $h = 0.01$ m, $R = 2$ m, $R_0 = 0.5$ m, $E = 75,600$ MPa, $v = 0.3$, and $\rho_0 = 2640$ kg/m³. The stiffeners have the same tetragonal cross section, whose height $h_s = 4$ cm and width $\xi = 3.33$ cm. The plane pressure wave in the fluid was assigned in the following form:

$$q = p_0 \exp\left[-\alpha^{-1}\left(t + x_s a_0^{-1}\right)\right] H\left[\alpha^{-1}\left(t + x_s a_0^{-1}\right)\right], \tag{6.19}$$

where H is the Heaviside function; $p_0 = 8$ MPa and $\alpha = 10^{-3}$ second are constants defining the amplitude and duration of the pulse, respectively; $a_0 = 1500$ m/s is the speed of sound in the fluid; and $x_s = r \sin\theta - R$.

We investigated the interaction of the underwater wave and shell by the stepwise complication of the dome model. The dome was treated as an undeformable body, or a smooth and reinforced (annular or crossing stiffeners) shell.

Smooth shell. Calculations were performed under the assumption of the shell's absolute rigidity and deformability. The results of the calculation for time $t = 75 \times 10^{-6}$ second are presented in Figure 6.10.

In the case of the absolutely rigid shell, the pressure field was compared with the test. The equations derived by Grigolyuk and Gorshkov [45] on the basis of the analytic solution of the wave equation for a fluid were used to calculate the test data. As it is apparent, the computational results are in good agreement with the test results in the central portion of the shell where its surface is nearly flat. The significant difference between these data near the edges is explained by the fact that in this zone, the fluid flow is complex in nature and the analytic theory [45] does not provide for sufficient accuracy of pressure calculation.

Interaction of Extreme Underwater Waves

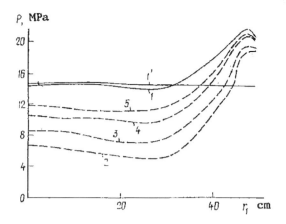

FIGURE 6.10 Pressure distribution along surface of absolutely rigid (1, 1') and deformable (2–5) shell: 1') test, 2) $h = 2.07$ cm, 3) $h = 3.14$ cm, 4) $h = 5$ cm, and 5) $h = 8$ cm.

Calculations performed with allowance for shell deformability indicated that the pressure on its surface varies rather smoothly and depends heavily on the thickness of the shell. The results of the calculations are approximated to those obtained for an absolutely rigid body as h increases.

Annular stiffeners. We investigated the SSS of domes reinforced by annular stiffeners while accounting for their interaction with the fluid. A comparative analysis of the computational results of stiffened and smooth shells of similar weight was performed to determine the thickness of an equivalent smooth shell. We examined cases of reinforcement with two, four, and nine stiffeners. The thickness of the corresponding smooth shell is determined from the following equation:

$$h_e = N h_s \xi R_0^{-1},$$

where h_s, ξ, and N are the height, width, and number of stiffeners, respectively. The stiffeners were located with the coordinates $r_1^{(1)} = 10$ cm and $r_1^{(2)} = 30$ cm when $N = 2$; $r_1^{(j)} = 10j$ cm when $N = 4$; and $r_1^{(j)} = 5j$ cm when $N = 9$.

The deflections at the pole of the stiffened and smooth shells for different times are made in Figure 6.11. As is apparent, the reinforcement exerts a significant influence on the magnitude of the deflection. The maximum deflection of the shell having the reinforcement is nearly half of the maximum deflection of the smooth shell if $N = 2$. At the same time the maximum deflection of the shell having the reinforcement is larger than the maximum deflection of the corresponding equivalent smooth shell when $N = 4$ and $N = 9$.

It follows from the calculations that for the shell having the reinforcement, the strain state (SS) may both approach (when $N = 4$) the SS of a smooth shell of equal weight, and differ significantly from it (when $N = 2$ and 9), depending on the number of stiffeners.

Laws governing the distribution of force and bending waves along the surface during hydropulsed immersion were established as a result of the numeric calculations.

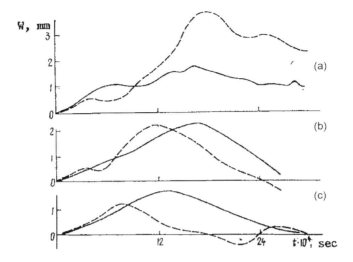

FIGURE 6.11 Comparison of deflections at pole of stiffened (solid lines) and smooth (broken lines) shells of similar weight: (a) $N = 2$, (b) $N = 4$, and (c) $N = 9$.

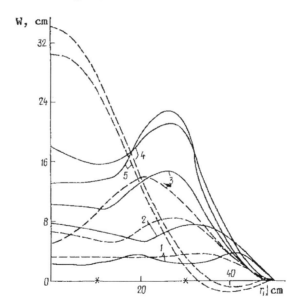

FIGURE 6.12 Distribution of deflection along radius of support perimeter of stiffened (solid lines) and smooth (broken lines) shells at various times: 1) $t = 2.25 \times 10^{-4}$ second, 2) $t = 4.5 \times 10^{-4}$ second, 3) $t = 7.5 \times 10^{-4}$ second, 4) $t = 15.7 \times 10^{-4}$ second, and 5) $t = 21 \times 10^{-4}$ second. (Here and in Figure 6.13, the location of stiffener arrangement is denoted by points.)

The distribution of deflections along the radius ribbed and smooth shells of similar weight for different times is shown in Figure 6.12. As it is apparent, one inflection zone, which subsequently moves toward the center, appears in the initial load step

Interaction of Extreme Underwater Waves 173

of the smooth shell near the edge, while several inflection zones appear in the case of the stiffened shell, depending on the number of stiffeners and their location. The pole part of the ribbed shell, which is surrounded by the first annular stiffener, moves as a rigid body, i.e., without appreciable bends. A region of vigorous bending may develop, however, near the edges.

A representation of the influence exerted by annular stiffeners on the character of the stress distribution along the surface of a stiffened shell is given in Figure 6.13. It is apparent that the presence of stiffeners leads to an increase in the forces N_1 and N_2 and in the moment M_2; and to a decrease in the moment M_1 and intersecting force Q as compared to the case of a smooth shell. The values of N_2 and M_2 vary in a jump-wise manner at points of stiffener deployment.

The reinforcement of a shell with annular stiffeners may therefore be irrational; that is, the bearing capacity of a stiffened shell is lower than the bearing capacity of a smooth shell of the same weight.

Crossing system of stiffeners. We examined stiffeners arranged similarly with respect to the X and Y axes, i.e., the stiffener coordinates $X_j = Y_i$ when $i = j$ and $N = M$. The effect of the number of stiffeners $N_s = N + M$ on the dome's dynamics, disregarding the effect of the fluid, was investigated initially. The parameters of the external load were assigned in the form (6.19), where $p_0 = 5$ MPa, $\alpha = 10^{-3}$ seconds, and $X_s = 0$. We examined the cases $N_s = 4, 6$, and 8. Each computational variant was repeated for a smooth dome with a weight equal to that of the stiffened dome using one-dimensional equations of shell motion.

Analysis of the results obtained indicated that for the infrequent deployment of stiffeners in the crossing system, the bearing capacity of the stiffened shell increases significantly as compared to that of the smooth shell (Figure 6.14).

It was therefore established that when $N_s \geq 6$, one-dimensional equations of motion of a smooth shell can be used for an approximate determination of the SS parameters of a shell of equal weight, reinforced by a crossing system of stiffeners.

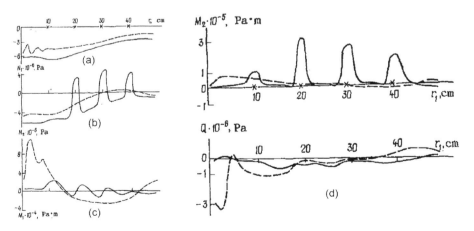

FIGURE 6.13 Distribution of forces (a and b), moments, (c and d), and intersecting forces (e) along surface of stiffened (solid lines) and smooth (broken lines) shells when $t = 15 \times 10^{-4}$ second.

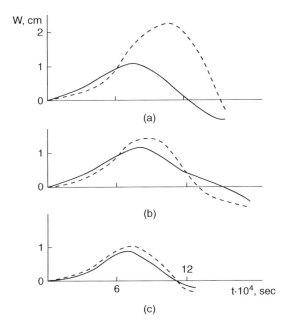

FIGURE 6.14 Curves of displacements at pole of shell reinforced by crossing system of stiffeners (solid lines) and smooth shell of equal weight (broken lines) versus time: (a) $N_s = 4$; (b) $N_s = 6$; (c) $N_s = 8$.

Conclusions

1. Reinforcement of a shell with annular stiffeners does not always lead to an increase in bearing capacity.
2. When $N_s \geq 6$, one-dimensional equations of motion of a smooth shell can be used for an approximate determination of the SS parameters of a shell of equal weight, reinforced with a crossing system of stiffeners; the greater the number of stiffeners, the better the accuracy of the solution using the smooth shell model.
3. An algorithm for calculating the SSS of shells, which are reinforced with a crossing system of stiffeners and which interact with a fluid, is developed.

6.5 EXTREME AMPLIFICATION OF WAVES AT VICINITY OF THE STIFFENING RIB

To increase the rigidity and stability of structures under the action of extreme waves, they are often supported by stiffeners made of a more durable material. Examples of calculations of such structures are given above. However, shell models do not always make it possible to evaluate the strength of a structure, especially in places where stresses are concentrated. This can be done more accurately using the equations of elasticity theory.

Interaction of Extreme Underwater Waves

Examples of such calculations are given in monographs [17–19,27]. To illustrate the dynamic stresses arising in the vicinity of the stiffeners, we give some calculated results without description of the calculation method itself.

It is assumed that the shell and the stiffener are made of different materials, namely, glass and steel. Calculation parameters are as follows: $R = 1$ м, $d = 0.9$ м, $e = 0.8$ м, $L = 2.2$ м and $L_1 = 1$ м (Figure 6.15). Let us consider the stresses arising in the vicinity of the rib (stiffener) at the first moments of time, when perturbations from the cylinder ends have not yet reached reinforcements. The stresses are related to the amplitude of the underwater wave.

A long wave. With the action of a long wave, a gradual increase in the stresses σ_{rr} in time is observed (Figure 6.16). At the corner point, there is a more intensive surge of stresses.

Stretching stresses σ_{xx} and $\sigma_{\varphi\varphi}$ occur only near the ends of the cylinder. Stresses σ_{rr} within the considered site do not exceed the dynamic pressure of the liquid.

A comparison is made of the last results with those obtained for entirely glass cylinder. The nature of the change in the stressed-deformed state of the body did not change qualitatively. The quantitative results differed up to two times.

Short wave. The action of a short wave is accompanied by a completely different picture of stress distribution. We illustrate this with an example σ_{rr} at the most characteristic moments of the interaction time (Figure 6.17). At $t = 0.03$ second, the compression wave approaches the contact surface of the cylinder with the rib and partially passes into the rib. The stress σ_{rr} at the angular point increases and at $t = 0.075$ second is 6 times greater than the maximum pressure in the underwater wave.

The wave reflects from the cylindrical surface of the rib at $t = 0.1$ second. As a result the wave of the tensile stresses forms. In the subsequent, there is a decrease in stress along the entire section of the rib (Figure 6.17). Wave interactions smooth of the wave fronts.

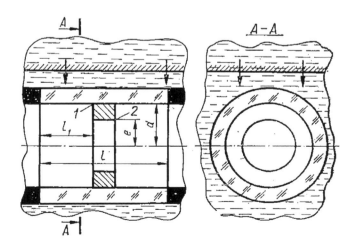

FIGURE 6.15 A rigidly fixed glass cylinder reinforced by a steel stiffener under the action of an extreme underwater wave. The wave front is parallel to the surface of the cylinder.

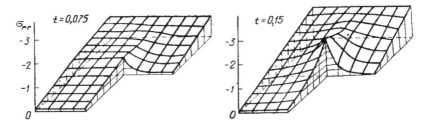

FIGURE 6.16 The action of a long wave.

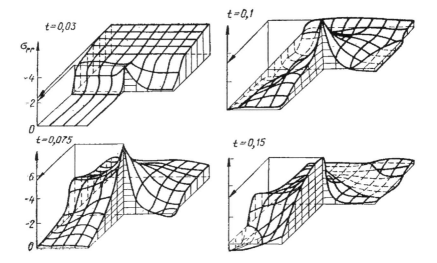

FIGURE 6.17 The action of a short wave and an evolution of the wave σ_{rr}.

The remaining components of the stress tensor change more smoothly with time. The stress distribution for cases of long (a) and short (b) waves for the instant $t = 0.15$ second is shown in Figure 6.18. The pattern of distribution σ_{rx} and $\sigma_{\varphi\varphi}$ under the action of long and short waves is approximately the same.

On the outer surface of the cylinder, over the lateral surfaces of the rib, in the case of a short wave, tensile stresses σ_{xx} occur, which are two times greater than the amplitude pressure in the incident wave. There can be a fracture of the material of the cylinder within the tension wave. The fractures are symmetrical with respect to the middle of the rib, and the split surface in the section $\varphi = 0$ has a shape close to the ellipse.

Conclusion. In this part of the book, we pay special attention to the destruction of liquid within rarefaction waves.

We found that in the wide range of variations of the pressure amplitude and the structural elements, the results of calculations for the models of cavitation and bubbly liquids are much more in line with each other than with results obtained using the ideal elastic liquid. With regard to the applicability and accuracy of the models

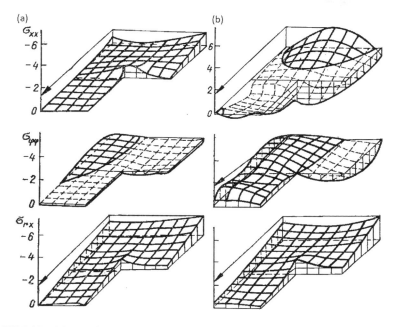

FIGURE 6.18 Waves of stresses σ_{xx}, σ_{rx} and $\sigma_{\varphi\varphi}$.

of destructible (cavitating) and bubbly liquids, it is difficult to say something as certain. Note that the programming and calculation of problems based on the bubbly liquid model takes much longer than the other model. In addition, let the growth time of the bubbles before their merging and the complete destruction of the liquid in a certain volume is very small and much shorter than the time of deformation of the structure. In this case apparently this growth of the bubbles has little effect on the deformation and more reasonably uses the instantaneously destructible (cavitation) model of liquid. Moreover, we note that the used model of bubbly liquid also has its drawbacks, since it does not take into account the mutual influence of the bubbles. Apparently, the faster the destruction of the liquid, the more applicable the cavitation liquid model. The last conclusion agrees with the conclusion made in the theory of dynamic destruction of solids about the applicability of the instantaneous split criterion (see Section 2.1).

REFERENCES

1. Knapp RT, Daili JW, Hammit FG (1970) *Cavitation.* McGraw Hill Book Company, New York (1970).
2. Brennen C. *Cavitation and Bubble Dynamics.* Oxford University Press, Oxford (1995).
3. Galiev ShU et al. Method of calculating the load from nonlinear waves on a tank cover. *Strength Mat* 5: 663–669 (1984).
4. Anuchin MA et al. *Theoretical Principles of Stamping by Explosion.* Mashinostroenle, Moscow (1972) (in Russian).
5. Enzra AA. *Principles and Practice of Explosive Metalworking.* Garden City Press Limited, London (1973).

6. Mansour AE, Ertekin RC (Eds.) *Proceedings of the 15th International Ship and Offshore Structures Congress.* Elsevier (2003).
7. Frieze PA, Shenoi RA (eds.). *Proceedings of the 16th International Ship and Offshore Structures Congress.* Volumes 1-2, Southampton, UK. University of Southampton Press (2006).
8. Galiev SU, Flay RGJ. Interaction of breaking waves with plates: The effect of hull cavitation. *Ocean Eng* 88: 27–33 (2014).
9. Liu G-Z, Jian-hu L, Jun Wang J, Jian-qiang P, Hai-bin M. A numerical method for double-plated structure completely filled with liquid subjected to underwater explosion. *Marine Struct* 53: 164–180 (2017).
10. Zhang Z-H, Wang Y, Zhang L-J, Yuan J-H, Zhao H-F. Similarity research of anomalous dynamic response of ship girder subjected to near field underwater explosion. *Appl Math Mech-Engl Ed* 32 (12): 1491–1504 (2011).
11. Cole RH. *Underwater Explosions.* Princeton University Press, Princeton, NJ (1948).
12. Zamislaev BV, Iakovlev YS. *Dynamical Loads during Underwater Explosions.* Sudostroenie, Leningrad (1967).
13. Felippa CA, DeRuntz JA. Finite element analysis of shock-induced hull cavitation. *Comput Methods Appl Mech Eng* 44: 297–337 (1984).
14. Sandberg G. A new finite – element formulation of shock – induced hull cavitation. *Comput Methods Appl Mech Eng* 120 (1–2): 33–44 (1995).
15. Felippa CA, Park KC, Farhat C. Partitional analysis of coupled mechanical systems. Report CU-CAS-99-06, Center for Aerospace Structures, College of Engineering, University of Colorado (1999).
16. Shauer HM. The after flow theory of the reloading of air-backed plate at under-water explosions. *Proceedings of first U.S. National Congress of Applied Mechanics*, Illinois Institute of Technology, Chicago, 1951, pp. 887–892; Shin YS. Ship shock modelling and simulation for far-field underwater explosion. *Comput Struct* 82: 2211–2219 (2004).
17. Galiyev ShU. *Dynamics of Structure Element Interaction with a Pressure Wave in a Fluid.* Naukova Dumka, Kiev (1977) (in Russian) (translated in English: Dep. Navy, Off. Naval Res., Arlington, USA, 1980).
18. Galiyev ShU. *Dynamics of Hydroelastoplastic Systems.* Naukova Dumka, Kiev (1981) (in Russian).
19. Galiyev ShU. *Nonlinear Waves in Bounded Continua.* Naukova Dumka, Kiev (1988) (in Russian).
20. Galiyev ShU., Pavlov AA. Experimental investigation of the cavitation interaction of a compression wave with a plate in liquid. *Strength Mat* 8: 988–993 (1977).
21. Galiyev ShU, Demina VM, Shelom VK. Impulse expansion of cylindrical shells by a liquid. *Strength Mat* 12: 1379–1383 (1979).
22. Galiyev ShU. Influence of cavitation on transient deformation plates and shells by the liquid. *Proceedings of IUTAM Symposium on Impact Dynamics*, Peking University Press, Beijing, 1994.
23. Galiyev ShU. *Charles Darwin's Geophysical Reports as Models of the Theory of Catastrophic Waves.* Center of Modern Education. Moscow (2011) (in Russian)
24. Galiyev ShU. *Darwin, Geodynamics and Extreme Waves.* Springer, Cham (2015).
25. Galiyev ShU. Numerical investigation of the deformation of plates by a liquid in press guns. *Strength Mat* 11: 1293–1299 (1979).
26. Pertsev AK, Platonov EG. *Dynamics of Plates and Shells.* Sudostroenie, Leningrad (1987).
27. Galiyev ShU, Babish UN, Zhurakhovskii SV, Nechitailo NB, Romashchenko VA. *Numerical Modeling of Wave Processes in Bounded Media.* Naukova Dumka, Kiev (1989) (in Russian).

28. Galiyev ShU. Influence of cavitation upon anomalous behavior of a plate/liquid/underwater explosion system. *Int J Impact Eng* 19 (4): 345–359 (1997); Johnson W. Comments on 'Influence of cavitation upon anomalous behavior of a plate/underwater explosion systems'. *Int J Impact Eng* 21(1): 113–115 (1998).
29. Galiyev ShU, Abdirashitov A, Uliahshiev AU. Local pulsed loading of coaxial cylindrical shells interacting through a liquid. *Strength Mat* 2: 234–238 (1988).
30. Abdirashitov A, Galiyev ShU. Effect of the bubbling-up of a liquid on pulsed deformation of the tank walls. *Strength Mat* 2: 259–264 (1988).
31. Abdirashidov A, Galiyev ShU. Effect of the bubbling-up of liquids in low-pressure waves on the transient deformation of hydroelastic systems. *Strength Mat* 20 (10): 1367–1370, 1988.
32. Sychev VV et al. *Thermodynamic Properties of a Nitrogen.* Moscow (1977); Sychev, BB et al. Thermodynamic properties of oxygen. Moscow (1981); Sychev, BB et al. *Thermodynamic Properties of Helium.* Moscow (1984).
33. Galiyev ShU, Romashchenko VA. A method of solving nonstationary three-dimensional problems of hydroelasticity with allowance for fluid failure. *Int J Impact Eng* 22: 469–483 (1999).
34. Bleich HH, Sandler IS. Interaction between structures and bilinear fluid. *Int J Solid Struct* 6: 617–639 (1970).
35. Newton RE. Effects of cavitation on underwater shock loading-plane problem. Final Report NPS-69-81-001, Naval Postgraduate School, Monterey, CA (1981).
36. Abdirashidov A, Karshiev AB. Nonsteady deformation of an infinite elastoplastic plate by an explosion in a liquid. *Strength Mat* 6: 89–92 (1989).
37. Abdirashidov A, Galiyev ShU. Dynamics of interaction of a disintegrating solid layer, lying on a liquid, with an expanding gas bubble. *Strength Mat* 19 (12):1668–1673 (1987).
38. Wilkins ML. Calculation of elastic-plastic flows. University of California, UCRL 7322 (1963).
39. Abdirashidov A, Galiyev ShU. Nonsteady interaction of a spherical panel with an underwater bubble. *Strength Mat* 20 (10):1386–1390 (1988).
40. Galiyev ShU, Karshiev AB. Numeric analysis of the accuracy of models of stiffened spherical domes subjected to pulsed loading. *Strength Mat* 22 (5): 726–730 (1990).
41. Vol'mir AS. *Nonlinear Dynamics of Plates and Shells.* Nauka, Moscow (1972) (in Russian).
42. Amiro IYA et al. *Oscillations of Stiffened Shells of Revolution.* Naukova Dumka. Kiev (1988) (in Russian).
43. Il'in VP, Karpov VV. *Stability of Stiffened Shells Subjected to Large Displacements.* Stroiizdat, Leningrad (1986) (in Russian).
44. Galiyev ShU, Karshiev AB, Abdirashidov A. Nonstationary interaction between a stiffered spherical done and a fluid. *Strength Mat* 11: 1650–1659 (1990).
45. Grigolyuk EI, Gorshkov AG. *Nonstationary Hydroelasticity of Shells.* Sudostroenie, Leningrad (1974) (in Russian).

Part III

Counterintuitive Behavior of Structural Elements after Impact Loads

Everything should be as simple as possible, but not simpler.

Albert Einstein

It is known that the solids can demonstrate different kinds of nonlinearity in different situations. In some cases, it is elastic nonlinearity. In other cases, nonlinearity manifests itself in the form of plasticity [1–7]. Often the nonlinear elastic–plastic properties can manifest themselves together (see Sections 1.2, 4.4.4, and 6.1). In particular, during the impulse load, the thin-walled constructions can lead self-some as a plastic body, but after removing the load, it turns out that the body conserves enough elastic forces which can manifest itself. As a result, the structure does not always receive the expected form. This is so-called counterintuitive behavior (CIB). CIB means an appearance of final deflections of these elements which are contrary to the direction of the impulsive loading. The CIB effect can be accompanied by a failure of structural elements.

Here the material published in [8–10] is presented.

7 Experimental Data

7.1 INTRODUCTION AND METHOD OF IMPACT LOADING

CIB of circular plates was discovered during studies of action of underwater [11,12] and air [3] explosion on the plates. In [11, pp. 377–378], plates of different thicknesses of iron, copper, and lead were examined. The plates covered a round hole in the wall of a thick-walled cistern filled by water. Explosions of different power were carried out within the cistern. The plates usually showed a final deflection directed outside of the cistern. After each explosion, the deformed plate was replaced by a new one.

However in one series of the experiments, the following paradoxical result was found. The final deflection of the plates, which was normally directed outside of the cistern, increased to a definite value together with a rise of the explosion power. When the explosion power was increased still further through, the direction of the final deflection was changed so that it pointed toward the inside of the cistern. It was noted [11] that this CIB was not fully understood and no theory was put forward to explain it.

CIB sometimes occurred as a result of loading of metal plates by an electrical discharge in water [12]. The circular plates of aluminum alloy AMΓ and ductile steel (steel 3) were used. Sometimes, a layer of rubber was placed on the metal plate. In some cases, a layer of air was left between the rubber and the metal. The thickness, h, of the metal plates varied from 1 to 3 mm, and the radius, R, varied from 150 to 500 mm. The electrical discharge took place at a distance from 100 to 200 mm away from the plates. In [12], figures of plates after CIB occurred were given. The center part of the plates was deflected contrary to the direction of the underwater electrical discharge loading. There were no theoretical explanations for these figures.

Thin circular metal plates ($h/R \approx 0.01$) of aluminum alloy AMΓ-6 sometimes exhibited CIB after blast loading if a plastic deformation was small [13, p. 162]. This behavior was explained in [13] in the following manner. The plate bends after loading and oscillates as the shallow shell. In the case of CIB, vibration energy of this shell is enough for only one reverse snap-buckling.

Some figures illustrating the possible occurrence of CIB of the plates after loading by the underwater explosion were presented in the theoretical investigations [14,15].

Ross et al. [16] described tests of aluminum plates and beams subjected to blast loading and showed the final shape of one beam specimen as having a counterintuitive deflection. Moessner [17] carried out tests on clamped steel beams, using explosive pressure, and found one case of the counterintuitive deflection about 1 mm in a specimen of length 200 mm.

The first theoretical investigations into CIB of elastic–plastic structures were published in 1985 [14,18,19]. Moreover, two directions of investigation appeared simultaneously. The first direction [18,20–25] considers different models of a relatively simple structural element (i.e., a beam), subjected to a uniformly distributed rectangular pressure pulse. The dynamics of those models was studied very thoroughly using currently available mathematical methods. The second direction considers circular plates and shallow shells subjected to a uniformly distributed rectangular pressure pulse, and an air or underwater explosion using more traditional (i.e., numerical) methods of study [14,19,26–33].

It was shown that the pulse deforms the beam into a shallow elastic–plastic arch. This arch can sometimes lose its stability by snap-buckling under the action of unloading forces. This behavior occurred within some narrow ranges of structure and loading parameters during a transition from elastic to moderately small plastic deformations. Particularly in the realm of CIB, the character of the beam response depends strongly on initial conditions, and very small perturbations of them can change the magnitude and the sign of the residual plastic deflection. These phenomena are associated with chaotic vibrations [20–25]. The above theoretical investigations opened the way for purposive experiments [34,35], which confirmed the theoretical results.

Generally speaking, the results for the circular plates and shallow shells subjected to uniformly distributed rectangular pressure pulse [28,29,32] agree with results obtained for the beam. In particular, these structural elements exhibit a behavior in certain ranges of their initial parameters and pulse loads which cannot be predicted. Within these ranges, nonlinear equations of motion for the above elements are so sensitive to perturbations of problem parameters that it is impossible to predict the dynamic behavior of real structural elements by solving the equations, since the initial parameters and loads are always known approximately and any correction of them can change calculation results significantly. In flat plate, CIB due to reverse snap-buckling occurs only with elastic–plastic deformation, whereas for a shallow spherical cap or cylindrical panel, it can occur with viscoelastic [15,29], elastic, and elastic–plastic [28,30] material behavior.

The above theoretical results were obtained for the rectangular or exponential pressure pulse. These approximations of actual load shape simplify a theoretical analysis and are usually an acceptable idealization for many practical problems. At the same time, the final displacements of the plates can depend significantly upon the pulse shape [36]. The pulse shape especially is important for the phenomenon of CIB, which is very sensitive to the changes of the parameters of loading.

CIB of some structural elements loaded by a blast was studied in [37], where the influence on this behavior of a depression wave moving after a compression shock wave was emphasized. The structure deforms into a new shape during loading by the compression shock wave. This shape may lose its stability as a result of unloading of the structural material and of the action of the depression wave. In case of an underwater explosion, the load strongly depends on the structure/liquid interaction and the appearance of cavitation in the liquid. These phenomena as well as unloading of the material and a static pressure in order to study CIB of structures in contact with liquid must be taken into account [26,31,23].

On the whole, the most publications considered are concerned with the results of theoretical investigations. There are only a few publications devoted to results of experiments, which are limited to cases of beams and circular plates.

Method. Here data of experiments carried out in Russia from 1972 till 1985 are presented. The main purpose of the experiments was to study the strength of thin-walled constructions under the action of an air explosion. Every test was repeated three or more times. Phenomenon of CIB occasionally occurred during the experiments with different single- and double-layer plates, and circular single- and double-layer membranes made from a wire net, and shallow spherical shells (caps). The possibility of a failure of the double-layer plates and cups after reverse snap-buckling under the action of unloading forces was also discovered. The masses of trotyl charges, distances L (Figure 7.1) from charges to center of the tested structural elements, and amplitudes of loads had been defined under which CIB of the specimens occurred. The experimental results will be discussed below using various models and results of aerodynamics and structural dynamics.

First, we describe a method of loading and a profile of the pressure acting on the specimens. The load was created by an explosion of a spherical trotyl charge in the atmosphere. It is known [38] that after an air explosion of moderate power, a discontinuous increase in pressure occurs at a distance equal to 10–11 radii of the spherical charge. The compression pressure behind the discontinuous jump is much more than the atmospheric pressure p_0. The spherical compression shock wave diverges, and the discontinuous increase in pressure becomes weaker. The shock wave changes its shape in time. Particularly, a depression wave arises traveling behind the compression shock wave [39]. At first, the depression wave has a modulus of an amplitude much smaller than the modulus of an amplitude of the compression shock wave. Therefore in case of the action of such spherical blast on a structure, the influence of the depression wave on its response is very small. However, in the next moment

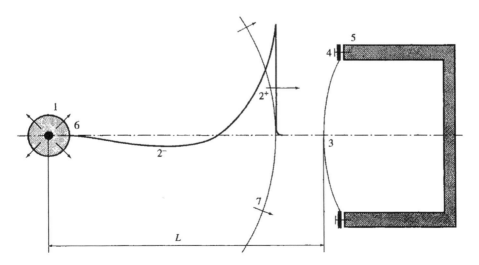

FIGURE 7.1 A scheme of experiments. 1 is a charge, 2^+ is a compression shock wave (positive pressure wave), 2^- is a depression wave (negative pressure wave), 3 is a specimen (plate or cap), 4 is a bolt, 5 is a motionless support, 6 is an explosion product, and 7 is a spherical surface of a discontinuous pressure jump.

of time the amplitude of the compression shock wave reduces and the effect of the depression wave on the structure response may be important.

We assume [39] that the depression wave must be taken into account if a relation of the amplitude p_2, of the compression shock wave, to the pressure p_1 before the discontinuous jump, is less than $(\gamma+1)/(\gamma-1)$, where γ is the adiabatic coefficient of the gas. For atmospheric air, $\gamma = 1.4$ and $p_2 = p_0$, so the influence of the depression wave will be important if the amplitude of the compression wave acting on the specimen is less than 0.6 MPa. Therefore, we will take into account the depression wave only if

$$p_2 < 0.6\,\text{MPa}. \tag{7.1}$$

In the experiments, the trotyl mass q was varied from 0.05 to 1 kg. Assuming that the density of trotyl is 1.65 g/cm^3, we can find the resulting distances at which the discontinuous jump arises. For the above variation of q, these distances change from 25 to 70 cm. In the experiments, the distance L (Figure 7.1) was approximately two times larger than the distance of the discontinuous jump arose. The amplitude p_2 of the compression shock wave pressure acting on the specimen as a result of the above charge explosion varied from 0.04 to 1 MPa.

After the explosion, the pressure on the specimen underwent a discontinuous jump up to the largest value $P > p_2$ (see Eq. (7.3)), then decreased below atmospheric (negative pressure wave on Figure 7.1), and finally returned to its initial value. As a result, the total pressure (dynamic pressure plus atmospheric pressure) on the deformed structural element could change direction during loading.

Different structural elements were examined [13] on the action of the blast loading, but CIB demonstrated only initially flat plates and membranes, and shallow caps. The plates were either circular or rectangular.

The dynamic displacements were measured by electromagnetic sensors. They allowed us to measure a displacement amplitude of up to 6 mm. Larger displacements were measured with the help of a light sensor which recorded the change of a beam of light reflected by the test specimen. The jump pressure was obtained using piezoelectric transducer. A description of the electrical and optical schemes of the above sensors can be found in [13, pp. 147–148].

Typical oscillograms of counterintuitive displacement of the center of specimens are given in Figure 7.2.

At first, the displacement directs from the charge. Then, as a result of reverse snap-buckling, the specimen moves contrary to the direction of the blast loading and begins to oscillate near a new equilibrium position. The amplitude of these oscillations reduces and an average displacement increases a little in course of time. One can see that CIB occurred if the rate of deforming under the shock wave loading was approximately four times less than the period of free oscillations of the specimens. It is interesting to note that an amplitude of the free oscillations may be comparable with the maximum displacement of the specimen during blast loading. So, the specimen can have considerable elastic energy after reverse snap-buckling to the new equilibrium position. Later, this energy dissipates due to cyclic plastic deformations connected with the concentric bending waves traveling from the edge to the

Experimental Data

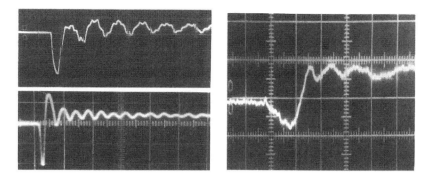

FIGURE 7.2 Three typical oscillograms showing counterintuitive displacement at the center of specimens.

center of the specimen and back again. The final displacement of the specimen may be comparable to the maximum displacement during loading by the compression shock wave.

7.2 CIB OF CIRCULAR PLATES: RESULTS AND DISCUSSION

These plates sometimes exhibited CIB after blast loading if a plastic deformation was small. This behavior was explained in [13] in the following manner. The plate bends after loading and oscillates as a shallow shell. In the case of CIB, vibration energy of this shell is enough for only one reverse snap-buckling. Two specimens after test are presented in Figure 7.3.

Single-layer and double-layer circular plates were tested. Below some mechanical and geometrical parameters of specimens which demonstrated CIB are given, and some elements of the experimental equipment are described.

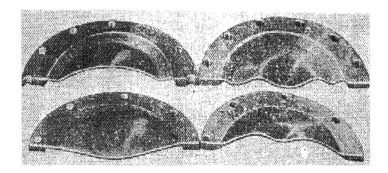

FIGURE 7.3 Typical forms of plates after reverse snap-buckling. The radius R of the clamped specimens was 90 mm; the thickness h varied from 0.4 to 5 mm. The thin specimens ($h/R = 0.01$) showed a variety of behavior under loading. Sometimes, bending waves were observed. They usually focused on the center of the specimens.

Single-layer circular plates and caps were produced from an aluminum–magnesium alloy AMΓ-6 (density $\rho = 2700\,\text{kg/m}^3$, Young's modulus $E = 70{,}000\,\text{MPa}$, Poisson's ratio $\nu = 0.3$, yield stress $\sigma_0 = 190\,\text{MPa}$, strain-hardening exponent $E_1 = 3500\,\text{MPa}$) together with thick edges (circular stiffening frame). These edges had holes. The holes were used in order to bolt the thick edges of the specimens to the motionless equipment (see Figure 7.3 and 7.4). The thickness of the above specimens varied from 0.5 to 4 mm.

The double-layer circular plates were produced by gluing to the metal plates a layer of an elastic material (polymethylmethacrylate: $\rho = 1200\,\text{kg/m}^3$; $E = 5350\,\text{MPa}$; $\nu = 0.35$). The thickness h_1 of the elastic layer was varied from 2 to 6 mm. The double-layer plates were fronted toward the trotyl charge by the elastic layer.

The circular round-meshed plates were also tested. The perforated circular holes formed a square mesh on the plate surface. Figure 7.4 shows the round-meshed plate after its failure by an underwater explosion. The locations of the holes and the circular stiffening frame are visible. The minimum distance between the hole centers was 10 mm, which was the same for all the plates. The diameter, d, of the holes (perforations) was constant for each plate but varied from 2 to 7 mm for the different plates.

Results. Some data of the tests for which CIB of the circular plane specimens occurred are presented in Table 7.1.

In Table 7.1, the data of each line correspond to an individual test. N is the number of the test, t_1 is the time of loading of the plate by the compression shock wave, and other notations of Table 7.1 were defined above. Number 1 corresponds to the

FIGURE 7.4 Photograph of a failed round-meshed (perforated) plate.

TABLE 7.1
Data of the Explosions Loads and Specimens

N	L(m)	q(kg)	P(MPa)	t_1(s)	h(mm)	h_1(mm)	d_1(mm)	p_2(MPa)
1	0.6	0.053	1.28	0.0006	1	0	0	0.41
2	1.3	0%	1.95	0.0013	2	0	5	0.53
3	1.1	0%	3.37	0.0011	2	0	5	0.77
4	1	0%	4.6	0.0010	2	2	7	0.95
5	1	0%	4.6	0.0010	1	6	0	0.95

single-layer plate; numbers 2, 3, and 4 correspond to the round-meshed (perforated) plates; numbers 4 and 5 correspond to the double-layer plates.

The shallow shells formed by the explosion from the circular plates usually had the shape of a cupola or a cone with a spherical center section. However, sometimes axisymmetric shells having more complex profiles were formed. One of the above profiles corresponding to test 1 of Table 7.1 is shown in Figure 7.5. The profile is plotted schematically without an adhesion to ratios of real geometrical dimensions. The profile approximately corresponds to the very shallow cap for which

$$k = H/2R = 0.028. \tag{7.2}$$

The shape of the profile is apparently a result of the oscillations of the plate, with a large amplitude, after the reverse snap-buckling (see Figure 7.2). The reverse snap-buckling of the double-layer plate having large perforations (test 4 of Table 7.1) was accompanied by the failure of polymethylmethacrylate. The amplitude and shape of oscillations of this plate changed instantly, after the above failure (see the oscillogram in Figure 7.6 and compare this oscillogram with that in Figure 7.2). The failure of the elastic layer occurred in the center and near the edge of the plate.

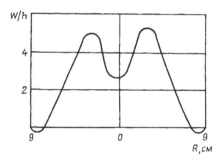

FIGURE 7.5 Variation of the dimensionless final displacement (w/h) of a circular plate after reverse snap-buckling.

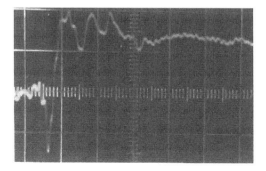

FIGURE 7.6 The oscillogram of the counterintuitive displacement of the center of the double-layer perforated plate. After failure, the oscillations practically end as a result of an instantaneous decrease of the total (elastic plus kinetic) energy of the plate.

The values of p_2 of Table 7.1 were calculated using the formula [40, pp. 152–154]:

$$P = p_2 \left[p_2(3\gamma - 1) - p_1(\gamma - 1) \right] / \left[p_2(\gamma - 1) + p_1(\gamma + 1) \right]. \tag{7.3}$$

In our tests, $p_1 = p_0$. This formula is for a reflection of a shock wave on a rigid wall. The plates can be considered motionless at the first moment of the reflection. We did not take into account the influence of the permeability of the plate. A comparison of tests 4 and 5 shows that the influence of the permeability on P was small. Knowing p_2 will allow us to discuss the influence of the depression wave on the behavior of the tested plates.

Discussion. 1. First, we shall consider test 1. In order to simplify the analysis, we start with the assumption that the stress state existing before the reverse snap-buckling of the tested plate was similar to a stress state of the same plate under the static uniform load. The influence of membrane forces and plastic deformations are also ignored. We take into account that the experimental boundary conditions do not correspond perfectly to the conditions of a fully clamped plate. For the analysis of the stress state of the tested plate, we used the curves in Figure 7.7 [41].

According to Figure 7.7, plastic deformations can first appear either in the center or near the edge of the plate. Consequently, if these plastic deformations are moderately small, then the main volume of the plate material undergoes only elastic deformation and thus accumulates elastic energy. At the moment $t = t_1$, when the action of the compression shock wave is ending, the plate can begin its reverse motion and start subsequent oscillations. If plastic deformation occurs during the reverse motion of the plate, then we can have the phenomenon of CIB (see Figure 7.2). If the plastic deformations of the reverse motion and subsequent oscillations are very small, then the oscillations can become chaotic [22–24].

Apparently, the occurrence of CIB discussed above is possible only if the time of loading, by the positive pressure, is less than the period T of the fundamental frequency of the plate oscillations. We found T assuming that the plate was fully clamped [42]: $T = 0.0032\,\text{s}^{-1}$. So the time of loading of the plate by the compression shock wave was a little less than $T/4$ ($t_1 = 0.0006\,\text{s}$).

The depression wave acted on the plate after the compression shock wave. An amplitude of the depression wave A cannot be large since $|A| < p_0$. However, the

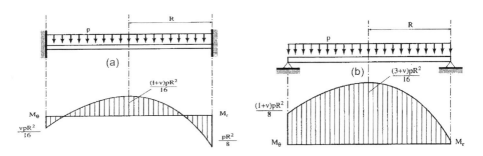

FIGURE 7.7 The profiles of radial M_r and tangential M_θ moments of a circular plate: (a) fully clamped edge and (b) simply supported edge.

influence of the depression wave on the result of test 1 was important because the amplitude of the compression wave ($p_2 = 0.41$ MPa) for this test was not large; thus, the condition (7.1) fulfills. The effect of the depression wave increases because the time and direction of its action coincides with the time and direction of the reverse snap-buckling of the plate.

Figure 7.8 shows schemes of loading and motion of two points on the plate, which are summarized above. One can see that CIB of the plate is explained by the simultaneous appearances of unloading of the material of the plate and the negative pressure on the plate. Of course, the occurrence of CIB is possible only if the essential material of the plate deforms elastically before its reverse snap-buckling.

Discussion. 2. Now, we consider the perforated plates (tests 2–4). The perforation resulted in the permeability of the plates. We could change the permeability of the plate by the variation of, d, in order to vary a plate/blast wave interaction. The table shows that the perforated plates were subjected to more intensive and longer duration blast loading than the plate of test 1.

Because of the increase of p_2, we can ignore the influence of the depression wave acting on the perforated plates (see Eq. (7.1)). For tests 2–4, aerodynamic processes (connected with transition of the discontinuous jump of pressure and transient air flow through the permeable plate) were of primary importance. In the first moments of loading, the air flows into the box (Figure 7.9(a)). As a result, the pressure in the box becomes larger than atmospheric pressure outside the box. When the loading

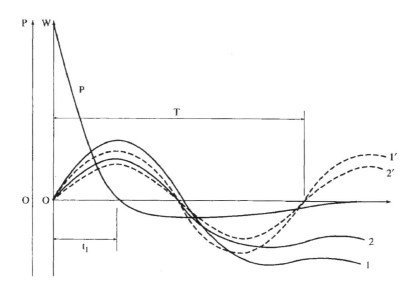

FIGURE 7.8 Scheme of loading and the counterintuitive displacements of points on a circular plate. P is the curve of the pressure variation. 1; 2 and 1'; 2' are the curves for the elastic–plastic and elastic plates, respectively. 1 is the displacement of the plate point where the material obtained plastic strains. 1' is the same for the elastic plate. 2 is the displacement of the plate point where the material has not obtained plastic strains. 2' is the same for the elastic plate.

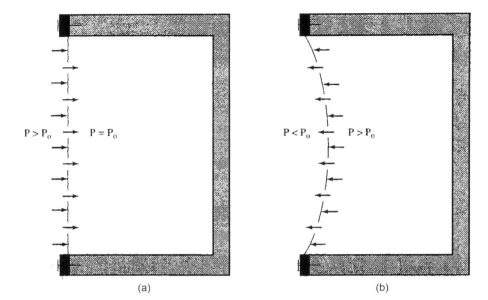

FIGURE 7.9 Scheme for the interaction of the blast with the perforated plate: (a) the direction of the pressure and air flow at the first moments of the action of the compression shock wave on the plate; and (b) the direction of the pressure and air counterflow during of the action of the depression wave on the plate.

of the plate by the compression shock wave ends, the plate begins reverse motion under: the action of unloading of its material, the pressure inside the box, and the air counterflow (Figure 7.9(b)).

Apparently, CIB of the plates for tests 2; 3; 4 are associated both with unloading of the plate material and the transient aerodynamical processes of the air wave flow through the permeable surface of the plates.

Discussion. 3. Now, we shall consider test 5. Reasons for CIB of the thick $((h + h_1)/R = 0.077)$ two-layer plate used in test 5 may only be due to the mechanical properties of the plate. The thicker layer of the plate was elastic (polymethylmethacrylate), and the other was elastic–plastic (metal). During the blast loading, the elastic layer accumulated elastic strain energy. In the next moments during unloading, reverse snap-buckling, and the following oscillations of the plate, this energy was absorbed by the elastic–plastic deformed metal layer. As a result the double-layer plate fixed, the final displacement directed to the charge.

Remark 1. The example of the plate for test 5 (Table 7.1) shows that the structures (systems) containing both elastic and elastic–plastic elements can apparently demonstrate the anomalous response on an impulse load. In particular, they can demonstrate CIB in cases when large plastic deformations occur if the energy accumulated by the elastic elements is much larger than the other elastic–plastic elements of the structure (system), which are able to absorb elastically. Therefore, the energy of the elastic elements can be dissipated by the large plastic deformations of the elastic–plastic elements.

Experimental Data

Remark 2. Test 1 (Figure 7.5) and the peculiarities of the failure of the plate of test 4 show that the most intensive bending stresses existed in the center and near the edge of the plate. Therefore, the stress state of the tested plates was closer to the stress state of the fully clamped plate (Figure 7.7(a)) than to the simply supported plate (Figure 7.6(b)). Figure 7.5 shows that the amplitude of the concentric bending wave of the displacement was near the center of the plate much more than the amplitude of this wave near the edge. Apparently, the experimental boundary conditions for the tested plates can be simulated by the conditions of the fully clamped edge more accurately than the simple support conditions.

Remark 3. The maximum displacement of the double-layer plate (test 4 of the table) after the reverse snap-buckling was more than its maximum displacement during loading by the compression shock wave (Figure 7.6). This is apparently due to the focusing of bending waves traveling from the edge to the center of the plate and with the action of the box pressure and the air counterflow.

Remark 4. Test 4 of Table 7.1 (Figure 7.6) shows that the failure is accompanied by the simultaneous dissipation of the total energy and ending of the large oscillations of the specimen center.

7.3 CIB OF RECTANGULAR PLATES AND SHALLOW CAPS

Rectangular plates. Plates made from alloy AMΓ-6 were tested. Their thickness h ranged from 1 to 4 mm. The length of the larger edge of the plate was 250 or 300 mm; the length of the other edge changed from 60 to 250 mm. Plates were examined with fully clamped edges or combined boundary conditions (i.e., fully clamped short edges and free long edges). Different shapes of explosion loading were used. About 150 fully clamped plates were tested [13].

In the case of the combined boundary conditions, 45 plates were tested [13]. The time of the blast loading was less than $T/4$ (T is the fundamental period of elastic oscillations of the plate). The test results agreed with the data of the tests of the other researchers [43,44]. The final displacements of these plates were less than $10h$, and the plastic deformations were not large. The experimental results [13, pp. 153–156] did not therefore coincide with those of the calculations according to the theory [45].

CIB of the rectangular plate occurred after an explosion of 0.18 kg trotyl at a distance of 0.65 m from the fully clamped plate. The sides of the plate were 150 and 300 mm, and the thickness was 4 mm. The maximum final displacement of this plate obtained as a result of reverse snap-buckling was 5.41 mm.

Discussion. According to [39] (see also Table 7.1 and Eq. (7.1)), an explosion of 0.18 kg trotyl did not form an appreciable depression wave at a distance of 0.65 m. Therefore, CIB of the rectangular plate was only associated with unloading of its material.

Spherical shallow shells. The shallow caps used in the test were made from alloy AMΓ-6. The radius R of the cap's edge varied from 30 to 150 mm, the thickness h from 0.3 to 5 mm, and the rise height H from 2 to 40 mm. The geometry of the caps was defined by the dimensionless ratio $K = R/\sqrt{ah}$ where a is the radius of the initial cap curvature. Various methods for causing the impulse load were used. Under explosive loading, the caps demonstrated either a dynamic axisymmetric

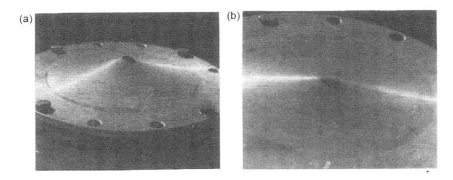

FIGURE 7.10 The cap after the blast loading: (a) the final shape of the cap after reverse snap-buckling and (b) the type of failure of the cap near its pole.

snap-buckling if $K \leq 6.3$, or a dynamic non-axisymmetric snap-buckling if $K > 6.3$ [13, pp. 168–169]. The cap behavior was similar to flat plate behavior when $K < 3$. The results of many tests are presented in [13].

The numerical investigations showed the possibility of CIB of shallow spherical caps subjected to an impact loading on the concave side [15,28,32]. The analogous result was found experimentally for the caps loaded by a blast.

The dimensions of the caps were as follows: $h = 0.55$ mm, $H = 7.5$ mm, $R = 90$ mm, and $a = 535$ mm. Therefore, $K = 1.65$ and $k = 0.042$. The trotyl mass was 0.053 kg. Loading was realized by an explosion at a distance of 0.8 m (when $P = 3$ MPa) or 1.15 m (when $P = 2.2$ MPa) from the cap. At first, after shock loading, the caps lost stability and deformed in the direction of the blast; but then reverse snap-buckling of the caps occurred. The caps turned out completely and after CIB formed a cone shape (Figure 7.10). When $L = 0.8$ m, the metal became thinner up to failure around the specimen pole. The hole caused by the failure was star-shaped. So through CIB, the caps can lose their strength completely.

7.3.1 Discussion of CIB of Shallow Caps

1. **Reasons for CIB of the caps.** The caps were very shallow, and their k did not differ strongly from (7.2). The trotyl mass q and the distances L did not also differ significantly from test 1 in the table. The amplitude p_2 of the compression shock wave which acted on the caps was 0.78 MPa, when $L = 0.8$ m and 0.44 MPa when $L = 1.15$ m (p_2 was calculated using (7.3)). The time of the positive pressure loading was approximately 0.7 ms. Then, the caps were subjected to the depression wave loading. The influence of the depression wave was important for the case $p_2 = 0.44$ Mpa (see Eq. (7.1)) and was not very significant for $p_2 = 0.78$ MPa. The fundamental frequency of the cap did not differ noticeably from the fundamental frequency of the flat plate for test 1 in the table. The increase in the fundamental frequency caused by the cap curvature was suppressed by the reduction in the cap thickness.

Experimental Data

The parameters of the caps and loading were similar to the parameters in the test of the above plate. The reasons noted above allow us to think that CIB of the caps was associated with the simultaneous action of the material unloading and the depression wave.

2. **Reasons for the failure of the cap pole.** Now, we consider the reasons for the failure of the metal in the cap pole after the explosion at a distance 0.8 m from the cap.

   2.1. **Linear elastic analysis.** The influence of the curvature on the stress state of the cap is very important. We shall evaluate this influence using the data of paper [46] (these data can also be found in [47, pp. 124–125]) where a cap with $h = 0.057R$ was studied. The linear equations of the cap motion were solved, and the data from these calculations were presented in [46] with the help of dimensionless values. In particular, the stresses and pressure were divided into $E/(1 - \nu^2)$. Some stresses for the fully clamped caps are given in Figure 7.11. The calculations were produced for various values of relation $k = H/2R$. One can see that for the small value k, the bending stresses are more important than the membrane stresses. Particularly, if $k = 0$ (flat plate), the bending stresses reach a maximum and the membrane stresses equal zero. When the value of k increases, the bending stresses reduce rapidly and the membrane stresses first increase (if $k \leq 0.03$) and then reduce (if $k > 0.03$).

   Figure 7.11 shows that some particulars of the dynamic stress state of the cap can be similar to its static stress state. It agrees with the

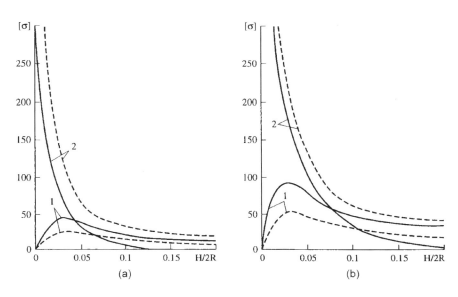

FIGURE 7.11 The maximum values of modulus of the stresses in the caps subjected to the uniformly distributed static pressure $p = 1$ (a) or the dynamic step pressure with an amplitude equal 1 (b). 1 is the membrane stresses. 2 is the bending stresses. Solid lines (———) are the stresses in the pole, whereas dashed lines (-----) are the stresses in the edge.

assumption used for the analysis of the circular plate tests. However, the dynamic load is more stressful for the structural element than the static load. The stress state depends very strongly on the curvature of the cap. It is interesting to note that the tested caps had a rise height unfavorable for their strength ($k = 0.042$). The tested caps had simultaneously both high membrane stresses and high bending stresses in the pole. The stress state near the edge of the tested caps was also complex, but the membrane stresses were less than that in the pole. It is known that the membrane stresses are often more dangerous than the bending stresses because they are uniformly distributed through the thickness of a wall. If the maximum membrane stress is σ_0, then the structure may fail, while a maximum bending stress equal to σ_0 does not cause failure.

Thus, linear analysis has demonstrated the simultaneous existence in the pole of the tested caps of unfavorable membrane and bending stresses.

2.2. **Nonlinear elastic analysis.** The influence on the cap's behavior of geometric nonlinearity is more than its influence on the plates' behavior. The snap-buckling of the shells is always a dynamic process even for static loading. These peculiarities are very interesting for the problem of CIB of structural elements.

We give below the results illustrating the importance of the above features. The calculations of the shallow cap subjected to uniformly distributed quasi-static load presented in [13, p. 61] are used (Figure 7.12). The cap from alloy AMΓ-6 with parameters – $a = 400$ mm, $R = 55$ mm, $H = 3.8$ mm, $h = 1$ mm – was fully clamped on the edge.

The numerical solution of the nonlinear equation of motion showed that in the first instance, the deflection of the cap pole increased proportionally to the load (section AB of Figure 7.12), next the snap-buckling of the cap, and the transient process of cap oscillations in the new equilibrium state (section BC) took place. When the load began to reduce, the deflection also began to reduce (section CD) and reverse snap-buckling (section DE) occurred. It is interesting and important for the cap strength that the pole transient oscillations after reverse snap-buckling were much larger than the oscillations after snap-buckling. The reasons for the pole oscillations are considered below.

Remark 5. Snap-buckling is a strong nonlinear and rapid process (Figure 7.12). As a result of snap-buckling, the cap can obtain a large elastic deflection. In particular, the snap-bucklings, following one after another, generate (in the fully clamped cap) intensive bending stresses waves. The amplitude of these waves is more than that of bending stresses wave caused by the deformation and then reverse snap-buckling of the fully clamped plates. Thus, in the caps, bending stresses were generating, as a result of CIB, more dangerous for the strength than bending stresses generated by CIB of the plate under similar loading. It corrects the results of the linear analysis (Figure 7.11).

Experimental Data

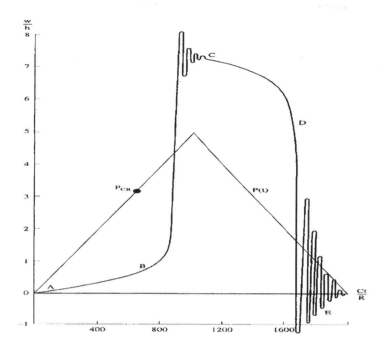

FIGURE 7.12 Variation of the pressure and the deflection of the cap pole in time. P_{CR} is the static buckling (critical) load for the cap. C in the expression Ct/R is the sound velocity in air.

2.3. **Nonlinear elastic–plastic analysis.** The data in Figures 7.11 and 7.12 were calculated for the elastic cap. However, the tested caps had plastic deformations. Figure 7.10 shows that near the pole, the cap deformed to a conical shape. So there are no small plastic deformations. Finite dynamic displacements of the elastic–plastic caps were studied in [48, pp. 122–123]. Below are some results from [48] for the case $h/a = 0.02$.

In Figure 7.13, the dimensionless displacement in the pole of the fully clamped semispherical cap subjected to a dynamic uniformly distributed step pressure is presented. W_0 in Figure 6.13 is the maximum displacement of a complete spherical shell subjected to the same step pressure. The displacement in the pole changes and increases very quickly and has an alternating character. Consequently in the pole, the transient alternative oscillations of the bending stresses also exist. These stresses are very dangerous for the strength of the cap material. The oscillations of the pole are excited by the concentric bending waves traveling from the edge. The waves focus on the pole and largely increase the stresses there. Figure 7.14 shows the profiles of the cap displacement. In [48], it is emphasized that the above noted transient alternative bending stresses increase for the shallow caps. Really, the

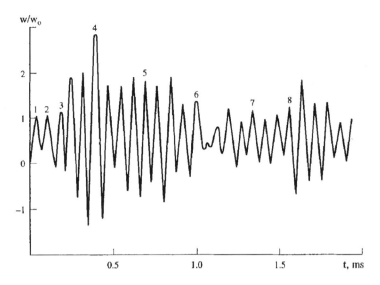

FIGURE 7.13 Variation of the displacement of the cap pole in time. 1–8 are the peaks of displacement.

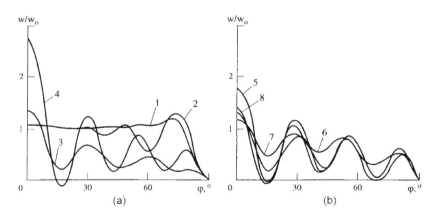

FIGURE 7.14 The profiles 1–4 (a) and 5–8 (b) of the displacements corresponding to the peaks 1–8 of Figure 12.

amplitude of the bending stresses increases as the rise height reduces (see Figure 7.11).

2.4. **Reasons for the failure.** The above analysis allows us to connect the tested cap failure with both the appearance of the plastic membrane strain near the pole during loading of the cap by the compression shock wave and the high-frequency oscillating alternating intensive bending stresses. The maximum membrane stresses were approximately equal to the maximum bending stresses. The concentric bending waves generated near the edge of the cap and focused on the pole are a result of the snap-buckling and reverse snap-buckling of the cap which followed one after another. The bending stresses caused an increase in plastic deformation and fatigue failure of the metal in the pole. Amplitudes of the bending and membrane stresses were considerably higher in the pole than those near the edge, and therefore, the failure occurred at the cap pole.

There are four stages in the deformation for the pole of the cap shown in Figure 7.15. They are as follows:

1. The displacement and snap-buckling of the cap under the action of the compression shock wave (section OB).
2. The deformation of the cap under the action of the compression shock wave and the appearance in the plate pole of plastic strains connected with membrane forces (section BC).

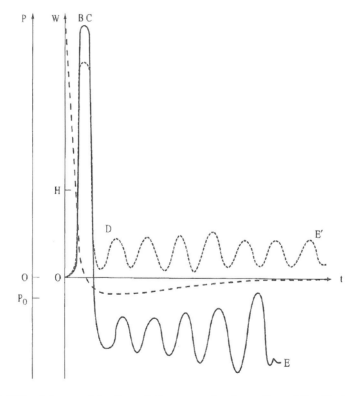

FIGURE 7.15 Scheme of loading of the cap under the action of the blast pressure P (interrupted line), in the case of large displacement of the pole and the failure of the metal (solid line), in the case of small plastic displacement and the damped oscillations of the pole (points).

3. The reverse snap-buckling of the cap under the action of unloading of the cap material and the depression wave (section CD).
4. Intensive high-frequency oscillations of the cap pole which can be accompanied by the rise of plastic strains because of bending stresses, metal fatigue and the failure (section DE, the case of the explosion in the distance $L = 0.8\,\text{m}$) or damped oscillations (section DE', the case of the explosion in the distance $L = 1.15\,\text{m}$).

Remark 6. It is necessary to note that the oscillogram of the pole displacement (it is not presented here) showed that the failure of the cap took place after CIB.

Remark 7. Of course, our consideration is only qualitative because the curves of Figures 10–13 were calculated for loads and caps different from those of the tests.

7.3.2 Cap/Permeable Membrane System

We have considered (above) CIB of the mechanical system consisting of a box closed by a perforated plate. CIB of this system was connected with the aerodynamic processes of the transient air flow through the perforated plate. The system consisting of a cap closed by a permeable membrane was also tested. The permeable membranes were used as a protective screen which reduced the effect of blast loading on the concave side of the cap (Figure 7.16).

We will not discuss in detail the experiments with the cap. They are considered more carefully in [10] using the results presented in [49–57].

Results. Single- and double-layer permeable metal circular membranes ($R = 90\,\text{mm}$) were used. They were made of steel wire net (steel alloy 12X18H10T:

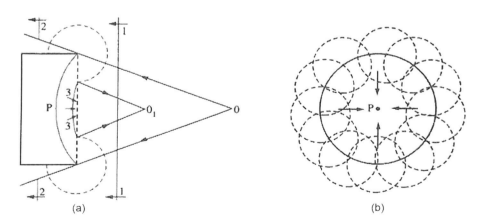

FIGURE 7.16 Scheme of loading of the cap (a), scheme of converging of the rarefaction waves from support to the center of the cap/membrane system (b). O is the point of the explosion, O_1 is the focus of the spherical shock wave 33 reflected by the cap, 11 and 22 are the locations of the blast wave before and after of the cap loading, P is the cap pole, dashed line (------) is the surface of the spherical wave generated by the points of the support after loading by the blast wave or the surface of the membrane.

$\rho = 7800$ kg/cm^3, $E = 200000$ MPa, $\nu = 0.3$, $\sigma_0 = 260$ MPa, $E_1 = 1130$ MPa) having square cells. The area of each square cell of the net was 0.004 cm^2. The thickness of the wire was 0.5 mm. The double-layer membranes were made by positioning two wire nets side by side. The wires of the different layers formed an angle of 45°. The radius of the cap was 90 mm, the rise height varied from 5 to 40 mm, and the thickness was 0.5 or 1 mm. The edges of the membrane and cap were bolted on to the motionless support of the experimental equipment with the help of a grip ring (see Figures 7.1 and 7.10).

First, the caps were tested without the protective permeable membrane. Explosions of different power were carried out at a distance of 1 m from the cap (Figure 7.16). After the explosion of 0.8 kg trotyl, the amplitude of the pressure measured on the concave side of the cap was 4.6 MPa ($p_2 = 0.95$ MPa, see the above table). The cap failed owing to tearing of the cap material at the supports. Next, the same caps with a protective membrane were tested. The protected caps remained intact. Behind the membrane, on the cap surface, the amplitude P of the blast load was 2.3 MPa.

It was discovered that CIB of membranes from the wire net was possible. The deflection of the single-layer membrane was smaller than the deflection of the double-layer membrane for the same blast load. The final deflection of the double-layer membrane after CIB was large and reached a value of 0.6 R. It was surprising that CIB was exhibited only in the external layer of the double-layer membrane, whereas the internal layer did not show a noticeable plastic deflection.

Figure 7.17 shows the final shape of the double-layer membrane after an explosion of 0.8 kg trotyl at a distance of 1 m from the cap/permeable membrane system. The center the membrane obtained a parabola shape as a result of CIB. The displacement near the edge is much less than that in the center of the membrane.

Discussion. First, we consider the process of blast loading of the concave surface of the cap. The influence of the membrane and failure of the cap are not taken into account. In this case, a part of the blast wave (its amplitude is 0.95 MPa) loads and reflects from the specimen (Figure 7.16). The amplitude of the reflected shock wave is 4.6 MPa. Its curvature approximately equals to that of the cap. The spherical

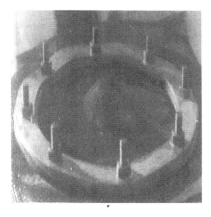

FIGURE 7.17 Final shape of a double-layer membrane after CIB.

reflected wave converges to point O_1 shown in Figure 7.16. The other part of the blast wave passes by the experimental equipment. As a result, the pressure near the support reduces very rapidly and the spherical rarefaction waves begin to propagate from the support. These waves expand to the cap pole (Figure 7.16) and reduce the pressure acting on the cap.

Thus, after the blast loading, a complex wave process occurs when the new shock and the rarefaction waves appear near the cap. Generally speaking, the mathematical investigation of this process needs the solution of the three-dimensional aerodynamic equations. However, for qualitative analysis, we will consider points of the cap edge as sources of radial spherical waves converging to the center (Figure 7.16). The amplitude of each pressure wave propagating from the edge point reduced as r^{-1} (see Chapter 9). Thus, the single spherical wave propagating from the cap edge (Figure 7.16) damps rapidly, just as the pressure in the pole of the cap instantly increases rapidly due the focus of fronts of converging spherical wave 33 (Figure 7.16).

In case of the cap preserved by the membrane, the above considerations became complicated by the reflected wave from the membrane and the oscillations of air between the cap and membrane. However, the influence of the last details is not significant. The distance between the membrane and cap near the support is small; therefore, the pressure in front of and behind the membrane becomes equal, and the membrane does not obtain a large displacement at the first moments of loading. Then, the rarefaction waves, propagating from the support to the center, unload the edges of the cap and membrane.

The large deflection of the membrane center (Figure 7.17) is because of loading by the reflected converging spherical shock wave 33 (Figure 7.16). The initial amplitude of this wave is 2.3 MPa. The amplitude has increased proportional to r^{-1} at the moment of its collision with membrane, which is, as a result, equal to 5.2 MPa for $H = 40$ mm. The rarefaction waves from the support did not influence the pressure near the center because the time for propagation of the waves from the membrane center to the cap pole and back again was $2H/c_0$, where c_0 is the sound velocity in air. This is less than the time of the wave propagation from the support to the center (R/c_0). As a result, the double-layer membrane achieved a large deflection at the center. The single-layer membrane had, after CIB, a center deflection smaller than the double-layer membrane, because the shock wave passed by the single-layer membrane and loaded it less than the double-layer membrane.

The large final deflection only in the external layer is explained by the interaction of the layers and peculiarities of the air flow through the double-layer net. The shock wave air flow through the double-layer net seems more complex than that through the single-layer net or the perforated plate. The paradoxical peculiarities of the air flow through the double-layer net were noted in [56].

Thus, CIB of the cap/permeable membrane system is associated with aerodynamic processes (both with the reflection of the blast wave 11 (Figure 7.16) from the concave side of the cap and loading of the membrane center by the reflected converging shock wave 33 (Figure 7.16) having the amplitude equal 5.2 MPa). The influence of unloading of the membrane material (steel wire) was small.

Experimental Data

7.3.3 CIB OF PANELS

The large volume of experiments was carried out in Russia from 1972 to 1985 [13,48]. The purpose of the experiments was not studying the anomalous and counterintuitive behavior (CIB) of impulsively loaded structural elements. Therefore, a lot of results have not been fully saved.

Here, we consider some data concerning the anomalous and CBIs of cylindrical and conical panels subjected to blast loading. A scheme of experiments and using equipment were described in [13,48]. The experimental data testify that the paradoxical results are possible for different structural elements and under various impulse loads. However, the structure must be flat or shallow. The time of positive pressure loading must not significantly exceed a quarter of the period of the fundamental frequency of the structural element. The influence of the amplitude and shape of the external pressure pulse, peculiarities of the structure – blast interaction, and material properties of the structure is very important. Sometimes, CIB can be accompanied by local plastic deformations. However, typically the plastic deformations were usually small during CIB.

Cylindrical panels. The thickness of the panels varied from 1 to 3 mm. One side of them was 14, 21, or 28 cm, and the other side was 14 cm and the radius $R = 41$ cm. The blast load moved along the panels (traveling load) or was normal to panel surface. CIB was usually observed for the traveling load. If a time loading was larger than the fundamental frequency of the panel, then a dent occurred. If this time was reduced, a few waves were generated dependent on the angle. CID had place along the central line and/or near the straight edges of the panels. Two specimens after the test are presented in Figure 7.18.

Conical panels. The length of the specimens was 20 cm and $h = 1$ mm. The radius of large side was 13.5 cm and that of the small side was 8.5 cm. The cone

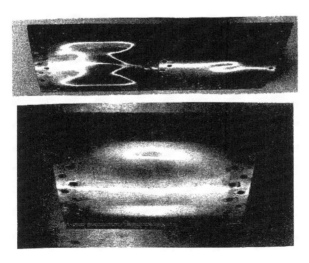

FIGURE 7.18 Cylindrical panels after the experiment.

angle was 30°, and the circumferential angle was 60°, 75°, or 90°. A small traveling load forms the one dent along the central line of the panels (Figures 7.19 and 7.20). Sometimes, small amplification of this load changed the deformation history dramatically; that is, the displacement strongly increased and counterintuitive displacement took place at the central line. Near the straight edges the dents were generated.

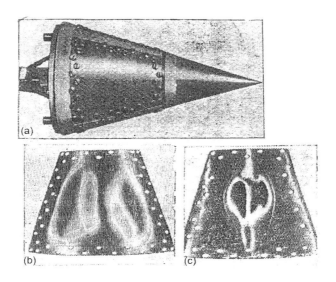

FIGURE 7.19 View of the conical panel (a) and results of certain experiments (b, c).

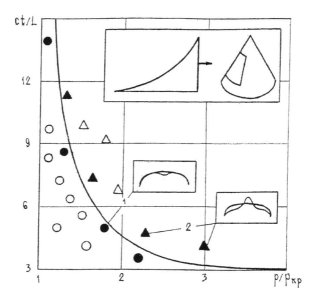

FIGURE 7.20 Results of experiments.

Some results of the experiments are presented in Figure 7.20: the solid line defines the dents of thickness h (1), the bright circles correspond to stable panels, and the dark circles correspond to the dents of 0.6–0.7 h. The dark triangle corresponds to the anomalous form (2), whereas the bright one corresponds to the snap-buckling with a large amplitude.

Dynamical buckling is a very complex process which is very sensitive to imperfections of structures. This process can sometimes be accompanied by elastic unloading of the material and formation of waves. In particular, the part of the loaded structural element can obtain a final displacement to the direction of the initial loading. A "crater" (cylindrical shell) and a "peak" (conical shell) forms of the structure after impulsive loading which are the partly directed contrary to the direction of the impulsive loading are, for example, shown below.

The above processes are complicated in the cases of panels and finite shells. For example, if we have a shell under an internal pressure.

The above process complicates if there are static loads. For example, if we have a shell under an internal pressure, then the shell can obtain the final displacement opposite to the direction of the dynamic external pressure as a result acting of the internal static pressure!

Conclusion. The survey of the publications and the experimental data in this chapter testify that the occurrence of CIB is possible for different structural elements and under various impulse loads. However, the structure must be flat or shallow. The time of positive pressure loading must not significantly exceed a quarter of the period of the fundamental frequency of the structural element. The influence on CIB of the amplitude and shape of the external pressure pulse, peculiarities of the structure/blast interaction, and material properties of the structure are very important. CIB can be accompanied by local plastic deformations. This can lead to local failure of structural elements.

The examples of the failure of the double-layer plate (in the center and near the edge) and the cap (in the pole) have been given. In the first case, only the elastic layer has failed. The plastic (metal) layer remained intact. The failure was caused by the intensive bending waves of stress. They were generated near the edge and focused on the center of the plate because of one oscillation, which was excited first by the shock wave loading and then reverse snap-buckling. The failure of the cap was also caused by the bending waves traveling from the edge to the pole and vice versa. These waves were generated by snap-bucklings of the cap which followed one after another and had a large amplitude. The plastic membrane strains appeared near the pole as a result of the shock wave pressure loading, and then the high-frequency oscillating bending waves arrived. They caused the fatigue failure of the metal in the pole.

The geometry and loading of the failed cap did not strongly differ from those of the plate (test 1 of the table). However, the plate kept intact because the bending waves generated by CIB of the plate are weaker than those generated by CIB of the cap under the action of the similar load. The membrane stresses in the plate are also smaller than those in the cap.

The discussion of the experimental data has had been entirely qualitative. The complete analysis of the experimental results requires numerical calculations, theories and data from aerodynamics, structural dynamic mechanics, and the mechanics of dynamic plastic failure.

8 CIB of Plates and Shallow Shells
Theory and Calculations

8.1 DISTINCTIVE FEATURES OF CIB OF PLATES AND SHALLOW SHELLS

Notation: r and z – coordinates; ρ – density; S_{ij} – deviatoric stress tensor components; U, W – displacements; ε^e_{ij} – elastic strain tensor components; $\sigma_e = (3\, S_{ij}\, S_{ij}/2)^{1/2}$ – equivalent stress; $\varepsilon_e = (2\, \varepsilon_{ij}\, \varepsilon_{ij}/3)^{1/2}$ – equivalent strain; N_1, N_2, N, and Q – forces; $\Delta_n \varepsilon^p_{ij}$ – incremental change of the plastic components of the strain tensor for the time step n; $\Delta_n \varepsilon^p_e$ – incremental change of the plastic component of the equivalent strain for the time step n; P – load; M_1 and M_2 – moments; ν – Poisson's ratio; R_0 – radius of the plate supporting contour; R – radius of the curvature; f – rise height of the shell; ψ – rotation angle; E_1 - strain-hardening exponent; ε_{ij} – strain tensor components; σ_{ij} – stress tensor components; h – thickness; t – time; σ_0 – yield stress; E – Young's modulus; $(.)' = \dfrac{\partial(.)}{\partial r}$; $(\dot{.}) = \dfrac{\partial(.)}{\partial t}$.

8.1.1 INVESTIGATION TECHNIQUES

This section discusses the influences of small perturbations of the parameters of the loading, structure (including shell form and curvature), and the numerical scheme on the deformation within the ranges where reverse snap-buckling occurs. The analysis is based on the numerical calculations performed on the one-dimensional equations of the various structures.

Transient motions of a plate, a thin shell in the form of a spherical cap (or cupola), and a cylindrical panel of infinite length in one direction were studied. Timoshenko equations of motion (see Section 1.5.1 and [47–57]) were used.

Nonlinear terms in equations of motion were taken into account according to [50,62]. In particular, it is supposed that all the components of the strain tensor in any shell point are small as compared to unity ($\varepsilon_{ij} \ll 1$), and $\left(\dfrac{\partial W}{\partial r}\right)^2 \ll 1$. Under these conditions, normal displacement W of a shell can be of the order of thickness and even larger, but it should be small as compared to the main geometrical dimension of the shell. Shallow shells are considered for which the ratio of the main geometrical dimensions to the curvature radius is much less than unity (e.g., for a spherical cap $R_0/R \ll 1$). The theory of shallow shells enables studying displacement comparable with the shell rise value. It is widely used in studies of shells response to a dynamic load [36,43,47,50,62].

For the spherical cap under axisymmetric loading, the equations of the noted above nonlinear theory of the shallow shells take the following form (see, also, sections 1.5, 6.1 and 6.3):

$$(rN_1)' - N_2 = r\rho h \ddot{U};$$

$$(rQ)' + \left[rN_1(W' + r/R)\right]' = r\rho h \ddot{W} - rP; \quad (8.1)$$

$$(rM_1)' - M_2 - rQ = r\rho h^3 \ddot{\psi}/12.$$

For a cylindrical panel, the following one-dimensional equations are used:

$$N' = \rho h \ddot{U};$$

$$Q' + N/R + (NW')' + P = \rho h \ddot{W}; \quad (8.2)$$

$$M' - Q = \rho h \ddot{\psi}/2.$$

The equations for a plate and a beam are obtained, respectively, from (8.1) and (8.2) by putting the factor $1/R$ equal to zero. By neglecting in Eqs. (8.1) and (8.2) the terms that consider the influence of transverse shear and rotatory inertia, or of finite displacements, or of the shell curvature, we get different equations of motion for beams and plates which are used in [1,36]. The expressions for the forces and moments in (8.1) and (8.2) have the usual forms:

$$(N_1, N_2, Q) = \int_{-h/2}^{h/2} (\sigma_{11}, \sigma_{22}, \sigma_{13}) dz;$$

$$(M_1, M_2) = \int_{-h/2}^{h/2} (\sigma_{11}, \sigma_{22}) z \, dz. \quad (8.3)$$

We shall assume that plastic strain increments may be calculated in accordance with standard theory governing plastic deformations. The total strain increment is the sum of elastic and plastic increments, which is given as follows:

$$\varepsilon_{ij} = \varepsilon_{ij}^e + \sum_{n=1}^{N} \Delta_n \varepsilon_{ij}^p. \quad (8.4)$$

Since the elastic strain increments are calculable from the stress increments by Hook's law, the stress components of thin plate or shell theory are written as

$$\sigma_{11} = E\left[\varepsilon_{11} + v\varepsilon_{22} - \sum_{n=1}^{N}\left(\Delta_n\varepsilon_{11}^p + v\Delta_n\varepsilon_{22}^p\right)\right]\bigg/\left(1-v^2\right);$$

$$\sigma_{22} = E\left[\varepsilon_{22} + v\varepsilon_{11} - \sum_{n=1}^{N}\left(\Delta_n\varepsilon_{22}^p + v\Delta_n\varepsilon_{11}^p\right)\right]\bigg/\left(1-v^2\right); \quad (8.5)$$

$$\sigma_{13} = E\left[\varepsilon_{13} - \sum_{n=1}^{N}\left(\Delta_n\varepsilon_{13}^p\right)\right]\bigg/\left[2\left(1+v^2\right)\right].$$

The radial and tangential strains ε_{11}, ε_{22} and the shear strain ε_{13} of the layer at a distance z from the middle surface of the structural element are expressed in terms of the radial and transverse displacements U, W and the rotation angle ψ. For example, in the case of the spherical cap, the forms are

$$\varepsilon_{11} = U' - W/R + (W')^2/2 + z\psi'; \quad \varepsilon_{22} = U/r - W/R + z\psi/r; \quad \varepsilon_{13} = (W' + \psi)f(z).$$

Here, the function $f(z)$ expresses the shear stress distribution over the thickness of the plate or shell [47,61]: $f(z) = 6[1/4 - (z/h)^2]$.

Expressions (8.4) and (8.5) differ from those written for an elastic body in terms of accumulated plastic strains. These terms can be found using the constitutive equations for the stress–strain behavior of materials. Some important properties of this behavior for some materials are defined by the σ_e–ε_e relation (Figure 8.1):

$$G(\sigma_e, \varepsilon_e) = 0. \quad (8.6)$$

The plastic strain appears if $\sigma_e > \varepsilon_0$. The value σ_0 depends on the deformation history. Equation (8.6) can be defined, for example, using experimental data for the uniaxial tension of the material. However, Eq. (8.6) does not generally provide a relation

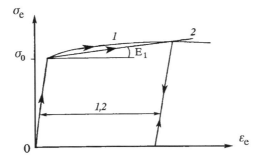

FIGURE 8.1 Schematic representation of a stress–strain curve taking into account the influence of the material strain hardening: curve 1 corresponds to Eq. (8.6); curve 2 corresponds to bilinear approximation of curve 1.

between the individual stress and strain components. Therefore, the Prandtl–Reuss constitutive equations will be used for defining σ_{ij} and ε_{ij} : $\Delta_n\varepsilon^p{}_{ij} = 3\Delta_n\varepsilon^p{}_e s_{ij}/(2\sigma_e)$.

The above equations are sufficient for study of the elastic–plastic strain of structural elements [1]. They contain both the finite and infinitesimal quantities and are valid only for small plastic segments of the stress–strain curve. An iteration step-by-step algorithm is constructed to solve the plastic problems. For each step n in accordance with this algorithm, the calculation proceeds by first regarding the accumulated plastic strain increments as known ones (see Section 1.2 and Figure 1.3). Assuming wholly elastic behavior, Eqs. (8.1)–(8.5) are solved to obtain increments of stress resultants, total strains, and displacements and rotation angle. If the yield condition is violated, the increments of stress and strain are recalculated using the iteration algorithm so that the yield condition is satisfied. The iterative algorithm used has been described in Ref. [62]. An explicit difference scheme was used for the solution of the equations of motion, according to which the displacements and rotation angle are obtained at the mesh nodes, whereas forces, moments, and strains were found at the centers of the mesh elements. The number K of mesh elements defined over the surface of the body was varied from 15 to 25 in the calculations.

The plates and shells studied were assumed fully clamped along the boundary contour and to have zero initial displacements and velocities. Uniformly distributed pressure pulse over the surface was applied. The load instantaneously reached the value P and was removed after reaching the maximum positive deflection. The plates considered were made of aluminum alloy D16AT for which $\rho = 2.8$ g/cm^3, $E = 720{,}000$ MPa, and $\nu = 0.3$. The bilinear approximation of equation (8.6) in accordance with Figure 8.1 were also used ($\sigma_0 = 282$ MPa, $E_1 = 1500$ MPa).

8.1.2 Results and Discussion: Plates, Spherical Caps, and Cylindrical Panels

Extensive numerical calculations both within and outside the ranges in which counterintuitive behavior (CIB) was observed were included in this study. The ranges are typically characterized by quite small plastic deformations and stress states approaching the momentless (membrane) case.

Plates. We first show results of numerical studies of Eq. (8.1) with $r/R = 0$ and plate radius $R_0 = 55$ mm, showing the insensitivity to small changes of the plate thickness h and of the mesh number K.

Figure 8.2 shows the center deflection–time curves for five values of the pulse load parameter P. The solid and dash-dotted curves are for $h = 1$ mm and $K = 15$ and 25, respectively; the dashed lines are for $h = 1.05$ mm and $K = 25$. For the smallest value, $P = 0.4$ MPa, the behavior is essentially elastic, whereas for $P = 1.0$ MPa, the plastic deformations are large enough so that the plate remains bent in the direction of the loading. These cases indicate the intuitively expected behaviors: small changes in the parameters of the structure, the loading, or the numerical scheme produce correspondingly small changes in the observed response. The range of pulse magnitudes between these extremes illustrates the range in which the behavior is counterintuitive in two respects: first, there is an abnormal sensitivity to small changes in the two

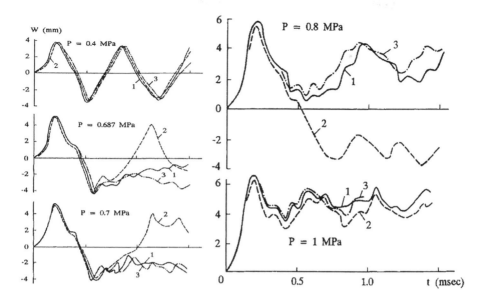

FIGURE 8.2 The influence of small disturbances upon the stability of the numerical calculations the center deflection–time curves for five values of the pulse load parameter P. The solid and dash-dotted curves are for $h = 1.00$ mm and $K = 15$ and 25, respectively; the dashed lines are for $h = 1.05$ mm and $K = 25$.

parameters; and second, the final vibration takes place about a mean deflection that may be either positive or negative. After deflecting in the direction of the loading, the structure snaps through to the opposite side. In some cases, the deflection remains on that side (with negative deflections), whereas in others, it snaps back to and remains on the positive side. The curves for $P = 1$ MPa in Figure 8.2 illustrate the fact that when the plastic strains are larger than a certain magnitude, the solution becomes stable with respect to small changes in the parameters.

Results of further calculations of this type are shown in Figures 8.3–8.5. The number of elements of numerical mesh K and the thickness h were fixed. In Figure 8.3 curves numbered 1–5 correspond to $P = 0.4, 0.5, 0.66, 0.687,$ and 0.689 MPa, respectively.

In Figure 8.4, curves numbered 1–5 correspond to $P = 0.75, 0.775, 0.777, 0.8,$ and 1 MPa, respectively. In each figure, there are groups of curves which apparently coincide up to a certain time and then bifurcate, i.e., visibly separate. For example in Figure 8.3, curves 3, 4, and 5 coincide up to a time t of approximately 0.58 msec when the curves 4 and 5 diverge from the curve 3; subsequently, the curves 4 and 5 appeared coincide up to a time of approximately 1.38 ms and then separate. The points where the bifurcation occurs are marked by large black circles. The smaller the difference in P, the longer the delay in the appearance of this point.

In Figure 8.5, a much narrower band of pulse magnitudes is studied: seven curves are plotted for P ranging from 0.800 to 0.811 MPa. Four points of bifurcation are marked: $t = 0.4$ ms (curves $P = 0.8$ and 0.81 MPa), $t = 0.55$ ms (curves $P = 0.81$ and

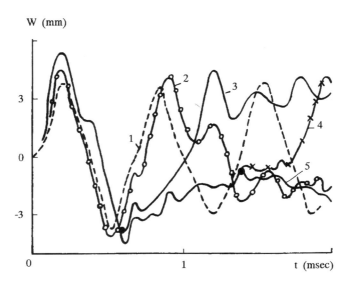

FIGURE 8.3 The change of the plate displacement character with the advancement in the range of the counterintuitive behavior: curves numbered 1–5 correspond to $P = 0.4, 0.5, 0.66, 0.687$, and $0.689\,\text{MPa}$, respectively.

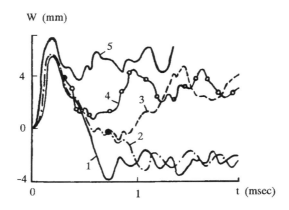

FIGURE 8.4 The change of the plate displacement character with the going out of the counterintuitive behavior range: curves numbered 1–5 correspond to $P = 0.75, 0.775, 0.777, 0.8$, and $1\,\text{MPa}$, respectively.

$0.811\,\text{MPa}$), $t = 0.8$ ms (curves $P = 0.811, 0.8105$, and $0.81025\,\text{MPa}$), and $t = 0.95$ ms (curves $P = 0.8105, 0.8103$, and $0.81027\,\text{MPa}$). The pattern suggests that the divergence is occurring exponentially, a characteristic feature of chaotic vibrations. It should be noted that the pulse force magnitudes have been chosen arbitrarily within certain ranges. (The term "bifurcation" is used here as a descriptive term to describe the appearance of visible separations, which are not bifurcations in the technical sense commonly used, namely of changes in the type of solution, for example from

CIB of Plates and Shallow Shells

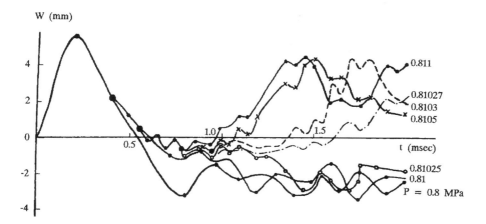

FIGURE 8.5 The results of the numerical investigations of the bifurcation possibility of the plate motion equations are plotted for P ranging from 0.800 to 0.811 MPa. Four points of bifurcation are marked: $t = 0.4$ ms (curves $P = 0.8$ and 0.81 MPa), $t = 0.55$ ms (curves $P = 0.81$ and 0.811 MPa), $t = 0.8$ ms (curves $P = 0.811$, 0.8105, and 0.81025 MPa), and $t = 0.95$ ms (curves $P = 0.8105$, 0.8103, and 0.81027 MPa).

oscillatory to nonoscillatory, which takes place at a particular value of a parameter. Here, they are due to a general property of the system of equations, i.e., an extreme sensitivity to initial conditions. Two solutions, starting from closely neighboring initial values, diverge exponentially rather than linearly. Hence, after a period in which they apparently coincide, they separate and may after sufficient time lose all resemblance to each other.)

Figure 8.6 shows for three values of P curves: 1 is the external work, 2 is the internal energy, 3 is the work done in plastic deformation, 4 is the kinetic energy, and 5 is the elastic strain energy.

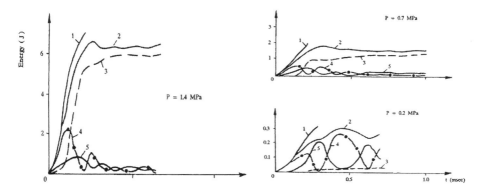

FIGURE 8.6 The changes for the time of the different kinds of the energy outside and inside of the range of the counterintuitive behavior plate: 1 is the external work, 2 is the internal energy, 3 is the work done in plastic deformation, 4 is the kinetic energy, and 5 is the elastic strain energy.

The curves 1, 2, and 4 were computed according to classical formulae [47,61,62], whereas the curves 3 and 5 were obtained from the stress–strain diagram. The curves for $P = 0.7\,\text{MPa}$ correspond to a case when CIB may occur. Figure 8.6 shows the require for CIB. CIB takes place if the energy dissipated in plastic deformation is approximately the same as the sum of the kinetic and elastic strain energies. The internal work does not change greatly after the load is removed. It characterizes the precision of the calculations.

Spherical cap. Figures 8.7 and 8.8 show results of calculations mainly assuming elastic behavior of a spherical shell in the form of a cap (or cupola). Taking $R_0 = 55\,\text{mm}$ and $h = 0.5\,\text{mm}$, two cases were considered: $f = R_0/11$ and $f = R_0/5.5$. The pulse load was applied from the *concave* side. Its type and amplitude were chosen from consideration of behavior of flat plates.

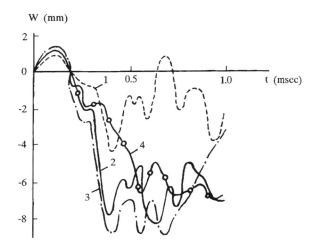

FIGURE 8.7 Deflection of a spherical shell for the various load amplitudes (the curves numbered 1–4 correspond to $P = 0.4$, 0.5, 0.6, and 0.7 MPa, respectively).

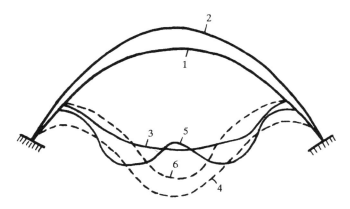

FIGURE 8.8 Time-dependent changes in the spherical shell shape. Profiles numbered 1–6 are calculated for $P = 0.6\,\text{MPa}$ at successive instants of time: $t = 0, 0.1, 0.34, 0.4, 0.55$, and 1.00 ms.

Figure 8.7 shows center point deflection–time curves calculated for $f = R_0/11$. The curves numbered 1–4 correspond to $P = 0.4, 0.5, 0.6$, and 0.7 MPa, respectively. Elastic behavior was assumed in all cases except curve 4, for which plastic deformation was considered. It is seen that snap-buckling can occur with wholly elastic and with elastic–plastic behavior.

Shell profiles numbered 1–6 are shown in Figure 8.8 for $P = 0.6$ MPa at successive instants of time: $t = 0, 0.1, 0.34, 0.4, 0.55$, and 1.00 ms. These profiles are plotted schematically, without adheres to ratios of real geometrical dimensions. It can be seen that a wrinkled or wave-like concentric shape appears during the passage from the positive to the negative region of displacements. The shell appears to prefer a snap-buckled state. Figure 8.9 shows the external work and internal energy, and demonstrates the satisfactory precision of the calculations.

Calculations made with rise $f = R_0/5.5$ indicate that when the curvature is increased, the shell stiffness increases. Plastic strains occur at small displacements. These strains hinder a snap-buckling after load removal.

Cylindrical panel. Shallow cylindrical panels were studied with span between supports 110 mm, rise $f = 4$ mm, and thickness $h = 0.5$ mm; the long dimension was assumed infinite. A pulse of uniform pressure pulse was applied on the *concave* side. P was taken in the range from 0.05 to 0.8 MPa. Results of numerical solutions of Eq. (8.2) are shown in Figures 8.10–8.12.

Figure 8.10 shows midpoint deflection–time plots for six pulse magnitudes ranging from 0.35 to 0.80 MPa. The curves labeled 1, 2, and 3 correspond to $P = 0.8, 0.6$, and 0.55 MPa, respectively. Substantial plastic deformation occurs in these cases. Curves 4, 5, and 6 are for $P = 0.40, 0.375$, and 0.35 MPa, respectively. As the plastic

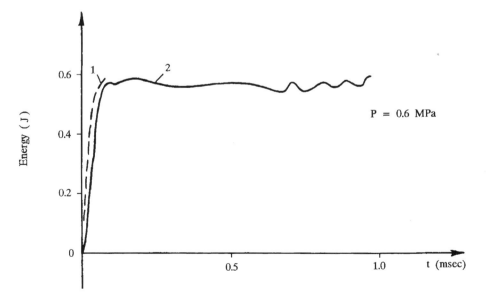

FIGURE 8.9 The change curves of the external work (1) and internal energy (2) inside of the range of the counterintuitive behavior of the cupola ($P = 0.6$ MPa).

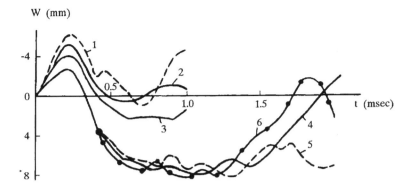

FIGURE 8.10 The transition from the elastic–plastic to the elastic deformation of the cylindrical panel. Curves labeled 1, 2, 3, 4, 5, and 6 correspond to P = 0.8, 0.6, 0.55, 0.40, 0.375, and 0.35 MPa, respectively.

deformations are reduced, the panel snaps through to a deflection in the reverse direction approaching twice the rise height (4 mm) of the undeformed shell.

Figure 8.11 shows additional plots for P in the neighborhood of and smaller than 0.375 MPa. Finally, Figure 8.12 shows response curves for eight values of P in the narrow band between 0.379 and 0.3801 MPa. These show that small perturbations of a fiducial curve lead quickly to solutions with no resemblance to the fiducial.

If the curve for P = 0.38 MPa is regarded as the fiducial curve, the other curves correspond to perturbations δP ranging from 0.000001 to + 0.001 MPa. As the increment δP decreases, the time of separation increases. (There is a rough proportionality between $-\log |\delta P|$ and the separation time, indicating an exponential rate of divergence between the fiducial curve and the curve for the perturbed pulse magnitude). It can be seen that a change in P from 0.37990 to 0.37991 MPa; or from 0.379910 to 0.379911 MPa leads to a large change in the response curve. This great a sensitivity to a parameter change is absent in normal response of panels.

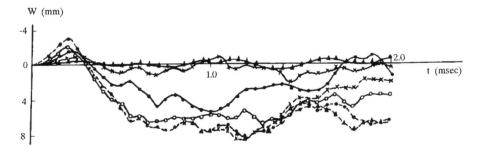

FIGURE 8.11 The transition from the elastic–plastic to the elastic deformation of the cylindrical panel: lines with black points are for P = 0.38 and P = 0.175 MPa, lines with triangles are for P = 0.375 and P = 0.05 MPa, lines with white points is for P = 0.25 MPa, and lines with crosses are for P = 0.37 and P = 0.1 MPa.

CIB of Plates and Shallow Shells

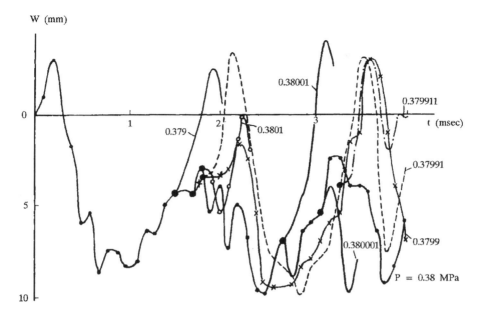

FIGURE 8.12 The results of the numerical investigations of the bifurcation possibility of the cylindrical panel motion (*P* rings from 0.379 to 0.3801 MPA; the points where the bifurcation occurs are marked by large black circles).

Conclusions. For some structural elements (bars, plates, and shallow shells), it is found that within certain ranges of the parameters of loading, it may be impossible to predict the response behavior. This is because for these structures, obeying nonlinear equations of motion, there may be an extreme sensitivity to changes of the parameters in these ranges. Actual values are known in practice only with a finite degree of uncertainty, and the behavior must be regarded as unpredictable. This phenomenon occurs typically in a range during the transition from purely elastic behavior to mixed elastic and plastic response. This phenomenon differs from classical snap-buckling of shells [1,47,61] because it occurs within small ranges of the problem parameters.

8.2 INFLUENCES OF ATMOSPHERE AND CAVITATION ON CIB

Thin circular plates are considered in this chapter with their edges being fixed in two different ways. In the first case, the plates are considered to be fully clamped in an absolutely rigid screen separating the liquid from the atmosphere. In the second case, they are fixed in a thick annular plate, the outer edge of which is clamped in the abovementioned screen. For the second case, the calculations will be made for a plate of variable thickness with a thin central circular and thick peripheral annular parts (Figure 8.13). In some instances, the calculation results will be compared for the above two cases in order to study the influence of clamp condition on the possibility of anomalous behavior. It should be noted that some deformability of the

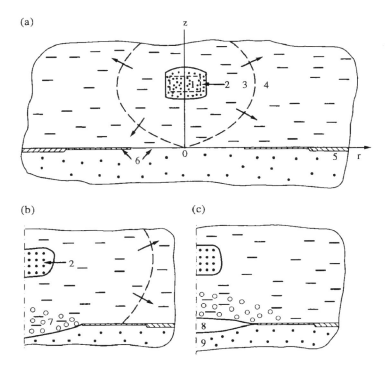

FIGURE 8.13 Dynamics of the plate/liquid/underwater explosion system: (a) dynamics of water/explosion interaction (1 – explosion product ($t = 0$); 2 – explosion product ($t = t_1$, $t_1 > 0$); 3 – liquid behind the pressure wave; 4 – undisturbed liquid; 5 – atmosphere; 6 – plate of variable thickness); (b) dynamics of the plate/water interaction (7 – cavitation zone); (c) anomalous and expected states of the system defined by different final deflections of the plate (curves 8 and 9, respectively).

edge of the plate fixed in the wall of a thick-walled tank took place in experiments (Chapter 7).

Deformations of the plate take place as a result of an instant explosion in the liquid near the plate center at the symmetric axis of the system (Figure 8.13).

The equations and boundary conditions describing the plate/liquid/underwater explosion interaction are presented below. The algorithm of the plastic strain calculation for the plate is outlined. Hydrodynamic equations are written according to different models of the generation, development, and disappearance of cavitation. Particularly, a model of the bubbly liquid is described. The expansion of the explosion products is described by the gasdynamic equations. Under the action of the underwater explosion, the plate deforms into a curved shape which may lose its stability as a result of both unloading of the structure material and the influence of the atmosphere. The anomalous behavior of the system is realized as snap-buckling of the plate toward the liquid. It is associated with the initiation of cavitation in the latter and can occur for both fully clamped and deformable edges of thin plates.

CIB of Plates and Shallow Shells   219

Notation. P_a – atmospheric pressure; n – bubbles quantity per unit volume; v_z – component of the velocity vector to z direction; t_k – moment of the occurrence of pressure $P < P_k$ in the continuous liquid; P_k – static "tensile strength" of the liquid; P_s – pressure in the cavitating liquid; P_{l0} – pressure in the undisturbed liquid; a_0 – sound velocity of the liquid; a_s – sound velocity in the cavitating liquid; P_k – static "tensile strength" of the liquid; Δt – time necessary to disrupt the liquid column by tension P_k; v – vector, liquid, or gas velocity; V – volume of the bubble; $(.)_g$ – gas parameter inside the bubble; and $(.)_0$ – undisturbed value.

8.2.1 THEORETICAL MODELS

Here, we remind the models described in Part I which will be used below.

1. **Timoshenko's model of thin plate.** Timoshenko's equations of motion are used (8.1) together with relations (8.3)–(8.6).

   The edge of the plate was fully clamped in the screen $W = U = \psi = 0$. For plates of variable thickness, it was also assumed that W, U, and ψ changed continuously along the coordinate r.

   The conditions of the plate/liquid interaction are described by $q = P$ and $\dot{W} = v_z$ ($z = h/2$). These conditions do not take into account the displacement of the plate/liquid contact surface because the effect of this displacement on the noted interaction is much less than that of nonlinearity of the plate equations and the cavitation in the liquid (Section 1.4.3 and Chapter 3).

2. **Models of liquid and gas.** We used the models described in Chapters 1 and 3. In the general case, the liquid may be considered either as keeping its continuity or as being ruptured. Four models of the liquid and the model of explosion products, which will be used for the calculations, are described below.

   2.1. **An ideal elastic liquid.** The following equations of motion and mass conservation occur for an inviscid liquid:

   $$\rho \frac{d}{dt}(\mathbf{v}) = -\operatorname{grad} P, \quad \frac{d\rho}{dt} + \rho \operatorname{div} \mathbf{v} = 0. \tag{8.7}$$

   The equation of state has the following form:

   $$(P + B)/(P_{l0} + B) = (\rho/\rho_0)^\kappa. \tag{8.8}$$

   This equation can be linearized if the pressure P is of the order of 100 MPa:

   $$\rho = \rho_0 [1 + K(P - P_{l0})], \quad K = 1/(\rho_0 a_0^2). \tag{8.9}$$

   Equations (8.8) and (8.9) are used outside of the cavitation zones. They are valid for water when $P < 3 \times 10^3$ MPa. For room temperature, $\kappa = 7.15$; $B = 304.5$ MPa.

2.2. Liquid models taking into account the cavitation phenomenon.

Transient cavitation has not been studied enough because physical and mathematical problems of the investigation of the initiation and the development of gas bubbles in the liquid are very complicated. Beforehand the site and time of the appearance (disappearance) of gas bubbles (cavitation zones) and the motion of them in the bulk of the liquid are unknown. Consequently, all approaches taking into account the initiation of cavitation during the transient structure/liquid interaction are based on a more or less simplifying statements.

Let us assume that the Eq. (8.7) is used for the medium in the cavitation zones. The property of the cavitating liquid in these zones is described by the equation of state.

a. Bilinear (constant pressure) model [62–64]. Experiments show that in cavitation zones, the pressure only varies slightly and hence it may be considered to be equal to the constant P_s. Then, the equation of state for the cavitating liquid can be given in the following form:

$$P = P_s. \tag{8.10}$$

The magnitude of P_s is of the order of 0.01 MPa. The condition for the initiation of cavitation at some point of a liquid (rupture criterion) may be written as (Chapter 2)

$$\int_{t_k}^{t} P\,dt \le P_k \Delta t. \tag{8.11}$$

The Δt value is of the order of 10 μs; for the sea water, P_k is $(-0.2) - (-0.35)$ MPa. In condition (8.11), $P < P_k$, i.e., the integration is conducted only when a negative pressure at the point liquid exists which is less than P_k. If the pressure does not drop very rapidly a less accurate but more convenient condition (8.12) is used instead of (8.11) to calculate the time and the site of the cavitation appearance:

$$P \le P_k. \tag{8.12}$$

Cavitation in the liquid vanishes if

$$\rho \ge \rho_0. \tag{8.13}$$

b. Bilinear (constant sound velocity) model [62]. The ruptured liquid in the cavitation zones contains gas bubbles. Due to this fact, this liquid is compressible although its density does not really change during compression. That is why the sound velocity in such liquid is considerably lower than that in the pure liquid or air. The sound velocity, a_s, varies slightly in the cavitation zones. Let a_s = const.:

$$\frac{dP}{d\rho} = a_s^2 \qquad (8.14)$$

Conditions for the initiation of cavitation (8.11), (8.12) and its disappearance (8.13) do not change. Equation (8.14) is similar to Eq. (8.9) for an ideally elastic liquid. The increase of liquid compressibility is taken into account in the cavitation zones since $a_s \ll a_0$. In the case of $a_s = 0$, Eq. (8.14) becomes analogous to Eq. (8.10).

Remark 1. In Ref. [9], it is emphasized that some analogy exists between the models of the behavior of an elastic–plastic body and the liquid capable of cavitation. In both cases, the use of different constitutive equations (equations of state) depends on the fulfillment of some criteria of the variation of the important mechanical properties of the medium. For example, the criterion $\sigma_e > \sigma_0$ of the plastic strain generation for an elastic–plastic body is similar to the criterion (8.12) for the cavitation initiation for a liquid. The sites of the zones where the liquid will exhibit the cavitation properties or the body will show plastic properties are not known in advance and are defined by the calculation.

   2.3. Bubbly liquid. Let us consider the motion of a liquid which contains the same spherical uniformly distributed gas bubbles. The volume of the gas is much smaller than that of the pure liquid. The exchange of the mass, momentum, and energy between the gas and the liquid is not taken into account. The length of the pressure wave in the mixture is much larger than the distance between the bubbles and their radius R. There is no interaction between the bubbles, which are under uniform pressure. The velocity of the bubbles and the liquid with respect to each other is equal to zero. The quantity of bubbles in the mixture does not change with time.

   If the initial gas concentration is low, the properties of the mixture in the compression zones do not differ appreciably from those of the pure liquid. In tension zones, however, gas bubbles begin to extend and change the properties of the mixture. An equation of state for the bubbly liquid can be presented (Chapter 3) in the following way:

$$\rho = \rho_0 \Big/ \big\{ (1 - nV_0)[1 - K(P - P_{l0})] + nV \big\}. \qquad (8.15)$$

The bubble wall motion is described by Rayleigh's equation (1.53):

$$R\ddot{R} + 3\dot{R}^2/2 = (P_g - P)/\rho_0. \qquad (8.16)$$

The pattern of the bubble motion depends upon the gas properties. If the pressure in the mixture changes slowly enough and appreciable heat exchange between the phases takes place, the process of the bubble oscillations may be considered as an isothermal one; if the pressure changes quickly, then heat exchange may be ignored and the adiabatic law is used. Thus,

$$P_g = P_{g0}\left(\rho_g/\rho_{0g}\right)^\gamma = P_{g0}\left(V_0/V\right)^\gamma; V = 4\pi R^3/3, \tag{8.17}$$

where $\gamma = 1.4$ in the case of adiabatic oscillations and $\gamma = 1$ in the case of isothermal ones.

Equations (8.7)–(8.15) suffice to determine all the hydrodynamic unknowns. These equations can be simplified if the pressure waves in the liquid are much larger than the radius of the bubbles R (the case of long waves). Then, we can neglect the terms on the left-hand side of Eq. (8.16) and obtain $P = P_g$. Taking into account the latter equation and Eq. (8.17), we rewrite (8.15) as

$$\rho_0/\rho = (1 - nV_0)\left[1 - K(P - P_{l0})\right] + nV_0\left(P_{l0}/P\right)^{1/\gamma}. \tag{8.18}$$

In compression zones, Eq. (8.18), in fact, coincides with Eq. (8.9), but in tension zones, it exhibits essentially new nonlinear properties. Equations (8.7) and (8.18) are much simpler than the initial ones ((8.7) and (8.15)–(8.17)). They are somewhat more complex than the equations for the pure liquid (8.7) and (8.8). However, they describe more correctly the behavior of a real liquid, where there is always some quantity of gas bubbles and where the appearance of high negative pressure is practically impossible. Its absence can be shown analytically, if $\gamma = 1$. From equation (8.18), it follows that

$$P = P_s + \frac{b + \left[b^2 + 4K(1 - nV_0)nV_0 P_{l0}\right]^{1/2}}{2K(1 - nV_0)}, \tag{8.19}$$

$$b = (1 - nV_0)(1 + K P_{l0}) - \rho_0/\rho.$$

Here, we take into account the pressure P_s of the saturated vapor of the liquid in the cavitation zones. From Eq. (8.19), it follows that the pressure in a bubbly liquid is always positive even when $\rho \to 0$ if $P_s > 0$ (see, also, Eq. (1.68)). Equations (8.7) and (8.19) form a simple system for studying a gas bubble/pure liquid mixture.

Figure 8.14 gives a scheme of the changes in the curves corresponding to the equations of state for some models of the liquid. Curves 1, 2, 3, and 4 correspond to the ideal elastic model (8.8), bilinear (constant pressure) model, bilinear (constant sound velocity) model, and the bubbly model (8.19), respectively. Curve 5 corresponds to Eqs. (8.15), (8.16), and (8.17) which take into account the inertia properties of gas bubbles in the liquid. It has been shown that in the compression zone, curves 1

CIB of Plates and Shallow Shells 223

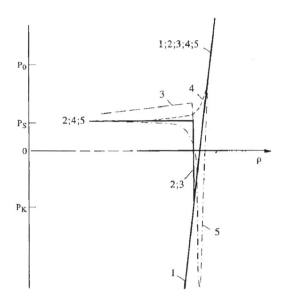

FIGURE 8.14 Relations P–ρ corresponding to different models of the liquid.

through 5 actually coincide, whereas in the tension zone $(P < P_{10})$, only curves 2 through 5 give nearly similar description of nonlinear properties of a real (cavitating) liquid.

Remark 2. It must be emphasized that a considerable negative pressure inside a very short tensile underwater waves can appear as a result of the inertia of the bubble extension. However, in present case, negative pressure in the liquid is small or absent because only long waves are considered here.

Remark 3. There is the analogy between the theories of transient cavitation of a liquid and the dynamic rupture of materials. For example, rupture criteria (8.11) and (8.12) were widely used to estimate the material strength inside of tensile waves. It is known that the development of voids inside of tensile waves is described by an equation of the type (8.16) (see Chapter 2).

    2.4. **A wide-range equation of state of water** (1.49)–(1.52). An equation of state for water that holds in a wide range of thermodynamic parameters was presented in Section 1.3. From Figures 8.14 and 8.15, we can see that the above models of the cavitating liquid, developed on the basis of various physical assumptions, are in good agreement (see, also, Section 1.4.3. The bubbly liquid model allows to study both the transient cavitation and the cavitation oscillations of a liquid.

    2.5. **The ideal gas model.** Let us assume that the temperature of the explosion products does not vary and Eq. (8.7) hold for these products. The equation of state is represented by the adiabatic law: $P = P_0 \, (\rho/\rho_0)^\gamma$, $\gamma = 3$ (Section 1.3).

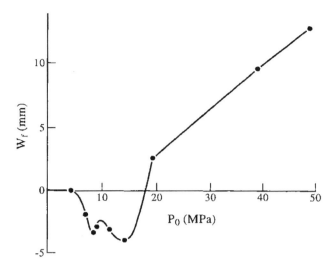

FIGURE 8.15 The location of the P_0 region where anomalous behavior of the plate interacting with liquid and underwater explosion occurs (W_f is the final deflection of the point $r = 0$, points on the curve indicate the P_0 values at which calculations were carried out).

8.2.2 CALCULATION DETAILS

Plate. Thin circular plates were considered which were either fully clamped in an absolutely rigid screen (the case of a smooth plate) or fixed in a thick annular plate (the case of a plate of variable thickness) (Figure 8.13). The radius of the thin plate was 55 mm, and its thickness was either 0.7 or 1 mm. The thickness of the thick annular plate was 8 mm, and its width was 55 mm. The plates considered were made of aluminum alloy D16AT for which $\rho = 2.8$ g/cm^3, $E = 72{,}000$ MPa, and $\nu = 0.3$. Bilinear approximation of Eq. (8.6) in accordance with Figure 8.1 was also used ($\sigma_0 = 282$ MPa, $E_1 = 1500$ MPa).

The calculations were made by a mesh-analysis method using the explicit procedure. The equations of motion (8.1) were integrated numerically. The displacements and rotation angle that were determined in the mesh nodes, whereas strains, forces, and moments were determined in the mesh center. For plates of variable thickness, the calculations were carried out at each time step in two stages. In the first stage, the unknown values U, W, and ψ were determined for the thick part of the plate using the cell of the thin part adjacent to the thick one. Then, we calculated the same unknown values for the thin part of the plate. In doing so, the U, W, and ψ values obtained in the first stage for the edge of the thick part of the plate were assumed to be boundary ones. The size of the mesh cell side was 3.67 mm.

Gas–liquid system. A computational scheme for treating the cavitating liquid has been devised by Galiev [50,62,64]. He used the velocity potential and the acoustic or nonlinear acoustic assumption. A highly efficient computational scheme for treating the cavitating liquid was also developed by Newton [65]. He used the displacement potential and the acoustic assumption. Here, as discussed in [15,27], we do not consider the liquid flow as irrotational. The above different models of the liquid are used. The equations of the liquid and gas motion were integrated using Wilkins's procedure [66].

The gas cylindrical volume (explosion product) was at a distance of 51.1 mm from the plate. Its length and radius were equal to 14.6 mm. The place for explosion (Figure 8.13) was chosen so that the accuracy of the calculations was provided. The numerical mesh was such that its boundaries in water do not influence the results of the calculations; that is, the numerical mesh used began to expand at the velocity of sound in the liquid from initial region ($0 < r < 367$ mm, $0 < z < 381$ mm) after the explosion.

In the region ($0 < r < 147$ mm, $0 < z < 161$ mm) adjacent to the plate, we used a uniform numerical mesh with a cell side 7.33 mm in length. Thus, the gas volume was divided into eight cells. Outside the boundaries of the abovementioned region, the distance between the mesh nodes changed nonuniformly. The cell sides along the coordinates increased three times in the range $147 < r < 367$ mm and $161 < z < 381$ mm, and then six times if $r \geq 367$ mm and $z \geq 381$ mm.

Plate/liquid contact condition. To satisfy the plate/liquid contact condition, we performed calculations at each time step in two stages. Initially, we calculated the liquid parameters using the plate velocity found previously. In the second stage, new displacements of the plate were calculated using the resultant pressure of the liquid.

Initial conditions. The initial velocity of the gas was equal to zero. Different initial values of P_0 inside the explosion products were used, but density ρ_0 of the explosion products was always assumed to be 1.65 g/cm^3. The plate was assumed to be motionless at $t = 0$. Other parameters of the system considered corresponded to undisturbed values for room atmospheric conditions. For example, the pressure of the liquid was equal to the room atmospheric pressure when $t = 0$.

A negative pressure ($P_s = P_k = -0.1$ MPa) was usually assumed to exist in cavitation zones. According to Ref. [30], anomalous behavior of the plate occurred when the final displacements were of the order of $(3-5)h$, which was used to determine the initial pressure of the explosion products in present calculations.

While calculating the gas and the liquid parameters, the time step ranged from 1 to 2 μs. The results showed that the latter time step provides a sufficiently high accuracy of calculations. The plate displacements were calculated with the time step equal to 0.2 μs ($h = 1$ mm) or 0.1 μs ($h = 0.7$ mm).

8.2.3 Results and Discussion

Numerical investigations of the possibility of the anomalous behavior of the plate/liquid/underwater explosion system were carried out by using different liquid models.

Some results of numerical experiments for a plate of variable thickness are plotted in Figure 8.15 for different values of P_0 inside of the explosion products. The thickness of the central part of the plate was 1 mm. It is shown in Figure 8.15 that the anomalous behavior of the plates is possible when the initial values of P_0 are between 10 and 15 MPa.

The displacement–time curves calculated at point $r = 0$ of the smooth plate ($h = 1$ mm) for different P_0 are plotted in Figure 8.16. The curves demonstrate the transition from elastic to plastic deformation within the region of the anomalous behavior. The calculations show that within this region, the maximum curvature of the plate has the order of the curvature which characterizes a shallow shell. The plastic strain energy was of the order of the elastic and kinetic energy [8].

The influence of the P_s value on the occurrence of the anomalous behavior of the plate was studied. For $P_s = 0$, the snap-buckling also occurred. However, when we excluded the action of the atmospheric pressure ($P_a = 0$), the reverse displacement of the plate was not observed in the case when $P_s = 0$.

In Figures 8.17 and 8.18, curves calculated using the different models of the liquid behavior are presented. The calculations were performed for $P_0 = 10$ MPa and $P_0 = 5$ MPa (curve 5 in Figure 8.18) in the region of the anomalous behavior illustrated in Figure 8.15.

The smooth plate displacements at the point $r = 0$ calculated for $h = 1$ mm are presented in Figure 8.17. Curve 1 was calculated using the equations of the ideal elastic model (8.8), curve 2 was calculated by the bilinear (constant pressure) model, curve 3 corresponds to the bilinear (constant sound velocity, $a_s = 10$ m/s) model, curve 4 corresponds to model (8.20), and curves 5 and 6 were plotted for the bubbly liquid model (8.19). The above models of the liquid behavior were defined by some constants. Most of them were presented above. When calculating curve 4, we assumed T being equal 293 K. For the bubbly liquid model, two cases were considered: $nV_0 = 10^{-4}$ (curve 5) and 9×10^{-6} (curve 6). Curves 2, 3, and 6 coincide. Calculations of curve 1 were carried out up to 423 µs. The deflection reached 10 mm and the arithmetic unit of the computer was overflowed because of the increasing pressure. For this reason, it was impossible to complete calculations using Eqs. (1.49)–(1.52) (curve 4).

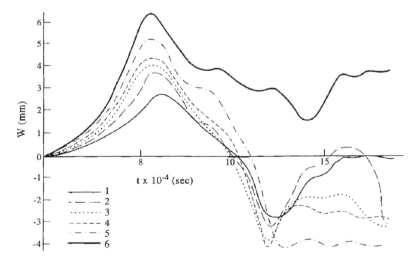

FIGURE 8.16 The displacement–time curves 1, 2, 3, 4, 5, and 6 corresponding to $P_0 = 5$, 8, 9.9, 12, 15, and 20 MPa, respectively.

CIB of Plates and Shallow Shells

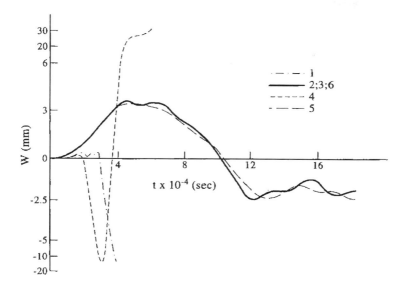

FIGURE 8.17 The displacement–time curves calculated using different liquid models.

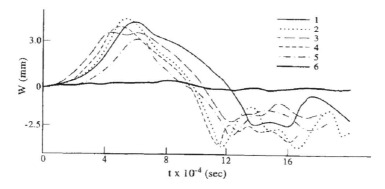

FIGURE 8.18 The displacement–time curves in the range of the anomalous behavior for different plates and P_0.

The above results show that the consideration of the cavitation and static (atmospheric) pressure is an important condition for studying the anomalous behavior of the plate loaded by an underwater wave.

The calculated results for displacements at points $r = 0$ (curves 1–5) and $r = 55$ mm (curve 6) for different plates are presented in Figure 8.18. Here, we used the bilinear (constant pressure) liquid model. Curves 1, 2, 5, and 6 were plotted for plates of variable thickness. In the central part, $h = 1$ mm (curves 1 and 6) or 0.7 mm (curves 2 and 5). Curves 3 and 4 correspond to smooth plates with $h = 1$ mm and $h = 0.7$ mm, respectively. We can observe snap-buckling of a thin plate ($h = 0.7$ mm, curve 5) even when $P_0 = 5$ MPa.

Thus, the reverse snap-through buckling appears to occur both in smooth and in stepwise varying plates. The influence of the peripheral part of the plate on the displacement of its central point is small. Figures 8.15–8.18 show that the possibility of anomalous behavior exists not only in the case of an idealized fully clamped plate but in the case of realistic boundary conditions (not fully motionless edge).

The region of the anomalous behavior which exists in the plate/water/underwater explosion system is not very small. The amplitude of initial pressure P_0 can change approximately twofold, while the possibility of the snap-buckling holds.

Let us consider the changes in the surface pressure of the plate of the variable thickness. These changes corresponding to curve 2 in Figure 8.18 are shown in Figure 8.19. The analyses of the pressure curves and other results of the calculations reveal that the rate of the cavitation formation on the plate is very high ($t = 40$ μs). Cavitation disappears at the instant of the plate motion retardation ($t = 450$ μs). In the range of 450–600 μs, the plate changes the direction of its movement and the pressure on it rises rapidly. A pressure impulse is generated by the plate and then propagates into the depth of the liquid. The pressure upon the plate decreases rapidly again resulting in the cavitation process which does not stop until the end of calculations.

Figure 8.20 shows the curves of W variation along r corresponding to curve 2 shown in Figure 8.19 for different values of time. The central part of the plate deformed only and the mean rate of reverse snap-buckling was approximately two times larger than the mean rate of the positive deflection.

Conclusions. The anomalous behavior of the structure/liquid/underwater explosion system covered in Ref. [11] was studied by considering the initiation of the cavitation in the liquid. Thus, the above anomalous behavior is connected not only with unloading of the plate material but with the transient cavitation in the liquid together with the action of the atmospheric pressure. A necessary condition of the anomalous behavior of the described system is the coincidence of the appearance of unloading and the transient cavitation in the liquid, which can be achieved only in a narrow range of the system parameters.

The negative pressure can occur on the surface of the plate when subjected to loading by the underwater explosion. The magnitude of this pressure is limited by the cavitation effect, and the negative pressure promotes the occurrence of anomalous

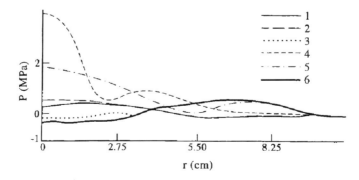

FIGURE 8.19 Variation of the liquid pressure across the plate surface with time: curves 1, 2, 3, 4, 5, and 6 correspond to $t = 30, 32, 40, 500, 600,$ and 900 μs, respectively.

CIB of Plates and Shallow Shells

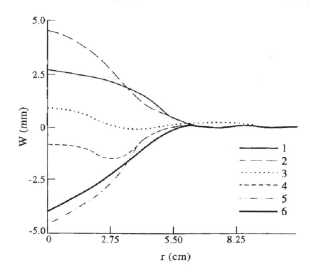

FIGURE 8.20 Dynamics of reversal of the plate interacting with the liquid: curves 1, 2, 3, 4, 5, and 6 correspond to $t = 0.4$, 0.6, 0.9, 1, 1.15, and 1.7 ms, respectively.

behavior. The similar effect occurs in structures under blast loading, when the depression wave loads the structure after a shock wave pressure load. The pressure in the depression wave can drop to zero and promotes to the anomalous behavior of the structure.

Anomalous behavior exists not only in the case of a fixed-end structure but in the case of the flexibly supporting condition.

Within these ranges of loading, the nonlinear equations of motion for the above elements are so sensitive to the perturbations of the parameters that it is impossible to predict the dynamic behavior of real structural elements by solving the equations, since the initial parameters and loads are always known approximately and any corrections of them can significantly change the result of calculations. The anomalous behavior is associated only with those structural elements which deform into shallow shells during loading phase and then lose their stability under the action pf unloading forces. In a flat plate, the anomalous final deflection due to reverse snap-buckling appears only with elastic–plastic deformation. In contrast, this effect for spherical and cylindrical panels may be for viscoelastic, elastic, and elastic–plastic material behavior.

In this part, a nonlinear (elastic–plastic) model of material behavior is used. Nonlinear problems are always difficult to understand. Even in the case of the simple models [18,20–25], they are able to demonstrate very complex chaotic behavior. The butterfly effect has become a well-known symbol of the dynamic behavior of real nonlinear systems. (The combination of words butterfly effect gained popularity thanks to the article: Lorenz, E. Predictability: Does the Flap of a Butterfly's Wings in Brazil Set off a Tornado in Texas? 1972. http://eaps4.mit.edu/research/Lorenz/Butterfly_1972.pdf)

The idea is that, in many strongly nonlinear systems, effects of even small initial perturbations may be very large. These effects increase according to the exponential law, which makes it impossible to make long-term forecasts.

Thus, the considered CIB is part of a huge problem in which predictability over a short period of time due to the use of deterministic equations is replaced by unpredictability over a long period of time because of the effect of a small disturbance in the initial data. Strictly speaking, the results of the numerical calculations are the extension of the classical results of Henri Poincaré and Edward Lorenz to elastic–plastic systems.

REFERENCES

1. Jones N. *Structural Impact*. Cambridge University Press, NY, USA (2012).
2. Yu TX, Qui XM. *Elastic-Plastic Waves*. John Wiley & Sons, New Jersey, USA (2018).
3. Karagiozova D, Alves M. Dynamic elastic-plastic buckling of structural elements: A Review. *Appl Mech Rev* 61 (4): 040803 (2008).
4. Zhao YG et al. Counterintuitive dynamic behavior in elastic-plastic structures. *Adv Mat Res* 428: 47–51 (2012).
5. Gu Y, Hu BY, Dong Q. Counter-intuitive behavior in elastic-plastic spherical shells and the application to the design of containment vessels. *ASME 2012 Pressure Vessels and Piping Conference*. High-Pressure Technology; ASME NDE Division. Toronto, Ontario, Canada, 15–19 July 2012, Volume 5. doi:10.1115/PVP2012-78106.
6. Aune V et al. Numerical study on the structural response of blast-loaded thin aluminium and steel plates. *Int J Impact Eng* 99: 131–144 (2017).
7. Xu Z, Liu Y, Huang F. Experimental study on counterintuitive behavior of thin aluminium plates under free airblast loading. *EPJ Web of Conferences* 183: 02059 (2018) doi:10.1051/epjconf/201818302059.
8. Galiev ShU. Distinctive features of counterintuitive behavior of plates and shells after removal of impulse load. *Int J Impact Eng* 19: 175–187 (1997).
9. Galiev ShU. The influence of cavitation upon anomalous behavior of a plate/liquid/underwater explosion system. *Int J Impact Eng* 19: 345–359 (1997).
10. Galiev ShU. Experimental observations and discussion of counterintuitive behavior of plate and shallow shells subjected to blast loading. *Int J Impact Eng* 18 (7–8): 783–802 (1996).
11. Lavrentiev MA, Shabat BV. *Problems of Hydrodynamics and their Models*. Nauka, Moscow (1977) (in Russian).
12. Mazyrovskii BJa, Sizev AN. *Electrodynamic Effect of Explosive Pressing of Sheets*. Naukova Dumka. Kiev (1983) (in Russian).
13. Karmishin AV, Skurlatov ED, Startsev VG, Felshtein VA. *Non-Stationary Aeroelasticity of Thin-Walled Structures*. Mashinostroenie, Moscow (1982) (in Russian).
14. Galiev ShU, Nechitailo NV. *Dynamics of Shape Changes of Thin Plates at Shells of Revolution*. Reprint of Institute for Problems of Strength, Academy of Sciences of the Ukrainian SSR, Kiev (1985) (in Russian).
15. Galiev ShU, Babich YuN, Zhurakhovsky SV, Nechitailo NV, Romashchenko VA. *Numerical Modeling of Wave Processes in Bounded Media*. Naukova Dumka, Kiev (1989) (in Russian).
16. Ross CA, Strickland WS, Sierakowski RL. Response and failure of simple structural elements subjected to blast loading. *Shock Vib Dig* 9 (12): 15–26 (1977).
17. Moessner JR. Snap-back of a clamped beam loaded impulsively. Laboratory Project No.45, Department of Mechanical Engineering, University of Cape Town, South Africa (1984).
18. Symonds PS, Yu TX. Counterintuitive behavior in a problem of elastic-plastic beam dynamics. *ASME J of Appl Mech* 52 (3): 517–522 (1985).

19. Galiev ShU. Nonlinear waves of different physico-mechanical nature in finite continua. *Strength Mat* 17 (12): 1633–1648 (1985).
20. Genna F, Symonds PS. Dynamics of plastic instabilities in response to short pulse excitation - Effects of slenderness ratio and damping. *Proc Royal Soc London A* 417: 31–44 (1988).
21. Borino G, Perego U, Symonds PS. An energy approach to anomalous damped elastic-plastic response to short pulse loading. *ASME J Appl Mech* 56 (2): 430–438 (1989).
22. Lee J-Y, Symonds PS, Borino G. Chaotic responses of a two degree-of-freedom elastic-plastic beam model to short pulse loading. *ASME J Appl Mech* 59 (4): 711–721 (1992).
23. Lee J-Y, Symonds PS. Extended energy approach to chaotic elastic-plastic response to impulsive loading. *Int J Mech Sci* 34: 139–157 (1992).
24. Qian Y, Symonds PS. Anomalous dynamic elastic-plastic response of a Galerkin beam model. Brown University, Providence, USF (National science foundation grant No.MSS-9024607) (1994).
25. Symonds PS, Lee J-Y. Fractal dimensions in elastic-plastic beam dynamics. *ASME J Appl Mech* 62 (2): 523–526 (1995).
26. Galiev ShU, Nechitailo NV. Unexpected behavior of plates during shock and hydrodynamics loading. *Strength Mat.* 18 (12): 1652–1663 (1986).
27. Galiev ShU. *Nonlinear Waves in Bounded Continua.* Naukova Dumka, Kiev (1988) (in Russian).
28. Galiev ShU, Abdirashidov A, Karshiev AB. Special features of deformation of spherical panels under pulsating load. *Strength Mat* 21 (3): 384–388 (1989).
29. Galiev ShU, Karshiev AV. Peculiarities of the counterintuitive behavior of flat and curved plates after compression under a pulse load. *Strength Mat* 22 (5): 744–747 (1990).
30. Galiev ShU. *Counterintuitive Behavior of the Structure Elements under Impulse Loading.* Preprint of Institute for Problem of Strength, Academy of Sciences of the Ukrainian SSR, Kiev (1990) (in Russian).
31. Galiev ShU. Numerical modeling unexpected behavior of sheets in experiments carried out by M.A.Lavrent'ev. *Strength Mat.* 25 (5): 381–386 (1993).
32. Galiev ShU, Blachut J, Skurlatov ED, Panova OP, Mpltschaniwskyj G, Cui Z. Experimental and theoretical design methodology of hemispherical shells under extreme static loading. *Strength Mat* 5: 98–108 (2004).
33. Galiev ShU, Romashchenko VA. A method of solving nonstationary three-dimensional problems of hydroelasticity with allowance for fluid failure. *Int J Impact Eng* 22: 469–483 (1999).
34. Kolsky H, Rush P, Symonds PS. Some experimental observations of anomalous response of fully clamped beams. *Int J Impact Eng* 11 (4): 445–456 (1991).
35. Li QM, Zhao LM, Yang GT. Experimental results on the counter-intuitive behavior of thin clamped beams subjected to projectile impact. *Int J Impact Eng* 11 (3): 341–348 (1991).
36. Jones N. *Structural Impact.* Cambridge University Press, NY, USA (1989).
37. Galiev ShU, Skurlatov ED, Panova OP. Anomalous behavior of structural elements under blast loading. *Strength Mat* 5: 98–109 (1997).
38. Sedov LI. *Similarity and Dimensional Methods in Mechanics.* Academic Press, Cambridge, MA, USA (1959).
39. Zel'dovich YaB, Paizer YuP. *Physics of Shock Waves and High-Temperature Hydrodynamic Phenomena.* Academic Press (1966).
40. Courant R, Friedrichs KO. *Supersonic Flow and Shock Waves.* Interscience Publishers, New York, USA (1948).
41. Pisarenko GS et al. *Strength of Material.* Visha Shcola, Kiev (1986) (in Russian).

42. Rocard Y. *General Dynamics of Vibrations*. Crosby Lockwood & Son Ltd, London, UK (1960).
43. Jones N, Uran TO, Tekin SA. The dynamic plastic behaviour of fully clamped rectangular plates. *Int J Solid and Struct* 6: 1499–1512 (1970).
44. Jones N, Griffin RN, van Duzer RE. An experimental study into the dynamic plastic behaviour of wide beams and rectangular plates. *Int J Mech Sci* 13 (8): 721–735 (1971).
45. Symonds PS, Mentel TJ. Impulsive loading of plastic beams with axial restraints. *J Mech Phys Solid* 6: 186–202 (1958).
46. Kurochkin VA. Interaction of shallow shell with acoustical shock wave. *Vzaimodeistvie obolochec s jidkost'yu* 12: 42–50 (1981) (in Russia).
47. Perchev AK, Platonov EG. *Dynamics of Shells and Plates (Transient Problems)*. Sydostroenie, Leningrad (1987) (in Russia).
48. Karmishin AV et al, *Methods of Dynamic Calculations and Tests of Thick-Wall Structures*. Mashinostroenie, Moscow (1989) (in Russian).
49. von Mises R. *Mathematical Theory of Compressible Fluid Flow*. Academic Press, Cambridge, MA, USA (1958).
50. Galiev ShU. *Dynamics of Structure Element Interaction with a Pressure Wave in a Fluid*. Naukova Dumka, Kiev (1977) (in Russian) (translated in English: Dep. Navy, Off. Naval Res., Arlington, USA, 1980).
51. Kelly SG, Nayfe AH. Non-linear propagation of general directional spherical waves. *J Sound Vib* 79: 145–156 (1981).
52. Galiev ShU. Nonlinear one-dimensional oscillations of a viscous diathermic gas in a spherical layer. *Trudi seminara teorii obolochek, Kazan*, 2: 240–253 (1971) (in Russian).
53. Galiev ShU. *Cavitational Resonant Oscillations of Fluid in Deformable Pipes and Containers*. Reprint of Institute for Problems of Strength, Academy of Sciences of the Ukrainian SSR, Kiev (1983) (in Russian).
54. Chester W. Acoustic resonance in spherically symmetric waves. *Proc R Soc Lond A* 434: 459–463 (1991).
55. Galiev ShU, Panova PO. Periodical shock waves in spherical resonators (survey). *Strength Math* 27 (10): 729–746 (1995).
56. Idel'chik IE. *Some Interesting Effects and Paradoxes in Aerodynamics and Hydraulics*. Mashinostroenie, Moscow (1982) (in Russian).
57. Timoshenko SP. On the correction for shear of the differential equation for transverse vibrations of prismatic bar. *Philos Mag* Series 6 41 (245): 744–746 (1921).
58. Mindlin RD. Influence of rotatory inertia, shear on flexural motions of isotropic, elastic plates. *ASME J Appl Mech* 18 (1): 31–38 (1951).
59. Reissner E. Stress strain relations in the theory of thin elastic shells. *J Math Phys* 31 (2): 109–119 (1952).
60. Naghdy PM. On the theory of thin elastic shells. *Quart App Math* 14 (4): 369–380 (1957).
61. Vol'mir AS. *Nonlinear Dynamics of Plates and Shells*. Nauka, Moscow (1972) (in Russian).
62. Galiev ShU. *Dynamics of Hydroelastoplastic Systems*. Naukova Dymka, Kiev (1981) (in Russian).
63. Bleich HH, Sandler IS. Interaction between structures and bilinear fluid. *Int J Solid and Struct* 6: 617–639 (1970).
64. Galiev ShU. *Stress-Strain State of a Hollow Cylinder Subjected to an Underwater Shock Wave*. Naukova Dumka, Kiev (1975) (in Russian).
65. Newton RE. Effects of cavitation on underwater shock loading. Report NPS 69-78-013, Naval Postgraduate School, Monterey, CA (1978).
66. Wilkins ML. Calculation of elastic-plastic flows. University of California, UCRL 7322 (1963).

Part IV

Extreme Waves Excited by Impact of Heat, Radiation, or Mass

> Imagination is no less important for a geometrican
>
> than for a poet in his moments of inspiration.
>
> **Jean D' Alembert**

It is shown in Chapters 7 and 8 that behavior of material of structure elements can strongly depend on parameters of shock loading. In this part, we continue to study the similar effects in material of structure elements in cases of very short and intense loads leading to the appearance of extreme waves of heat, destruction, melting, and evaporation. It is known that thermomechanical behavior of the material during the impact action by the extreme wave, in many respects, is determined by its length and amplitude. These parameters determine the temperature and pressure arising in the medium and consequently, the phase state of the matter. It is important to remember that in a very wide range of temperature and pressure, the matter does not exhibit strength properties. At temperatures of the order of $10^5 - 10^6$ K (K is Kelvin temperature), the matter exists as plasma, and at $10^4 - 10^5$ K, it is in the gaseous state. Only in a condensed (liquid or solid) state, which can take place up to temperatures of the order of 10^4 K, the matter has the property strength. The strength properties of the matter decrease with increasing pressure. As the pressure increases from values of the order of MPa, the properties of the matter are more accurately described by the models of liquid or gas.

The influence of various physical effects is determined by a time of existence of them. Despite the existence of "in-principle" mutual influence of many physical processes, there is often the main effect which is determined by the matter behavior (state) within various time or space intervals. Processes of the evaporation, the heat propagation, and the emergence of stress waves can be often considered independently.

9 Forming and Amplifying of Heat Waves

The classical parabolic model for heat conduction, for many choices of thermal conductivity, predicts an infinite speed of heat diffusion. The hyperbolic model of heat conduction considered here is more physically realistic as a finite speed of heat diffusion is predicted.

The relaxation model of heat transfer assumes the heat flux equation in the following form (1.13):

$$W + \tau W_t = -\kappa(T)T_z, \qquad (9.1)$$

where T is the temperature, W is the heat flux, and $\kappa(T)$ is the thermal conductivity. As earlier, letter subscripts denote differentiation with respect to the corresponding value. The relaxation time τ determines a time after which a heat flow forms when the temperature gradient is imposed. The first term in the left side of (9.1) differs this model from the classical Fourier law. In contrast to the latter, Eq. (9.1) describes the finite velocity of thermal waves.

Mathematical formulation of the problem includes also the energy conservation law (1.3) written in the following form:

$$E_t = -W_z, \qquad (9.2)$$

where $E = \rho C_V(T)T$ (E is the energy), $C(T)$ is the volumetric heat capacity. Now we can derive the nonlinear governing equation. First, we must apply the differential operator $\partial/\partial a$ to Eq. (9.1) and the differential operator $(-1 - \tau \partial/\partial t)$ to Eq. (9.2) obtained from $E_t = -W_z$ (9.2) after the substitution of $E = \rho C_V(T)T$ in it. After the summation of the resulting equations, we can write a nonlinear equation describing the propagation of heat waves:

$$\tau C_V T_{tt} + C_V T_t = \kappa T_{zz} + \kappa_T (T_z)^2, \qquad (9.3)$$

where $\kappa_T = \partial \kappa / \partial T$. Let us note that Eq. (9.3) is the quasi-linear hyperbolic partial differential equation.

9.1 LINEAR ANALYSIS – INFLUENCE OF HYPERBOLICITY

It is known that the propagation of heat is described by the parabolic-type equation. However, for nano- and picosecond range impacts, the rate of physical processes is such that the energy absorbed by the matter does not have time to distribute between electrons and more inertial atoms of the atomic lattice. Thus, for intervals of the order of 10^{-11} s in the microvolume of the matter the thermal flux W is not established instantly as it is predicted by the classical Fourier law. In this case, the heat flux will increase gradually from zero to maximum, and it is necessary to use the modified Fourier equation (1.13) so that the process of thermal conductivity can be described. This equation follows from the energy equation after discarding thermal and electron terms, and also terms related to the deformation of the medium. Thus, there appears the possibility of formation and existence in matter of a heat wave. This wave is the result of the hyperbolicity of the modified linearized Fourier equation (1.13). The front of this wave may be steep enough; however, the shock front cannot form because the equation used is linear. In this case, Eq. (9.3) becomes

$$\tau C_V T_{tt} + C_V T_t = \kappa T_{zz}. \tag{9.4}$$

To numerically study the influence of radiation frequency on the heat propagation, various presentations of the laws of the heat conduction are used below. Calculations were carried out with the following thermophysical constants: $C_V = 900\,\text{J/(kg K)}$, $\kappa = 250\,\text{W/m K}$, and $\tau = 10^{-10}$ s (W is a unit of power (Watt)).

The wavelength determines the depth absorption of the energy by the matter. The impulse is assumed to be instantaneous. For the description of the radiation energy, the Bouguer–Lambert law is used [1]. We considered the long wavelength and short wavelength range of the radiation frequency. For the absorption coefficient in the law, we used two values: $-L = 10^{-8}$ m^{-1} (long wavelength) and $L = 10^{-4}$ m^{-1} (short wavelength).The initial radiation energy was given exponential (see, for details, Chapter 10).

Figure 9.1 shows the fronts of the heat wave calculated for the long wavelength radiation according to a linear parabolic (curves 1a, 2a, 3a) and the linear hyperbolic

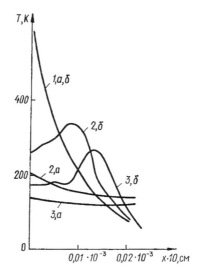

FIGURE 9.1 Fronts of a thermal wave for a linear parabolic (curves 1a, 2a, 3a) and linear hyperbolic (curves 1b, 2b, 3b) equations: (1) $t = 0$, (2) $t = 2.5 \times 10^{-12}$, and (3) $t = 4.5 \times 10^{-12}$ s.

(curves 1b, 2b, 3b) equations. Moments of time are (1) $t = 0$, (2) $t = 2.5 \times 10^{-12}$, and (3) $t = 4.5 \times 10^{-12}$ s. The absorption occurs in a very narrow layer. It can be seen that for a short time, the wave substantially changes its form and amplitude. However, if in the case of a parabolic equation, the wave is simply smooth, then for the hyperbolic equation there is a separation of the heat pulse from the loading surface and its advance into the depth of the material. The maximum temperature is not located on the loading surface. According to the hyperbolic model, the heat produced directly on this surface is greater than the heat calculated according to the classical heat equation.

It should also be noted that the time required for these processes is less than the characteristic time of mechanical phenomena, and therefore, the corresponding thermal dynamics in an essential way affects the formation of the compression pulse in the material.

Another picture is observed in the case of short-wave absorption. In this case, the absorption zone has a thickness of the order of 10^{-4} m. The results of calculations according to the parabolic and hyperbolic equations were practically the same. There is a slight change in shape and amplitude of thermal wave for the characteristic time of mechanical processes, which we consider the order of 10^{-3} s.

Thus, we can conclude that the thermal processes, that occur in the material, strongly depend on the wavelength of the radiation. It is impossible to neglect the processes of thermal conductivity during of the action of a long-wave radiation. If we are interested in the emerging and spreading impulse stress, then the hyperbolic equation shows large amplitudes for the temperature, and consequently, we have the large compression wave. The effect of the thermal conductivity on the stress can be understated, and the strength of the structural element may be overestimated if we use the classical heat equation.

Heat pulses quickly decay, for example, in metals at a distance of $L < 10^{-7}$ m and in porous materials of $-L < 10^{-4}$ m. At the same time, experiments show [2] that for temperatures close to absolute zero, the thermal impulses propagate slightly changing their form. In this case, they are analogous to acoustic waves. The heat pulses can very weakly decay. In particular, in very thin multilayered materials even their amplification is possible [1].

Amplification of thermal waves – Numerical analysis. The above analytic study of the possibility of amplification of thermal waves on the contact surface is based on a number of assumptions about the character of the solution of the equations [1] (see for details Chapter 10). Let us check this possibility solving numerically the hyperbolic equation for two-layer material. Initially, the wave is excited in the first layer. In accordance with theoretical analysis [1], the amplification may be done if the second layer has smaller conductivity and a large coefficient of thermal relaxation than the first layer. The characteristics of the first layer were presented above, and for the second layer, we take $C_V = 900$ J/(kg K), $\kappa = 25$ W/(m K), and $\tau = 10^{-9}$ s . We assume that at $t = 0$, the temperature varies as the stair. The initial and final profiles of waves calculated in two-layer structure are shown in Figure 9.2.

The wave profiles for a single layer are given in Figure 9.2 too. In the latter case, the formation of a wave profile of temperature and its propagation with the damping are shown. In the case of a two-layer material near the interface, a peak of

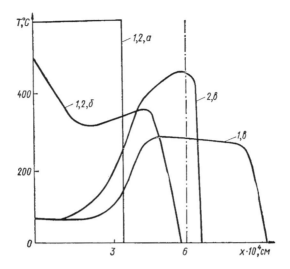

FIGURE 9.2 Heat wave profiles: 1 – the one layer; 1 – the two layers for $t = 0$ (а), $t = 0.8 \cdot 10^{-10}$ s (б), $t = 1.4 \times 10^{-10}$ s (в) (the dash-dotted line is a boundary of layers).

temperatures is formed, the amplitude of which is approximately twice as high as that calculated for single layer. Thus, a numerical calculation performed without any assumptions about the nature of the solution of the equations describing the propagation of heat fully confirms the result of an approximate analytical study [1].

The amplification of thermal waves in thin-layer structures can cause danger to their durability. On the one hand, some inner layer can melt if its critical melt temperature was appreciated according to the classical heat equation. On the other hand, the growth rate of the temperature amplitude might be quite large to exert a significant influence on the stress–strain state and lead to the formation of dangerous stresses.

9.2 FORMING AND AMPLIFYING NONLINEAR HEAT WAVES

Let us turn to the Fourier equation with nonlinear heat conductivity (9.1) and (9.2) or (9.3). In this case, the steep wave fronts can be formed in the matter. Therefore, the heat flow will be hindered due to the equality to zero the temperature just before the front. At the same time, the temperature on the heat front is maximum, so the heat flux is also maximum. As a result the steep heat front is conserved. However, there are specific heat conditions for realization of this scheme: special form of the coefficient of thermal conductivity and the requirement that the amplitude of the heat wave was significantly higher than the initial background heat along which the front moves.

Two-speed wave front in liquid helium. Thermal waves were studied both experimentally and theoretically [2–5]. It turned out that these waves can manifest themselves only in very extreme exotic conditions, i.e., metals at short-term energy

Forming and Amplifying of Heat Waves

impacts or cryogenic liquids at temperatures close to absolute zero. Above the generation and propagation of linear thermal waves in multilayer metal structures were studied [1,6,7]. The below is a numerical analysis of the propagation of such waves in liquid helium.

The purpose of this section is to provide the proof of suitability of the nonlinear relaxation model of heat conductivity for describing the main features of formation and propagation of waves of the second sound in liquid helium. This section discusses the possibility of the amplifying such waves. Experimental recording of second sound in He II [8] together with the detection of other specific effects caused the intense researches of quantum cryogenic liquids. Theoretical explanations [5] were related to the existence of thermal waves having a clearly pronounced front and a finite speed. This gave an impulse to extensive researches in the field of the so-called generalized thermodynamics created for a description of this range of temperature [9]. Thanks to the development of experimental techniques, additional details of the formation and the propagation of heat waves in liquid helium were discovered [4,10]. The necessity of interpretation of the received information required the development of an adequate mathematical model. It was found that the accounting for the relaxation and nonlinear thermal properties of cryogenic liquids is sufficient for a detailed description of the experimental data.

Two-stage waveform of the second sound. An initial research item is the results of the studying of waves of the second sound obtained in [4]. Dynamics of temperature arising in the volume of liquid helium due to pulsed heat generation in the form of a rectangle was recorded. In Figure 9.3, it is clearly shown that for $W_0 = 4 \times 10^5$ W/m^2, the temperature profile has a clearly expressed two-step form. The velocities of the steps are different. Existence of the second step is explained by processes of local boiling of the overheated liquid which is discussed in [4]. However, the two-step structure of the second sound wave can be modeled without additional physical mechanisms [10].

The thermal behavior of cryogenic helium is expressed by Eq. (9.3). Additionally, we take into account the dependences of coefficients of thermal conductivity $\kappa(T)$ [3] and specific volume heat capacity C_V [8] from temperature T. These dependencies are displayed graphically because it is very difficult to describe them analytically [7,11] (Figure 9.4).

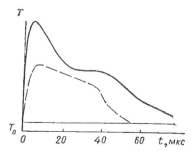

FIGURE 9.3 Oscillograms temperature in wave second sound at a distance 0.001 m from the emitter: $W_0 = 1.4 \times 10^5$ W/m^2 (the dashed line) and $W_0 = 4 \times 10^5$ W/m^2 (the solid line).

240   Extreme Waves and Shock-Excited Processes

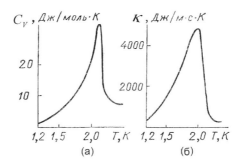

FIGURE 9.4 Dependencies of heat capacity (a) and heat conductivity (б) [3] of helium from the temperature.

The dependence $\kappa(T)$ was obtained on the basis of the following formula:

$$u_2 = \sqrt{\kappa/\rho C_V} \tag{9.5}$$

The measurements of the second sound velocities are given in [3]. The coefficient of thermal relaxation τ weakly depends on the temperature [3]. Therefore, it was assumed as constant and equal to 0.00472 sec. The density of liquid helium is 1500 kg/m². Thus, Eq. (9.1) combines two mechanisms for the formation of thermal waves [7,11], related to the relaxation term and the nonlinear term.

The system (9.1), (9.2) is supplemented by the initial conditions:

$$T(0,z) = T_0 = 0.5K, \quad E(0,z) = \rho C_V(T_0)T_0,$$

$$W(0,z) = 0, \quad 0 \le z \le H = 0. \tag{9.6}$$

Boundary conditions are

$$W(t,0) = W_0, \quad W(t,H) = 0. \tag{9.7}$$

The first of the conditions determines the heat flux W_0 at the left boundary of helium volume, and the second condition corresponds to the heat-insulated wall. The solution of the formulated boundary problem was sought numerically. The numerical method is described in [1].

The results of the calculations are shown in Figure 9.5. There is the correspondence of the calculations with experimental data. It should be noted that the lines shown in Figure 9.5 represent instantaneous distributions of the temperature over the volume of helium. On the contrary, the lines shown in Figure 9.3 display the evolution of temperature at a volume point. However, it is well known that in dynamic problems usually, these lines strongly correlate with each other.

A distinct two-wave form of the thermal wave is observed (the fronts a and the fronts б). Variations show that the speeds of both waves are independent of the load. The velocity can be estimated from formula (9.5). As a result, we obtain $u_2 = 19$ m/s.

Forming and Amplifying of Heat Waves

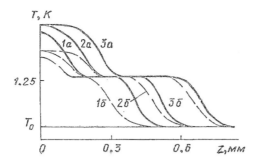

FIGURE 9.5 Spaces profiles of temperature fronts for various moments of time: the dashed curve corresponds to $W_0 = 1.96 \times 10^5$ W/m^2, solid curve $-W_0 = 7.84 \times 10^5$ W/m^2; (1) $t = 16 \times 10^{-6}$ s, (2) $t = 24 \times 10^{-6}$ s, (3) $t = 32 \times 10^{-6}$ s.

At the same time, numerical calculations give the value of $u_2 = 20$ m/s. Thus, the results are in good consent. The formation of a steep temperature front can be explained by joint action nonlinearity and hyperbolicity in Eq. (9.3).

The relaxation term in Eq. (9.1) takes into account the finite time for establishing a thermal flux in the microvolume of helium after action of temperature gradient. This effect hinders the passage of heat into the medium and increases the steepness of the thermal profile. The steepness growth in the case of a nonlinear mechanism is associated with the difficulty of passing of the heat from a heated and heat-conducting substance on the crest of the wave to a relatively colder and less heat-conducting helium before its front. Some of front blurring is explained by the fact that the areas, having different temperatures, move with various speeds.

Formation of the second thermal step in the framework of this model can be explained as a result of establishing an equilibrium between the heat arriving to the surface $z = 0$ and the heat flux from $z = 0$ due to thermal conductivity. The latter grows with increasing temperature, though not proportional to it. At the same time, the formation of a steep front and its speed are explained, as above, by the features of the liquid itself and therefore do not depend on the loading.

Reflection and amplification of wave steep front. The formation of a cooling wave during the heat wave reflection from the heat-conducting layer was shown in the experimental paper [4]. We will study this very interesting phenomena below.

Let us consider the reflection of a thermal wave from a non-heat-conducting surface. Some results of calculations are presented in Figure 9.6. It shows the growing of the wave amplitude as a result of the reflection.

Twenty-five percent of the amplification occurs. Then, with the approach of the second level (step) of the heat wave, the temperature stabilizes. More calculations allowed us to determine that the degree of amplification does not depend on the nonlinearity thermophysical parameters. This effect occurs more because of the relaxation properties of liquid helium.

As a result of the studies carried out in this section, we have described the features of the formation two-step (two-speed) experimentally observed wave of the second sound by the model that takes into account thermal relaxation of the medium

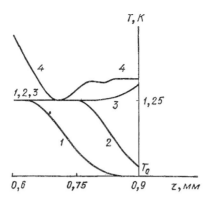

FIGURE 9.6 Heat wave amplification on heat-insulated surface: (1) $t = 32 \times 10^{-6}$ s; (2) $t = 39 \times 10^{-6}$ s; (3) $t = 47 \times 10^{-6}$ s; (4) $t = 66 \times 10^{-6}$ s.

and the nonlinear dependence of thermophysical parameters from temperature. The possibility of amplifying a thermal wave during its reflection from the heat-insulated surface was noted.

Comment. There is an analogy between propagation of the second sound waves and the waves propagating in weakly cohesive materials (see Chapter 3), bubbly liquids (see Figure 19.5), and elastoplastic materials [12]. In the last case, the processes may be described by the one-dimensional nonlinear wave equation:

$$u_{tt} = a^2 u_{zz}, \qquad (9.8)$$

where $a^2 = \rho_0^{-1} \partial \sigma / \partial e$ and $e = e(u_z)$. The last is determined by the stress–strain curve. According to (9.8), many solids can propagate two kinds of waves: elastic wave having the velocity a_e and the plastic wave having the velocity a_p. Since $a_p < a_e$, they form the two-step structure.

Thus, the elastic wave propagates faster than the plastic wave. Generally speaking, the speed of the elastic wave is constant. The speed of the plastic wave depends on the loading. As we see, this phenomenon is quite similar to particularity of thermal waves which was considered above.

9.3 STRONG NONLINEARITY OF THERMODYNAMIC FUNCTION AS A CAUSE OF FORMATION OF COOLING SHOCK WAVE

A one-dimensional model of relaxation heat transfer was investigated numerically for the case of helium II manifesting strongly nonlinear properties. A complicated structure of the heat wave of cooling generated by negative temperature jump on the boundary surface was studied. Particularly, two parts of the negative heat wave (cooling wave) were revealed. The first part has a smooth profile, but the second part has a steep (shock-like) profile. The latter can be formed from the initially smooth wave. The generation of the nonlinear wave of additional cooling after its reflection from thermally insulated wall was also studied.

Forming and Amplifying of Heat Waves

Introduction. Below the point of the phase λ-transition $(T_\lambda = 2.1768°\text{K})$, helium obtains properties of quantum liquid (He II) and exhibits an unusual behavior. The propagation of temperature waves with finite velocity and clearly manifested temperature profile is one of the physical effects appearing in helium at cryogenic temperatures that is referred to as "second sound". Peshkov [13] was the first who observed it experimentally.

The theoretical basis for the investigation of this phenomenon is the relaxation model of the wave-type heat transfer. Beginning with the pioneer work of Maxwell [14], the model was developed by a number of authors [15–17] and accepted for description of behavior of liquid helium. Theoretical analysis of the relaxation heat transfer was performed primarily within the framework of linear approximation [15] where the integral transformation is applicable.

However, near the temperature of phase transition the helium heat capacity tends to infinity [8]. Thus, the linear approach becomes inapplicable. The use of nonlinear equations was hindered by difficulties associated with the scarcity of the exact analytical solutions and the lack of generally accepted relations for thermal conductivity of liquid helium. The examples of the solution of nonlinear problems of heat conductivity are a few in number, and the numerical simulations dominate among them [18]. It was demonstrated, particularly, that the structure of thermal wave profile (second sound wave) generated by the energy pulse in liquid helium is very complex [13]. But the details of initiation and development of wave structure were not studied enough.

While the generation of compression shocks in continuum is a matter of common knowledge, the formation of thermal shocks during the heat propagation is a new issue in physics. The temperature jumps in this case may be considered as pressure jumps. In the light of relatively recent finding of rarefaction shock waves that were thought to be nonexistent for a long time [17], it is reasonable to suggest that thermal shocks of cooling (traveling negative temperature jumps) are also possible. But up to now, they have not been studied either experimentally or theoretically.

The objective of this section is to numerically demonstrate that the thermal shocks of cooling can be generated in liquid helium and these waves may be formed from smooth temperature profiles. Nonlinear reflection of the waves from thermally insulated boundaries with the generation of additional cooling is also studied.

Basic relations and governing equation. The nonlinear heat wave equation is written in the form (9.3). Let us note this is the quasi-linear hyperbolic partial differential equation. Taking this in mind, we can interpret the value (9.5) as the velocity of the heat wave. Solving (9.5) with respect to thermal conductivity, one can obtain

$$\kappa = \tau C_V u_2^2. \tag{9.9}$$

All the necessary temperature-dependent functions are plotted in Figure 9.7.

The system is completed with the uniform initial conditions for the functions $T(z,t)$:

$$T(z,0) = T, \tag{9.10}$$

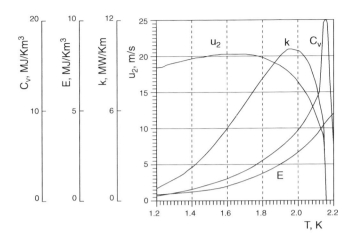

FIGURE 9.7 The temperature-dependent functions: heat capacity C_V, internal energy E, thermal conductivity κ, and second sound velocity u_2.

and the boundary conditions on the left boundary are

$$T(0,t) = \vartheta(t), \quad t > 0. \tag{9.11}$$

Equation (9.3) was solved numerically taking into account corresponding initial and boundary conditions.

Numerical results and discussion. The temperature evolution in the case of instantaneous cooling of the boundary (sudden imposing of negative temperature jump) was numerically simulated in the first series of calculations. The function ϑ in Eq. (9.11) has the form:

$$\vartheta = T_c = \text{const}, \quad \text{where} \quad T_c < T_0. \tag{9.12}$$

Corresponding temperature profiles are plotted for five instants in Figure 9.8.

The formation of a temperature jump (the front of the thermal wave) traveling with the velocity about 21 m/s can clearly be seen. During the considered time, the wave profile complicates and two parts of it can be distinguished. No profile change is manifested in the upper part, and its displacement is parallel to itself. At the same time, the lower portion of the profile changes its form during the propagation because of different velocities of points of the temperature profile.

The upper part of the temperature profile is the heat shock of cooling, i.e., the vertical jump of finite amplitude. The width of jump because of numerical representation is not zero. However, it reduced to almost zero when a size of the numerical mesh was reduced to zero. The lower portion of the profile represents the smooth nonlinear relaxation thermal wave. Its slope decreases during the wave evolution, and the obtuse angle at the point of the wave parts conjugation also decreases.

The thermal wave behavior described above can be explained qualitatively as follows. The thermal wave velocity is maximum (Figure 9.7) at the

Forming and Amplifying of Heat Waves

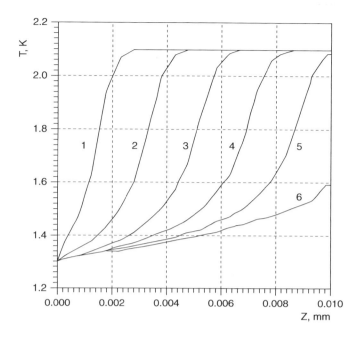

FIGURE 9.8 Profiles of the temperature wave for $t = 10^{-4}$ s (1), 2×10^{-4} s (2), 3×10^{-4} s (3), 4×10^{-4} s (4), and 5×10^{-4} s (5).

temperature near 1.8°K. Thus, the profile portions located either upper or lower, and then, this critical temperature travels more slowly than the point at which the temperature is just 2°K. The evolution of the initially straight temperature profile (1) into the wave of cooling (2) is sketched in Figure 9.9.

We can notice that the slope of the upper portion of the temperature profile becomes steeper due to the velocity growth together with temperature within the temperature interval considered. For the lower part of the curve, the trend of the form change is inverse.

The latter of the described processes has no physical limitations, but the steepening of the upper part of the profile can continue only while the slope is less than $\pi/2$ and the temperature-coordinate correspondence is one-to-one. After this, the cooling shock or traveling negative temperature jump is formed.

The formation of cooling shock due to the strong nonlinearity of thermodynamic properties of helium was additionally validated by the numerical simulation of the temperature evolution initiated by a linear decrease of the boundary temperature. In this case, we assume that

$$\vartheta = T_0 - tR \quad \text{if} \quad 0 \leq t \leq t^*; \quad \vartheta = T_c \quad \text{otherwise,} \tag{9.13}$$

where $R = (T_0 - T_c)/t^*$ and $T_0 > T_c$.

The time evolution of the temperature profiles for the two cases involved is compared in Figure 9.10. The difference between the couples of the matched profiles is

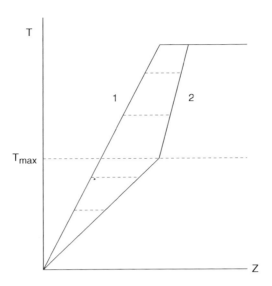

FIGURE 9.9 The sketch of the cooling shock formation: $t = t_0$ (1) and time $t = t_0 + \Delta t$ (2).

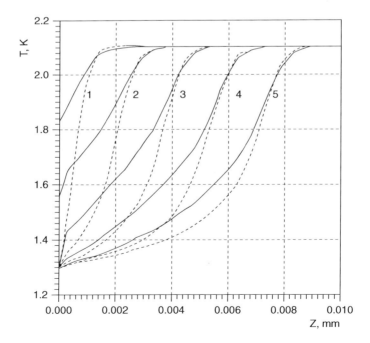

FIGURE 9.10 Profiles of the temperature waves caused by the linear time-dependent surface cooling with 3×10^{-4} (solid lines): $t = 10^{-4}$ s (1), 2×10^{-4} s (2), 3×10^{-4} s (3), 4×10^{-4} s (4), and 5×10^{-4} s (5), and the instant cooling (dashed lines).

Forming and Amplifying of Heat Waves

clearly manifested for all instants considered, but it is less for late instants. A gradual formation of the steep temperature curve from the gently sloping one can clearly be seen for the temperatures higher than T_{max}.

Beginning in 0.5 ms, the shock wave is formed completely and travels with a constant velocity (21 m/s), which is evident from comparison of the upper parts of two profiles considered in Figure 9.10. The coincidence of the lower parts of curves cannot ever be reached since their forms depend on the whole history of the wave formation.

The final part of the investigation involves the simulation of the cooled shock wave emerging at a thermally insulated boundary. The reflection of thermal waves in liquid helium from the solid surface was experimentally studied earlier in [19]. Theoretical analysis of this phenomenon was performed in [15] within the framework of the linear approach.

The linear model predicts that the amplitudes of the arriving and reflected temperature jumps are approximately identical, but in reality, the temperature is restricted by the absolute temperature zero. So for sufficiently big amplitudes of the reflected wave, the linear approach becomes inadequate. The temperature drop due to wave reflection in helium is limited by values of heat capacity and thermal conductivity when $T \to 0$. As a consequence, the thermal wave near zero temperature should generate the reflected wave with an appreciably smaller amplitude.

Figure 9.11 shows the results of the numerical simulation of the temperature wave reflected from a thermally insulated boundary. Due to a strong material nonlinearity,

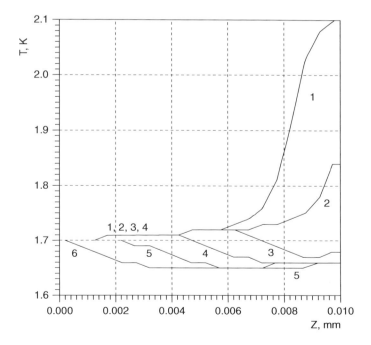

FIGURE 9.11 Reflection of the cooling waves from the isolated wall: $t = 5 \times 10^{-4}$ s (1), 6×10^{-4} s (2), 7×10^{-4} s (3), 8×10^{-4} s (4), 9×10^{-4} s (5), and 10^{-3} s (6).

the amplitude of the reflected wave is much lower than that of the initial wave. The further evolution of the thermal wave leads to gradual settlement of constant temperature background.

The phenomenon described may probably be used for the additional cooling of liquid helium by reflected thermal pulses [20].

CONCLUSIONS

1. The structure of the heat wave of cooling in helium II in the case of strong material nonlinearity was studied numerically. Two portions of the temperature profiles were revealed and investigated. The upper of them is cooling shock traveling with a constant velocity, and the lower (second) part is a smooth profile changing its shape during the propagation.
2. It was numerically demonstrated that the cooling shock can be formed from an initially smooth temperature profile and the causes and conditions of that transformation were found.
3. The generation of the nonlinear wave of additional cooling after thermal shock reflection from thermally insulated solid surface was studied numerically.
4. The phenomenon of amplification of thermal waves in thin-layer structures may present a certain danger to their strength. On the one hand, it is possible to reflow the internal fusible layer, calculated on the temperatures predicted by the classical the heat conduction equation. On the other hand, the growth rate amplitude of the temperature is large enough to have a significant influence on the stress–strain state and leads to the formation of dangerous stresses.

10 Extreme Waves Excited by Radiation Impact

1. This chapter contains the results published in the USSR at the end of the last century. It developed the approach in which the separation of the problem into a gas-dynamic and strength parts in a wide range of variation of the impact energy, frequencies, and time of action of radiation was not used from the very beginning [1,6,7].

2. The behavior of the material on which the radiation pulse falls is largely determined by the wavelength and the radiation power. Figure 10.1 shows the regions of influence of the most important physical and mechanical processes manifested during the action of radiation [6]. It follows from the figure that in the very wide temperature range (10^2–10^6 K), there is no effect of strength and material can be in the plasmous (10^5–10^6 K), gaseous (10^4–10^5 K), and condensed (10^2–10^3 K) states [21–31].

   Here, we study waves of pressure, stresses, and heat arising from the pulsed heating of material surface from 10^2 to 10^5 K. At low temperatures, when there is no melting, there are the processes of expansion of the near-surface layers of the material and their destruction in the form of a surface spalling. The presence of melting and evaporation at higher temperatures leads to the complication of the formation of stress and pressure waves, which are determined both by the reactive recoil of the evaporation products and by the expansion of the unevaporated part of the solid.

   The contribution of these two mechanisms to the formation of the waves depends on the intensity of the radiation incident on material. The effect

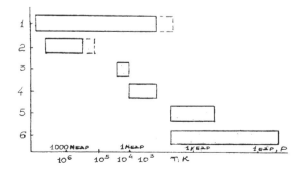

FIGURE 10.1 Fields of influence of some physical and mechanical processes: 1 – absent of strength, 2 – ionization and formation of plasma, 3 – evaporation, 4 – melting, 5 – phase transitions of the crystal lattice, and 6 – strength of the material.

of reactive recoil can be increased until the manifestation of the effect of screening radiation by flying pairs of material begins.
3. Certain additional results of Russian studies of the action of the radiation pulse on materials can be found in [21–32].

10.1 IMPULSIVE DEFORMATION AND DESTRUCTION OF BODIES AT TEMPERATURES BELOW THE MELTING POINT

In this section are considered thermoelastic waves caused by impact radiation. Results published in [6,33,34] are used. Destruction occurs in rarefaction waves in the form of a surface spalling. The considered models relate to the action of long-wave (thermal, laser) and short-wave radiation, acting during time less than 1 ns and more 10 ns.

10.1.1 Thermoelastic Waves Excited by Long-Wave Radiation

The long-wavelength radiation is absorbed in a very thin near-surface layer of material of approximately 10^{-8} m. Therefore, in the mathematical formulation of the problem, this layer can be assumed to be infinite thin, and the energy source can be introduced into the boundary conditions for the temperature.

In [6,33], the effect of the energy flux on metallic targets coated with transparent liquid dielectrics was studied. The formula for the shock-wave pressure is derived under the condition that the amplitudes of the shock wave in the liquid and the amplitude of the compression wave in the metal are equal:

$$p = \frac{(\gamma - 1)^{0.5} \rho_1 D_1 \rho_2 D_2}{\gamma^{0.5} (\rho_1 D_1 + \rho_2 D_2)} Q, \tag{10.1}$$

where ρ_1 and ρ_1 are the densities of the liquid and the metal, respectively, D_1 and D_1 shock-wave velocities, γ the adiabatic constant, and Q the absorbed radiation flux.

The next follows from a comparison of the pressure calculated according to (22.1) and the experiment (Figure 10.2) is discussed next. First, the theory and the experiment are in satisfactory agreement with the fact that the bulk of the energy is absorbed. Second, the energy of the recoil for copper and aluminum targets at about 0.5 J/m² is approximately of an order of magnitude, which is higher than that in the case of irradiation of the pure surface of a metal at the same densities of the radiation flux.

10.1.2 Thermoelastic Waves Excited by Short-Wave Radiation

A purely thermomechanical problem is being studied. A pulsed heat source is introduced in the thermal part of the problem, and then, the propagation of the thermoelastic wave is studied. Short pulse radiation is absorbed in a sufficiently thick layer of matter. This effect may be simulated assuming the heat sources appearing in the material. The study of the formation and propagation of heat waves and stress waves generated by the action of this radiation on copper mirrors was performed

Extreme Waves Excited by Radiation Impact

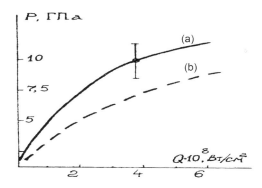

FIGURE 10.2 Dependence of pressure in a shock wave on the density of the incident radiation flux: (a) experiment and (b) theory (here ВТ is watt (W)), ГПа = GPa.

in [1,6,34]. The heat propagation in these mirrors was described by the relaxation model of the heat flux (the modified Fourier law) (1.13). Therefore, heat sources and the inertial term was introduced in the classical heat equation. As a result, we have

$$\rho C_V \left(\frac{\partial T}{\partial t} + \tau \frac{\partial^2 T}{\partial t^2} \right) = \kappa \frac{\partial^2 T}{\partial z^2} + \left(1 + \tau \frac{\partial}{\partial t} \right) Q(z,t), \tag{10.2}$$

where C_V is the heat capacity per unit volume, $\kappa(T)$ is the thermal conductivity, τ is the relaxation time of the heat flow, and $Q(z,t)$ is the density of volumetric heat sources. Moreover,

$$Q(z,t) = Q_0(t)\exp(-\mu z), \tag{10.3}$$

where $Q_0(t)$ is the heat release at the surface of a solid body (at $z = 0$), which is conveniently expressed in terms of the intensity of radiation $W(z,t)$, and μ is a constant of the material. We assume that

$$LW(t) = \int_0^\infty Q(z,t)dz = \mu^{-1}Q_0(t), \tag{10.4}$$

where L is the coefficient of absorption of radiation. The field of dynamic stresses in a half-space is determined by the equations for thermoelastic waves:

$$\frac{\partial^2 \sigma_{zz}}{\partial t^2} - c^2 \frac{\partial^2 \sigma_{zz}}{\partial z^2} = -\beta \frac{E}{1-2\nu} \frac{\partial^2 T}{\partial t^2}, \quad \sigma_{xx} = \sigma_{yy} = \frac{\nu}{1-\nu}\sigma_{zz} - E\frac{\alpha}{1-\nu}T, \tag{10.5}$$

where β is the coefficient of thermal expansion, E is the modulus of elasticity, ν is the Poisson's ratio, α is a constant, and c is the propagation velocity of the stress wave, which is given by

$$c^2 = (1-\nu)E\left[\rho(1+\nu)(1-2\nu)\right]^{-1}.$$

Initial and boundary conditions for temperature and stress are

$$T(z,0) = \partial T(z,0)/\partial t = 0,\ T(\infty,t) = 0,\ \partial T(0,t)/\partial z = 0, \qquad (10.6)$$

$$\sigma_{zz}(z,0) = \partial \sigma_{zz}(z,0)/\partial t = 0,\ \sigma_{zz}(0,t) = \sigma_{zz}(\infty,t) = 0. \qquad (10.7)$$

The problem can be solved by using analytical methods [34].

A case of a stepwise loading of the body surface was considered. It is assumed that the material breaks down when the stress reaches the yield point. It was shown that the change in the temperature field in a metal occurs both as a result of the volume heat sources and due to the heat wave moving into the material with a constant velocity. During the propagation, the amplitude of the thermal wave decreases exponentially.

In the material, there are two stress waves: a wave of compressive and tensile stresses. The last do not move with a velocity of elastic waves, but with the velocity of the thermal wave. The compression waves move ahead of thermal wave.

According to [34], the threshold of plastic deformations for the pulses longer than 0.1 ns is approximately from 20 to 50 times smaller than that for melting of the mirror. For pulse durations less than 0.1 ns, this discrepancy increased. In the last case, the difference between the thresholds was more than 100 times due to dynamic effects (Figure 10.3).

Figure 10.3a shows the threshold of fracture of copper mirrors by melting of the surface (1–4) and plastic deformation 5 [34]. Experimental points are shown for wavelengths of the radiation of 1.06×10^{-6}m (1), 3×10^{-6}m (2), and 10.6×10^{-6}m (3). Curves 4 and 5 are the results of calculations: four takes into account the melting, and five appreciates the influence of the strength of the material.

It must be borne in mind for understanding Figure 10.3a that the theoretical thresholds of destruction by melting by the pulsed radiation are substantially higher than those obtained in the experiment (see points noted as 1, 2, and 3). Therefore, one should expect a similar decrease of the real destructive thermoelastic stresses in comparison with the theoretical ones.

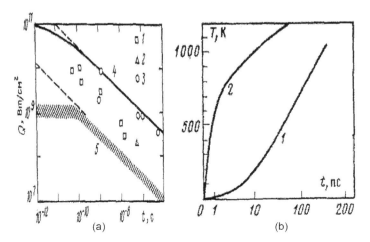

FIGURE 10.3 Threshold of destruction (a). Impulse duration effect (b).

Extreme Waves Excited by Radiation Impact

In Ref. [34], the influence of two dynamic effects on the allowable thermal load of mirrors was studied: the effect of the velocities of the propagation of the thermal and mechanical waves in the metal and the overheating of the electrons with respect to the crystal lattice. The first effect has a significant effect on the mechanism of mirror destruction and the second on the level of thermal stresses in the material (Figure 10.3). Figure 10.3b shows the effect of the duration of the laser pulse on the electron (curve 2) and lattice (curve 1) temperatures on the surface of a copper mirror at a radiation power density $W(t = 0) = 3 \times 10^{12}$ W/m² (10.4). It is shown in Figure 10.3 that there is an equalization of the temperatures of the electronic and lattice subsystems with an increase in the time.

10.1.3 Stress and Fracture Waves in Metals during Rapid Bulk Heating

The results of a calculation–experimental investigation of the main characteristics of stress wave propagation and fracture processes in copper specimens exposed to pulsed high-current electron beams are given in [31]. An integral fracture criterion was used for estimating strength in the calculations. The data obtained illustrate the possibility of using the method of numerical modeling of wave and spalling phenomena occurring during thermal shock.

The purpose of this chapter is a calculation-experimental investigation of the main characteristics of stress wave propagation and fracture processes in metals exposed to pulsed high-current electron beams. The maximum pulse energy of the electrons reached –3 MeV and pulse duration –10 ns. Copper specimens were used as the targets.

Figure 10.4 shows a schematic diagram of conducting the experiment and microstructure of the cross section of the copper specimen exposed to high-current electron beams *e*. Damages in the form of individual pores occur in the specimen.

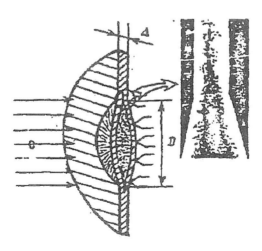

FIGURE 10.4 Diagram of conducting the experiment and microstructure of the cross section of the copper specimen as a result of thermal shock.

At the site of the most intense exposure (zone D), the pores merge into a main crack, which separates the specimen into parts. Furthermore, the temperature of heating the specimen was recorded in the experiments by means of copper-constantan thermocouples. These data were used further in calculations as the initial conditions for studying spalling phenomena and stress wave propagation.

Wave propagation processes in the specimens were studied by the method of numerical modeling on a computer. The problem was solved by means of a finite-difference scheme in Lagrange variables with the introduction of linear artificial toughness. We examined an approximate formulation of the problem. Digressing from the specific physical nature of the radiation pulses, we assigned the temperature field in time and as a function of the coordinates. The following variants were examined. First variant: a 0.04 cm thick copper plate was heated instantaneously to 400°C uniformly over the thickness. Second variant: the specimen was heated statically to 600°C, and then, the temperature was increased to 800°C. Such an approximate formulation of the problem is possible, since a practically uniform energy distribution is characteristic for thin specimens (the density of absorbed energy over the depth differs by no more than 10%–15%).

The physical and mechanical characteristics for copper used in the calculations and their change as a function of temperature are given in Table 10.1, where K and G are, respectively, the bulk and shear moduli, σ_y is the yield stress, and α is the coefficient of linear expansion. The melting point of copper is 1083°C.

The distribution of normal stresses over the thickness of the specimen for the aforementioned variants of loading is shown in Figures 10.5a and 10.6.

It is seen that for $t = 0.12\,\mu s$, tensile stresses are able to cause micro- and macrofractures form in the specimen. For $t = 0.23\,\mu s$, the formation of a pulse of compressive stresses occurs anew in the plate, and for $t = 0.34\,\mu s$, that of tensile stresses occurs; however, their value decreases (Figure 10.5). In the second variant, the wave picture is qualitatively similar to that in the first, but the value of the stresses is substantially less (Figure 10.6).

It is of interest to compare the results of calculations obtained in the second variant with the case when the temperature increases by the same value $\Delta T = 200°C$ from $T = 20°C$ Calculations show that in that case, the stresses exceed the stress σ_y by 50%.

TABLE 10.1
Physical and Mechanical Characteristics of Copper and Their Change as a Function of Temperature

T (°C)	K (GPa)	G (GPa)	σ_y (GPa)	$\alpha \times 10^6$ (/°C)
20	140	49.0	0.30	17.0
100	136	47.7	0.27	17.1
200	132	46.2	0.20	17.2
300	125	44.0	0.18	17.8
600	102	35.6	0.16	18.7
800	81.6	29.2	0.15	19.5

Extreme Waves Excited by Radiation Impact 255

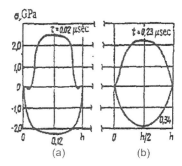

FIGURE 10.5 (a and b) Distribution of normal stresses over the thickness of the specimen with instantaneous heating to 400°C.

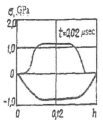

FIGURE 10.6 Distribution of normal stresses over the thickness of the specimen with an increase in temperature from 600°C to 800°C.

This can be explained by the temperature dependence of the physical and mechanical characteristics of the material.

The known results of investigating the temperature-time regularities of the fracture process under rapid bulk heating conditions showed that temperature is one of the important parameters affecting this process.

An integral fracture criterion (see Chapter 2) [31] was used in the calculations for estimating strength. As the measure of damage, we took the quantity covarying from 0 to 1 and determined from expression $\omega = \int_0^t \frac{(\sigma - \sigma_0)^2}{k} dt$, where σ_0 and k are certain constants of the material. The particular values of these quantities were selected from experiments on plane collision of plates, since the character of fracture and value of critical spalling stresses under the effect of electron beams and shock-wave loading agrees well [31].

The results of the calculations are given in Figures 10.7 and 10.8 in the form of the distribution of the measure of damage over the thickness of the plate and its change with time in the central part of the plate.

We note that the most pronounced damages are localized in the center of the target, although damage, especially for the first variant, occurs almost over the entire thickness of the plate. This agrees well with the experimental result (Figure 10.4).

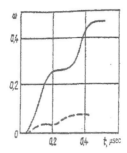

FIGURE 10.7 Change in the measure of damage with time in the central part of the plate. Here and in Figure 10.8: solid lines – heating by an increase from 20v to 400°C; dashed lines – static heating to 600°C and then instantaneous heating to 800°C.

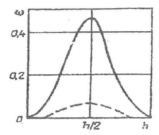

FIGURE 10.8 Distribution of the measure of damage over the thickness of the plate.

It is also seen from Figure 10.7 that the value of ω increases during the formation of tensile stresses. In experimental investigations, wave propagation processes in metals subjected to intense thermal heating are most often studied by recording stresses by a piezoelectric transducer located on the back side of the target exposed to radiation. Furthermore, information about the propagation process and interaction of the waves can be obtained by means of a Manganin transducer, making possible continuous recording of the change in the main parameters of the shock wave. One of the schemes for measuring stresses consists in the following: the back side of a specimen of the material exposed to radiation is fastened to a material with a smaller acoustic stiffness. The stress at the interface is recorded by means of a transducer, from which a conclusion about stresses in the specimen is made.

The results presented illustrate the possibility of using the method of numerical modeling of wave and spalling phenomena occurring in a specimen exposed to pulses of high-current electron beams.

10.1.4 Optimization of the Outer Laser-Induced Spalling

Thermomechanical waves were considered. They can be so strong that in the rarefaction zones, the material can fracture. Controlled dynamic fracture experiments can be conducted on a variety of platforms such as gas guns, split Hopkinson bars, and lasers. Typically, a plane-fronted compressive wave is generated. It propagates

Extreme Waves Excited by Radiation Impact

through the sample and reflects off a free surface. The compressive wave becomes tensile as it reflects off the surface. If the tension exceeds the material strength, fracture occurs. This kind of dynamic fracture is known as spallation. The problem of destruction of solid material, as shown in Chapter 2, is very complex. Therefore, simple models that illustrate well the main aspects of the process of the destruction are very important. Sometimes, they are very important for practice. One such model is presented in Ref. [35].

An analytic model of the outer spalling fracture caused by pulsed laser loading was developed. It was shown that our basic conclusions do not depend on the choice of a specific form of the model used to describe the absorption of pulse energy in the material.

The model of outer laser-induced spalling includes the equation of motion:

$$\rho \, \partial v / \partial t = \partial \sigma / \partial z, \tag{10.8}$$

and the state equation is

$$\partial \sigma / \partial t = \left(\lambda + \tfrac{4}{3} \mu \right) \partial v / \partial z. \tag{10.9}$$

Eqs. (10.8) and (10.9) yield

$$\partial^2 \sigma / \partial t^2 = \left(\lambda + \tfrac{4}{3} \mu \right) \rho^{-1} \partial^2 \sigma / \partial z^2. \tag{10.10}$$

Fracture criterion (2.1) (instant spalling) is

$$\sigma > \sigma_{sp}, \tag{10.11}$$

and three different laws of absorption of radiation in the material were used:

$$\sigma^{(1)}(t=0, z) = -\gamma C_V T_0, \; z < 1/L, \tag{10.12}$$

$$\sigma^{(2)}(t=0, z) = -\gamma C_V T_0 (1 - Lz/2), \; z < 2/L, \tag{10.13}$$

$$\sigma^{(3)}(t=0, z) = -\gamma C_V T_0 \exp(-Lz), \tag{10.14}$$

where v_ρ is the axial mass velocity, σ is the principal stress in the direction of radiation, λ and μ are the bulk and shear moduli of the material, respectively, σ_{sp} is its spalling resistance, T_0 is the surface temperature, L is the specific (per unit area) coefficient of attenuation of radiation in the material, C_V is the specific heat capacity per unit volume of the material, and γ is the Grüneisen coefficient.

The initial profile of temperature (the right-hand sides of Eqs. (10.12)–(10.14)) specifies the initial stress. We use three models of absorption, namely, the rectangular (version 1), triangular (version 2), and exponential (version 3) profiles (Figure 10.9). Note that the third profile appears when we use the classical Bouguer–Lambert law (10.44) [36] of absorption of the radiant energy in the material.

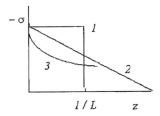

FIGURE 10.9 Initial profiles of thermal stresses: (1) rectangular, (2) triangular, and (3) Bouguer–Lambert law (exponent).

In deducing the system of Eqs. (10.10)–(10.14), it is assumed that the time of action of a pulse is much smaller than the time required for a wave to pass through the width of the absorption zone; that is, the pulse can be regarded as instantaneous. For pulses from the optical band, this corresponds to the duration of pulses varying within the range 10^{-8}–10^{-10} s. It is also assumed that the absorbed radiant pulse energy is insufficient for local evaporation and melting; that is, the material remains solid for the entire process of fracture.

The decomposition of the initial stress state $S(z) = -\sigma(0,z)$ leads to the formation of two waves. The first compression wave travels from the loading surface, and the second wave impinging upon this surface is reflected from it as an expansion wave. As a result, we have

$$\sigma(t,z) = -\tfrac{1}{2}\left[S(z-at) + S(z+at) - S(-z+at)\right]. \qquad (10.15)$$

These traveling stress waves also form the corresponding nonstationary field of mass velocities:

$$v_p(t,z) = \tfrac{1}{2}\left((\lambda + \tfrac{4}{3}\mu)/\rho\right)^{-1/2}\left[-S(z-at) + S(z+at) - S(-z+at)\right], \qquad (10.16)$$

where $\sqrt{(\lambda + \tfrac{4}{3}\mu)/\rho}$ is the velocity of plane stress waves.

In Figure 10.10, the profiles of stresses and mass velocities are presented for the triangular profile of absorption of radiation. It is easy to see that the maximum tensile stress attained on the moving boundary of two halfwaves gradually increases (for all initial profiles except rectangular). As soon as the maximum tensile stress exceeds the ultimate strength, we observe the formation of a plane cleavage crack perpendicular to the direction of loading.

The analysis of solution (10.15) shows that, for all three versions (10.12)–(10.14), the maximum of tensile stresses in the material is half as large as the level of compression stresses formed on the surface at the initial time and equal to $\gamma C_V T_0/2$. This means that there exists the following critical value of the specific radiant energy (per unit area of the irradiated surface):

$$E_{sp} = 2\sigma_{sp}/\gamma L. \qquad (10.17)$$

Extreme Waves Excited by Radiation Impact

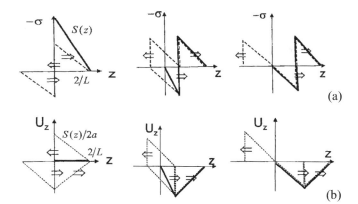

FIGURE 10.10 Evolution of the profile of stresses (a) and mass velocities (b) (the dashed lines correspond to the halfwaves).

As soon as this value is attained, the material suffers outer spalling. It should be emphasized that the critical value of specific radiant energy depends only on the thermodynamic properties of the material and does not depend on its elastic characteristics.

If the value of specific radiant energy exceeds E_{sp}, then, by using solution (10.15), one can find the thickness of spalled plates Z_{sp} for all three versions of numerical calculations as follows:

$$Z_{sp}^{(1)} = 1/2L, \quad Z_{sp}^{(2)} = \left(2\sigma_{sp}/\gamma EL\right)/L, \quad Z_{sp}^{(3)} = -\ln\left(-2\sigma_{sp}/\gamma EL\right)/2L. \quad (10.18)$$

The maximum possible thickness of spalled fragments computed by substituting relation (10.17) in relations (10.18) for all three versions of calculations is equal to

$$Z_{sp/\max}^{(1)} = 1/2L, \quad Z_{sp/\max}^{(2)} = 1/L, \quad Z_{sp/\max}^{(3)} = \infty. \quad (10.19)$$

Thus, as the rate of rise of the initial pulse of heating increases for a constant coefficient of attenuation of radiation, the thickness of the spalled plate decreases and approaches a limit independent of the radiation pulse energy.

If a cleavage crack is initiated, then a part of the mechanical energy of the wave is "locked" in the separated fragment. In this case, the momentum corresponding to the mass velocity of particles is also separated, which means that the spalled fragment acquires a certain mean recoil velocity and the main body gets a recoil momentum. For the indicated three versions of calculations, the mean velocity of the spalled plate $v_{\rho,sp} = \int_0^{Z_{sp}} v_\rho dz/Z_{sp}$ is equal to

$$v_{\rho,sp}^{(1)} = -\frac{\gamma EL\rho^{-1/2}}{\sqrt{\lambda + \frac{4}{3}\mu}}, \quad v_{\rho,sp}^{(2)} = -\frac{\gamma EL - \sigma_{sp}}{\sqrt{\left(\lambda + \frac{4}{3}\mu\right)\rho}}, \quad v_{\rho,sp}^{(3)} = \frac{\sigma_{sp}/Z_{sp}L}{\sqrt{\left(\lambda + \frac{4}{3}\mu\right)\rho}}. \quad (10.20)$$

and the corresponding specific recoil moment (per unit) to

$$I_{\rho,sp}^{(1)} = -\frac{1}{2}\frac{\gamma p^{1/2} E}{\sqrt{\lambda + \frac{4}{3}\mu}}, \quad I_{\rho,sp}^{(2)} = -\frac{(\gamma EL - 2\sigma_{sp})}{2\gamma E \sigma_{sp}}\sqrt{(\lambda + \tfrac{4}{3}\mu)\rho}, \quad I_{\rho,sp}^{(3)} = -\frac{\sigma_{sp}\rho^{1/2}/L}{\sqrt{\lambda + \frac{4}{3}\mu}}.$$

(10.21)

The analysis of the presented solutions enables us to conclude that the velocity of the spalled plate increases when the radiant energy exceeds its critical value. The same is true for the recoil momentum of the spalled plate with the sole exception that, in the exponential version, it is independent of the radiant energy.

The optimal depth of spalling is attained if we use pulses of radiation whose energy slightly exceeds the critical energy of outer spalling. Any subsequent increase in the radiant pulse energy leads to a decrease in the thickness of spalling.

10.2 EFFECTS OF MELTING OF MATERIAL UNDER IMPULSE LOADING

The elastic-plastic properties of the material, the possibility of its destruction and melting are introduced into consideration, but the transition of the material to the plasma state will be considered in Section 10.3.

10.2.1 MATHEMATICAL MODEL OF FRACTURE UNDER THERMAL FORCE LOADING

A more general approach other than those considered in Section 10.1 was formulated in [37]. It was based on the equations of the continuum mechanics describing the propagation of the coupled thermo-viscoelastoplastic waves of fracture initiated by radiation. The wide-range caloric and thermal equations describing the transformation of matter from solid to liquid state were used. However, the transition of the material to the plasma state is not considered. The conservation laws of continuum mechanics are written in Lagrangian coordinates for averaged quantities [1,7,32]:

$$\frac{1}{V}\frac{\partial V}{\partial t} = \frac{\partial v}{\partial z} \tag{10.22}$$

$$\frac{\rho_0}{V}\frac{\partial v}{\partial t} = \frac{\partial \sigma}{\partial z} \tag{10.23}$$

$$\frac{\partial E}{\partial t} = -(p+\omega)\frac{\partial V}{\partial t} + V\left(s\frac{\partial \varepsilon}{\partial t} - \frac{\partial W}{\partial z} - \frac{\partial Q}{\partial z}\right), \tag{10.24}$$

where $V = \rho_0/\rho$ is the relative volume, v is the velocity, E is the specific energy, W is the heat flux, σ and ε are the stress and strain, s is the deviator component of the stress tensor, and ω is artificial viscosity.

It is emphasized that we are considering the case $Q = 0$ (10.24) in this section. In this case, the system (10.22)–(10.24) describes also the temperature waves if the

Extreme Waves Excited by Radiation Impact

heat propagation is modeled by a relaxation model of heat flux (1.13). This model assumes the heat flux equation in the following form (9.1):

$$\tau \frac{\partial W}{\partial t} + W = -\kappa(T)\frac{\partial T}{\partial z}. \tag{10.25}$$

The connection between v and the strain rate ε_t is determined by the geometric relation $\varepsilon_t = v_z$.

Solid material. For the elements of material which conserve strength properties, the calculation is carried out by a step-by-step method, briefly described in Section 1.2. First, we consider the elastoplastic deformations and spalling of the material.

Full increments of the components of $d\varepsilon$ are a sum of the increments of the components of the elastic $d\varepsilon^e$ and plastic deformation $d\varepsilon^p$ (1.29):

$$d\varepsilon = d\varepsilon^e + d\varepsilon^p. \tag{10.26}$$

Accounting for changes in material damage. Equations (10.22)–(10.24) are written for the averaged values. Therefore, they are valid for porous and undamaged materials. Generally speaking, they can take into account the growth of pores in the material. Let us find the relationship of these quantities with the same values for an undamaged material. Now, we can consider the connection of these phenomena.

The increase in microscopic damage of the material is described by Eq. (2.49):

$$\frac{\partial \xi}{\partial t} = \frac{1}{\tau_k}\left[\exp\left(\frac{\sigma - \sigma_t}{\sigma_k}\right) - 1\right], \quad \sigma < \sigma_t, \xi < \xi^*, \tag{10.27}$$

where ξ^* is the critical level of microdamages. We must take into account the existence of pores and its dynamics writing the state equations for damaged materials.

1. *Relation of undamaged and damaged parameters of material.* For the rate of change in a undamaged material (without pores), we write Hooke's law in a differential form:

$$\partial \tilde{s}^e/\partial t = 4\mu(3\tilde{V})^{-1}\partial \tilde{V}/\partial t, \tag{10.28}$$

From Chapter 2 and taking into account that

$$\tilde{V} = V(1 - \xi), \tag{10.29}$$

and from (10.28),

$$\partial \tilde{s}^e/\partial t = 4\mu\left[\dot{V}V^{-1} - (1-\xi)^{-1}\dot{\xi}\right]/3. \tag{10.30}$$

Taking into account the pore dynamics and (10.22), we write for damaged material:

$$\partial s^e/\partial t = 2\mu\left[\dot{\varepsilon} - \dot{V}V^{-1}/3 - 2(1-\xi)^{-1}\dot{\xi}/3\right][1 - f(\xi)]. \tag{10.31}$$

2. *Accounting for plastic properties of the material.* The yield condition of von Mises can be used for the description of plastic properties of many metal materials (see (1.28) and (6.10)) [38]:

$$4\mu^2\left[\left(\dot{\varepsilon}-\frac{1}{3}\dot{V}V^{-1}\right)^2+\frac{2}{9}\dot{V}^2V^{-2}\right]=\frac{2}{3}\sigma_0^2. \tag{10.32}$$

The effects of microscopic damage of the material are described by the following expression (Section 2.4):

$$f(\xi)=\sqrt{\xi/\xi^*}. \tag{10.33}$$

In this case, (10.32) can be rewritten approximately in the following form [38]:

$$4\mu^2\left[\left(\dot{\varepsilon}-\frac{1}{3}\dot{V}V^{-1}\right)^2+\frac{2}{9}\dot{V}^2V^{-2}\right]=\frac{2}{3}\sigma_0^2[1-f(\xi)]. \tag{10.34}$$

Here, $d\varepsilon = d\varepsilon^e + d\varepsilon^p$ (10.26).

Melting material. Let us take into account the possibility of melting of materials. In this case, the system is closed by the equations of state (see Section 2.4):

$$p=p_\rho(\rho,\xi)[1-f(\xi)]+p_T, \quad E=E_\rho+E_T, \tag{10.35}$$

where

$$p_\rho=\sum_{k=1}^{3}A_k\left(\tilde{V}^{-1}-1\right)^k, \quad p_T=\gamma E_T V^{-1}, \tag{10.36}$$

$$E_\rho=\int_V^{\tilde{V}}p_\rho\,dV, \quad E_T=\int_{T_0}^{T}C_v(T)dT+E_0, \tag{10.37}$$

where p_ρ and E_ρ are nonlinear elastic components of the pressure and energy, respectively, p_T and E_T are thermal components of them, γ is the Grüneisen coefficient, and A_1, A_2, and A_3 are material constants. For an ideal gas $C_V = \frac{3}{2}R$ and a condensed state $C_V = 3R$.

The values of the coefficients of adiabatic curve (10.36) A_k are determined by expanding formulas for dynamic compression [39–41] into series in powers of $\tilde{V}^{-1}-1$:

$$p_\rho=-\rho_0\gamma_0^2(\tilde{V}-1)\left[\gamma_1(\tilde{V}-1)+1\right]^{-2} \tag{10.38}$$

and from the exponential relation between the velocity of the shock wave D and mass velocity beyond the shock wave D_1:

$$D=\gamma_0+\gamma_1 D_1. \tag{10.39}$$

As a result, the following are determined using (10.36):

$$A_1 = \rho_0 \gamma_0^2, \quad A_2 = \rho_0 \gamma_0^2 (2\gamma_1 - 1), \quad A_3 = \rho_0 \gamma_0^2 (\gamma_1 - 1)(3\gamma_1 - 1). \quad (10.40)$$

Boundary conditions. Boundary conditions are set to zero for stresses and heat influx due to thermal conductivity:

$$\sigma(0,t) = \sigma(H,t) = 0, \quad W(0,t) = W(H,t) = 0, \quad (10.41)$$

where H is the thickness of the loading plate.

Initial conditions. We assume that

$$v = \xi = 0, V = 1. \quad (10.42)$$

In this case, the temperature is given in the form of an initial distributed front:

$$T(z,t=0) = T_0(z,t=0). \quad (10.43)$$

The initial thermal stresses and the initial distribution of thermal energy arise in the material in this case.

It is emphasized again that we are considering in this section the case $Q = 0$ in (10.24). In this case, the system (10.22)–(10.42) describes the interaction of temperature, stress, and damage waves. The case when $Q \neq 0$ in (10.24) is discussed in Section 10.3.

10.2.2 Algorithm and Results

The essence of the above equations is that they describe a much wider range of phenomena than the thermoelastic approach presented in Section 10.1. However, one can study this range of phenomena only by using numerical methods.

The mathematical model presented above allows to take into account the influence of various mechanical and physical phenomena on the propagation of extreme waves. In the following calculations, all of them will not be simultaneously taken into account, since the goal of the study is to obtain an understanding of the influence of various factors on the formation of extreme waves.

Algorithm of calculation. Calculations begin with the satisfaction of the boundary conditions on the loaded surface [7,37]. In each numerical cell of the body, the equations of motion are solved, and thus, the velocities of cell boundaries are determined. The spherical part of the strain rate tensor (the rate of change volume) is calculated on the basis of the equation of continuity. The deviator components are found with the help of geometric relations. Now, the way to finding of the stress tensor opens. However, it is necessary first to solve the equation of microdamage growth, to take into account the effect of porosity on the stress state. If the value of the damage is larger than the critical level of microdamages ξ^*, the number of this cell is remembered.

Now, we can find the values of cold pressure p_ρ and deviator parts of stresses using the certain state equations. While considering the latter, various algorithms are implemented for cells of viscoelastic and viscoplastic layers. In the first case, the calculations are performed using an explicit numerical scheme. In the second case, an implicit iterative procedure is used to find variable radius of the fluidity circle after which the components of the deviator are determined.

The stress is the sum of the elastic pressure, thermal and electronic pressure, and deviator components of the stress tensor. Then, the energy of cold compression is calculated. The total energy is calculated as the sum of cold and thermal components. On the contact, boundaries of the layers satisfied conditions of ideal contact. Finally, on the free surface of a multilayer body boundary conditions for a free heat-insulated surface are calculated.

Some quantities, for example, temperature and the relative volume, are transferred from one phase to another. To reduce the error of this approximation, we used the integration of the task in general. As a rule, convergence is achieved by two iterations. The inertia of heat transfer requires a numerical solution of differential equations with a small parameter τ (10.25). It can significantly degrade the correctness of the numerical calculations; that is, in a number of calculations, three or four iterations were required at each step for the convergence of the calculations.

Thus, the numerical algorithm includes dynamical and thermal parts. The accuracy of the algorithms was verified by comparisons with known results. For the first part, we used the results described by M. Wilkins [38], and RI Nigmatulin and NA Akhmadeev [42,43] on the simulation of high-speed collision of plates without and with accounting for the destruction, respectively. The thermal part was checked by comparison with the analytical solution of the nonlinear heat equation (Chapter 9).

Results of calculations. In specific calculations, we studied a two-layer plate of materials A (iron) and B (aluminum). The mechanic and heat parameters for these materials are taken from [39–41,44]. The parameters of the kinetic equation (2. 27), defining the increase of the damage, are the same for materials A and B: $\xi^* = 0{,}075$, $\tau_k = 5 \times 10^{-8}$ second, $\sigma_t = 25 \times 10^8 \text{N/m}^2$, and $\sigma_k = 18 \times 10^8 \text{N/m}^2$ [42,43].

The absorption of energy and the release of heat were assumed to be instantaneous (see Eqs. (10.41)–(10.43)). The heating leads to the instant expansion of the near-surface material. Simultaneously with the compression wave propagating into the depth of the material, the material of loaded surface is stretched (the rarefaction wave), which leads to the damage of the material, to the external splitting of the material. With the advance into the interior of the body, the wave amplitude falls off lee nonlinear damping. As a result, the level of the damage decreases.

A slightly different picture can be observed in the case of a multilayered body. During the entrance of compression waves into a less dense medium is possible growth of the pores due to the reflection of the pulse from the boundary as the stretching wave. This fact was confirmed by the solution of the problem formulated above, and by analysis of its acoustic variant (the acoustic variant does not take into account the inverse influence of microdamages on the stress–strain state, $f(\xi) = 0$ (10.33)).

It was found that heat conduction and thermal coupling do not significantly change the wave amplitude. For example, Figure 10.11 shows the change in the form

Extreme Waves Excited by Radiation Impact

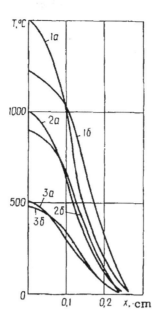

FIGURE 10.11 Motion of the temperature wave with different initial amplitudes: 1) $T = 500°C$, 2) $T = 1000°C$, and 3) $T = 1500°C$ (a and $б$ are the moments of time 0.40 and 0.54×10^{-6} second, respectively). Here, x corresponds z in Section 10.2.1.

of a thermal wave during the passage of a stress wave through the plate. It can be seen that this change is relatively small, but it occurs mainly at the moment of formation of the mechanical pulse and becomes more intensive with an increase in the curvature of the thermal front.

In Figures 10.12a and 10.13 for two moments of time are built the profiles of the stress waves σ and the damage ξ, arising when the body is loaded with a penetrating heat pulse calculated according to the full and acoustic approaches (notations in Figure 10.13 correspond to the notation of Figure 10.12). From this, we see that the acoustic approximation lowers the thickness of the external splitting off material. Faster development of damage (failure) when taking into account the interaction of pores and stresses leads to a loss of energy of the tension wave in the spalling material. Therefore, according to the full approach, the tension wave has a smaller amplitude than the acoustic approach determines. The interaction of pores and stress leads to the fact that the damage zone becomes narrowly localized.

The conclusion about insufficient accuracy of the acoustic approximation for the description the process of destruction is made from the analysis of curves shown in figures, which agrees with the data of other studies [39,40]. In the case of a two-layer material in addition to external splitting is possible internal splitting.

Separation of the heat pulse into two parts leads to deeper outside spalling than the one-time action of a pulse of double amplitude (Figure 10.14). This is explained by the rather slight dependence of spalling depth on load amplitude.

266   Extreme Waves and Shock-Excited Processes

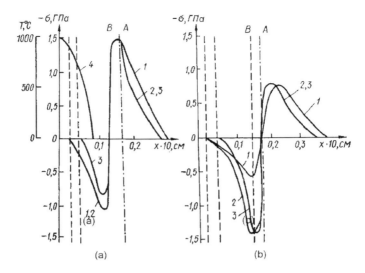

FIGURE 10.12 Initial temperature front and stress-wave propagation: 1) profile of stresses, acoustical approximation, single-layer plate; 2) acoustical approximation, multiple-layer plate; 3) nonacoustical approximation; and 4) initial temperature profile (a and b are the moments of time 0.40 and 0.54 × 10^{-6} second, respectively). Interrupt vertical lines are the spallation surfaces. Here, x corresponds z in Section 10.2.1.

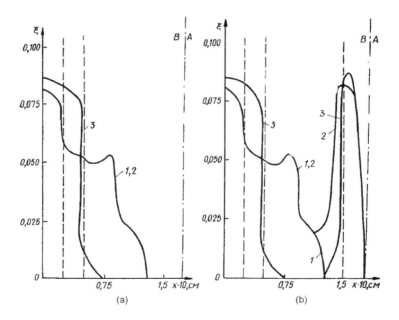

FIGURE 10.13 Propagation of microdamage wave: 1) damage content profile, acoustic approximation, single-layer plate; 2) acoustic approximation, multiple-layer plate; and 3) nonacoustical approximation (a and b are the moments of time 0.40 and 0.54 μs, respectively). Rupted vertical lines are the spallation surfaces (here x corresponds z).

Extreme Waves Excited by Radiation Impact

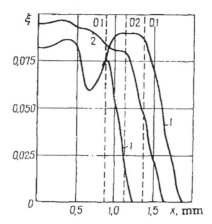

FIGURE 10.14 Increase in plate damage as a result of two successive thermal loadings (1) and with one loading of double amplitude (2). Rupted vertical lines are the spallation surfaces (*x* corresponds *z*).

10.3 MODELING OF FRACTURE, MELTING, VAPORIZATION, AND PHASE TRANSITION

So far we have considered waves of heat, stresses, and destruction arising from the action of relatively weak radiation pulses. Next, we take into account the possibility of the plasma appearance. In particular, two models are introduced for consideration: The first mode uses the wide-ranged equations of state (see Section 1.1), and the second model uses the separation of arising phases.

For the first (continuum) model, we continue to use the Eqs. (10.22)–(10.34). At the same time, the formulation of the problem changes because we replace the equations of state (10.35)–(10.37) by more general equations, which take into account the possibility of the material evaporation (the appearance of the plasma phase of material). When the second model is used, the surface of evaporation is determined at each step of the problem solution. The method of the previous part is used where material retains the strength characteristics. In the vapor zones, the equations of high-temperature gas are used. On the evaporation surface, the Knudsen conditions are written.

1. **The continuum model.** The possibility of the material evaporation is taken into account. Additionally, here we use the term $\partial Q/\partial z$ from (10.24). This term describes the radiation energy. The generalized Bouguer–Lambert law is written for this energy:

$$\partial Q/\partial z = -L(V,T)Q. \qquad (10.44)$$

Eq. (10.44) describes the dependence of the radiation attenuation coefficient $L(V,T)$ on the density and temperature of the material. If $L(V,T) = const$, then

$$Q(z,t) = Q_0(t)\exp(-Lz). \tag{10.45}$$

To close the above system of differential equations, we use wide-range equations of state (see Eqs. (1.10) and (2.50)) [44]:

$$p = p_\rho[1 - f(\xi)] + p_T + p_e, \; E = \tfrac{3}{2}\mu^{-1}RT_c\left(E_\rho + E_T + E_e\right). \tag{10.46}$$

The process of the transition of the solid to the liquid and gaseous states without explicit defining the phase boundary can be expressed using the above equations. Here, we assumed that the material is vaporized if the thermal component of pressure exceeds the greatest (negative) pressure of the cold component [44]:

$$p_{e,\min} = KA\frac{mn}{3(m-n)}\left[\left(\frac{n+3}{m+3}\right)^{\frac{n+3}{m-n}} - \left(\frac{n+3}{m+3}\right)^{\frac{m+3}{m-n}}\right]. \tag{10.47}$$

In this case,

$$\tilde{V}_{\min} = \left(\frac{n+3}{m+3}\right)^{\frac{3}{m-n}} = V_*. \tag{10.48}$$

For example, here $m = 6$, $n = 4$ for iron and aluminum. It is assumed that the substance is in the gaseous state in those regions, i.e., those finite-difference cells, in which $V > V_*$.

2. **The Knudsen model (conditions)**. When the second approach is used, the vaporization surface is identified in explicit form. Knudsen conditions are written on this surface, these being the boundary conditions of the problem. The part of the material, which has retained its strength properties, is described by (10.22)–(10.39), (10.44), and (10.45).

The motion of the gases (vaporization products) is described by equations that are valid for a high-temperature gas [18]:

$$\frac{\partial v}{\partial t} = -\frac{\partial p}{\partial z}, \; \frac{\partial}{\partial t}\left(\frac{1}{\rho}\right) = \frac{\partial v}{\partial z}, \; \frac{\partial E}{\partial t} = -p\frac{\partial v}{\partial z},$$
$$\frac{\partial z}{\partial t} = v, \; p = \rho RTN, \; E = RTN\left(C_p/C_V - 1\right)^{-1}. \tag{10.49}$$

where z is the Lagrangian coordinate, and N is the number of moles of gas.

The surface of contact of the solid and the gas is modeled by relations characterizing the kinetics of the phase transformation [45] (the Knudsen conditions):

$$\rho v = \rho_s\left(RT_s/2\pi\right)^{0.5} + \beta\rho(RT/2\pi)^{0.5}\left[\pi^{0.5}M\mathrm{erfc}(M) - \exp(-M^2)\right],$$

Extreme Waves Excited by Radiation Impact

$$\rho(v^2 + RT) = \tfrac{1}{2}\rho_s RT_s + \beta\rho RT\left[\left(M^2 + \tfrac{1}{2}\right)erfc(M) - M\pi^{-0.5}\exp(-M^2)\right],$$

$$\tfrac{1}{2}\rho v(5RT + v^2) = \rho_s R\left(RT_s/2\pi\right)^{0.5}$$

$$\times\left[2T_s + \tfrac{1}{2}(5 - 3C_p/C_v)(C_p/C_v - 1)(T_s - T)\right] + \beta\rho RT(RT/2\pi)^{0.5}$$

$$\times\left[\pi^{0.5}M\left(M^2 + \tfrac{5}{2}\right)erfc(M) - \left(M^2 + 2\right)\exp(-M^2)\right].$$

(10.50)

where $M = v(2RT)^{0.5}$, and *erfc* is the complementary error function. The quantities without indices correspond to the thermodynamic parameters for the gas at the discontinuity; the subscript s denotes the values of these quantities on the surface of the solid. It should be noted that the formulas from (10.50) for $C_p/C_v = 5/3$ are identical to those used in Ref. [46]. Thus, they constitute a special case of the equations for the Knudsen discontinuity [45]. Eqs. (10.50) correspond to the Hugoniot relations for a shock-wave discontinuity except for the fact that these equations presume a two-velocity gas model. These equations are analogous to the Hugoniot relations for an ideal gas when $C_p/C_v = 5/3$.

The presence of a nonequilibrium Knudsen layer is due to the flow of particle returned to the surface of discontinuity from the gaseous medium. The magnitude of this reverse flow depends on the conditions of gasdynamic dispersal, characterized by the Mach number (M). The latter determines the quantity of mass vaporized and the rebound pressure [47]. The maximum value of M is unity. Here, the dynamics of processes taking place in the solid and gas can be examined independently of one another.

The Knudsen model, in accordance with [45], is augmented by the equation of energy balance at the discontinuity:

$$-\kappa(T_s)\partial T/\partial z + \rho_l vL = q(t), \qquad (10.51)$$

and the mass conservation law is

$$\rho_l \dot{z} = \rho v, \qquad (10.52)$$

where ρ_l is the density of the substance ahead of the discontinuity, L is the specific heat of sublimation, and $\dot{z}$ is the velocity of the vaporization front. Eq. (10.51) describes the action of the external load, which is characterized by the specific power density associated with the incoming energy $q(t)$.

Thus, at $M = 1$ the second approach reduces to the solution of Eqs. (10.22)–(10.39) with the boundary conditions (10.50)–(10.52).

Boundary conditions [1,6,7]. Boundary conditions are set to zero for stresses and heat influx due to thermal conductivity:

$$\sigma(0,t) = \sigma(H,t) = 0, \; W(0,t) = W(H,t) = 0, \qquad (10.53)$$

where H is thickness of the loading plate.

1. **Distributed load**. On the loading surface,

$$Q(0,t) = Q_0(0,t). \tag{10.54}$$

2. **Instant load**. In this case, $Q_0(0,t) = 0$ in (10.53).
   Ideal contact conditions are assumed for a multilayer material on the surfaces of contacts of layers.

Initial conditions [1,6,7]. We assume that

$$v = \xi = 0, V = 1. \tag{10.55}$$

1. **Distributed load**. In this case, all unknown values are zero at $t = 0$.
2. **Instant load**. In this case, the temperature is given in the form of an initial distributed front:

$$T(z,t = 0) = T_0(z,t = 0). \tag{10.56}$$

The initial thermal stresses and the initial distribution of thermal energy arise in the material in this case. The energy flow $Q(z,t = 0) = 0$.

10.3.1 Calculations: Effects of Temperature

The complexity in studying the action of the energy pulse on solids requires special effort to obtain results of calculations. Ways to verify them experimentally are limited due to extreme conditions that arise in small body volumes during very small time intervals. On the other hand, numerical experiments with models make sense only if the models are reliable. The reliability of calculations increases if various mathematical models of the process under investigation and various numerical schemes of analysis are used [1,7].

Formation of waves near a loaded surface. We shall study the effect of the duration of the action of radiation on the formation of a stress wave near the loaded surface. The value of such study is that it becomes possible to compare numerical and analytical results. Here, the results from Ref. [7] are presented.

We fix the energy of the pulse:

$$\int_0^{\Delta t} Q(t)dt = const = \Delta E. \tag{10.57}$$

The stress profiles calculated for two different load durations ($\Delta t = 10^{-8}$ и 10^{-6} seconds) and the time $t = 10^{-6}$ second are shown in Figure 10.15.

There is a strong difference between the calculated curves 1 and 2, although a qualitative similarity of the curves is observed also. It is interesting that according to curve 2, there is an unstressed zone between compression and expansion pulses. This effect seems paradoxical at first glance. It is not easy to comprehend it on the basis of

Extreme Waves Excited by Radiation Impact

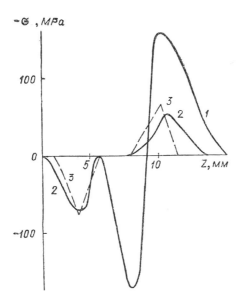

FIGURE 10.15 The stress profiles calculated for two durations of the load: $\Delta t = 10^{-8}$ second (1); $\Delta t = 10^{-6}$ second (2). Rupted curve 3 is analytic solution.

numerical calculations. At the same time, this feature also occurs in the linear case, which can be analytically investigated as discussed in Section 10.1.4.

To explain the results, we use the wave equation for the solution which is applicable to the case considered. Let us consider the following:

$$\partial^2 \sigma/\partial t^2 = c^2 \partial^2 \sigma/\partial z^2 + AQ(t,z) \ z \geq 0. \tag{10.58}$$

Initial and boundary conditions are

$$\sigma(0,z) = 0, \quad \partial \sigma(0,z)/\partial t = 0, \quad \partial \sigma(t,z=0)/\partial t = 0. \tag{10.59}$$

where $Q(t,z)$ is determined in (10.45). Further, we use the fact that the sum of two solutions of a linear homogeneous differential equation is also a solution. In addition, the general solution of the inhomogeneous equation is the sum of the particular solution of the inhomogeneous equation and the general solution of the homogeneous equation. Considering this, we study the initial stress pulse inside an elastic material ($t = 0$). Then, it breaks up into two pulses moving in different directions. Having reached the free boundary, one pulse is reflected by changing the sign.

According to the law (10.45), the initial temperature front is exponential. We replace the exponent with a triangle for simplicity. The dynamic of such impulse near the surface of the material is presented in Figure 10.16. It can be seen that this dynamics leads to the appearance of a wave that has areas of compression and tension.

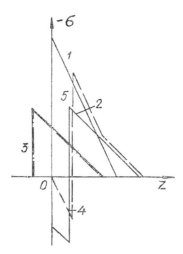

FIGURE 10.16 Scheme of evolution of the stress wave near free boundary: initial wave (1); a part of the wave moving to the right (2); a part of the wave moving to the left (3); the curve (4) is the profile of the wave (3) the partly reflected from the boundary; the curve (5) is the superposition of the waves (4) and (2).

It becomes more complicated when the energy loading is not instantaneous. In this case, the waveform emerging in the material can be more complicated than that shown in Figure 10.16. Let us consider the case of triangular heat release in a volume of a material. The profiles of the resulting stress waves depend on the duration. With instantaneous heat generation, the wave profile shown in Figure 10.17 (curve 1) is obtained (see, also, Figures 10.10 and 10.16). In the case of larger duration, a stress-free zone between the compression and stretching zones (curve 2) begins to form. With a further increase in the duration (curve 3), a stress-free zone arises similar to that shown in Figure 10.15.

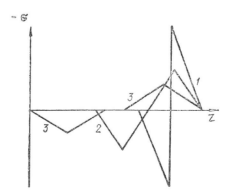

FIGURE 10.17 Three wave profiles calculated for different durations of radiation energy.

Interaction of heat waves and stresses. This issue was discussed already in Section 10.1. The results of analytical studies were used there. Here, we present the results of numerical studies presented in Ref. [7].

As is known (see Chapter 9), heat waves (heat pulses with pronounced steep front and rear fronts) can form when the heat is propagated in some absorbing materials. In particular, thermal wave amplification is possible in thin-film multilayer structure, if the thermal impedance of the next layer is greater.

In Figure 10.18 are presented the results of calculations for four cases. Calculations show a fairly significant decrease in the tension stresses (about 20%) due to thermal conductivity in a single-layer material. However, in the two-layer material, the effect can be reversed due to the thermal properties of the second material (for the effect of amplification of thermal waves, see Chapter 9).

Remark. However, it should be noted that the effect of thermal dynamics on the stress–strain state is rather weak and can be manifested only for small material thicknesses and for energy loading, which can be considered almost instantaneous.

10.3.2 Calculations: Effects of Vaporization

In realizing the equations of the first (continuum) model, we used a numerical scheme described briefly in Section 10.2 [1,6,35]. For the second (Knudsen) model, the equations of the solid were integrated in accordance with the Wilkins scheme [38], while models from (10.49), (10.50), (10.51), and (10.52) were satisfied by means of the Newton–Raphson iterative procedure.

Simple test of temperature. We performed simple calculations in order to compare the above-described algorithms in the solution of problems for a half-space filled with molten aluminum. An energy flux described by a step function and having

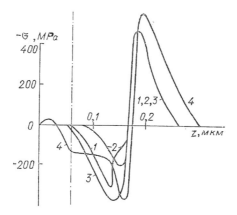

FIGURE 10.18 Effect of heat wave and material properties on the stress waves: single-layer non-heat-conducting material (1); single-layer heat-conducting material (2); two-layer, heat-conducting material, where the layers have the same mechanical and different thermal properties (3); and two-layer, heat-conducting material, where the layers have the same mechanical and thermal properties (4).

a duration of 10^{-6} second crosses the boundary of the half-space. The dynamics of dimensionless temperature $T^* = T_s/T_{vap}$ are illustrated in Figure 10.19, where curve 1 corresponds to the calculation in Ref. [45] based on the heat-conduction equation and Eqs. (10.50), (10.51), and (10.52). Curve 2 was obtained by solving Eqs. (10.22)–(10.34), (10.49), and (10.50) with known parameters from Ref. [45] and $\xi = 0$. Curve 3 shows the change in the Mach number over time and the presence of critical and subcritical flows in the Knudsen layer.

The curves obtained for temperature correspond to a point lying on the surface. The closeness of curves 1 and 2 indicates that the chosen model of the process and the algorithm used are sufficiently accurate. The general shape of the curves of T^* can be interpreted as follows. Temperature initially rises rapidly due to the supply of energy. However, this increase becomes slightly slow due to the effect of heat-conduction processes taking heat deeper into the material. The quantity of heat moved to the interior gradually becomes comparable to the amount of heat supplied externally, causing the temperature curve to reach a steady level. When the external supply of energy is cut off, the surface of the body abruptly cools. The surface then cools more gradually in connection with the equalization of temperature throughout the body.

Modeling of vaporization and fracture. To perform and compare the results of calculations by the two models, it is necessary to determine the location of the vaporization zone in the material.

As it was noted, in the case of the first model we assume that the material is vaporized if the sum of the thermal and electronic energies in it exceeds the maximum amount of elastic tension energy. Then, the elastic energy cannot resist further expansion of the body, and the substance is assumed to undergo transformation to the gaseous state.

In the case of the second model, the calculations were performed in accordance with the assumption that M = 1. Thus, since the gasdynamic equations were not used, the rate of displacement of the vaporization surface could be determined from Eq. (10.52) after the simultaneous solution of (10.50) and (10.51).

The second important aspect of the numerical calculations was connected with allowance for the permeability of the vaporization products relative to the radiant flux. In the case of the first model, we assumed that the gas is opaque at $V < V_0$ and that it becomes instantaneously transparent at $V \geq V_0$ (V_0 is the critical volume of clarification of the products). In the case of the second model, the question of clarification of the vaporization products is irrelevant; that is, the gas beyond the Knudsen layer is assumed to be transparent.

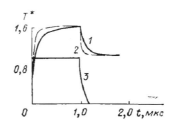

FIGURE 10.19 Temperature at the solid–gas boundary.

Extreme Waves Excited by Radiation Impact

Figure 10.20a shows the dependence of the mean rate of displacement of the vaporization front $\dot{z}$ on the energy flux q. The flow of energy starts at $t = 0$ and subsequently remained constant. The vaporization rate was determined after the vaporization process reached the steady state. The curve 1 is calculated according to model 2. The curves 2–4 are calculated according to model 2 for $V_0 = 1.1$ (curve 2), $V_0 = 1.2$ (curve 3), and $V_0 = 1.3$ (curve 4).

The curves show that there is a nearly linear relationship between the logarithm of the energy flux and the vaporization rate. The curves also show that there is a range of loading rates within which a correlation exists between data on vaporization thickness obtained by the two models. The best agreements between the calculated results are seen at $V_0 = 1.2$. With an increase in loading rate, the algorithm employing the Knudsen equations gives results which are somewhat higher than those obtained by the first model.

Thus, the data obtained according to the two mathematical models are in satisfactory agreement. At the same time, it seems to us that the first model is more general than the second, since it does not use the assumptions that $M = 1$ (concerning the velocity of the gas flow beyond the vaporization zone). The most of the results presented below were obtained on the basis of the first model.

Having examined the vaporization of near-surface layers of a material in the steady-state regime, we turn to the study of essentially nonsteady sublimation. It is reasonable to compare different loadings in this case if we assume a constant value of $E_* = q\tau$ – the total energy of the pulse transmitted through a unit area of the surface. In our calculations, we took $E_* = \text{const.} = 0.04$ J/cm^2.

Figure 10.20b shows the change in the mean vaporization rate calculated only for the time of action of the pulse. The curve 1 is calculated according to model 2, but curves 2 and 3 are calculated according to model 2 for $V_0 = 1.3$ and 1.2, correspondingly. An increase in loading time is initially accompanied by a sharp drop in

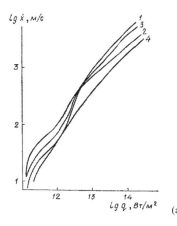
(a)

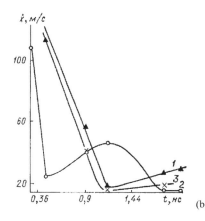
(b)

FIGURE 10.20 Dependences of the mean rate of displacement of the vaporization front $\dot{z}$ on the energy flux q (a) and on the time of action of the pulse (b) (in the figure BT is a watt (W) and $\dot{x} = \dot{z}$).

vaporization rate. This rate then stabilizes, which is interpreted as a transition from instantaneous vaporization to steady-state vaporization.

Another illustration of the above results is Figure 10.21a, which shows the dependence of the thickness of the vaporized material on the duration of the substantially nonlinear pulse. At $\tau = \Delta t = 10^{-9}$ second, there is a local minimum on the graph. Variation of the quantity V_0 shows that the value of this extreme changes but it does not disappear. In certain sense, it is independent of the clarification threshold of the material during expansion. Such complex behavior of vaporization can be explained on the basis of the competition between the mechanisms of propagation of energy into the depth of the material and the clarification of its vapors. Given sufficiently short loadings, the energy absorption (stress waves, heat condition) is absent. This leads to the formation of horizontal sections on the curves (Figure 10.21a).

With an increase in loading time, its duration approaches to the characteristic time of dynamic processes and becomes sufficient for the removal of energy from the absorbing cells due to the escape of the stress wave into the depth of the material. At the same time, the boundary cells expand. However, this expansion is insufficient for their clarification and, thus, for an increase in the number of absorbing cells. The flux of energy from the boundary region during the loading process leads to less intensive heating and a reduction in the volume of vaporized matter. A subsequent increase in the loading time results in the clarification effect. The radiation penetrates deeper into the matter and the volume of vaporized matter increases, as is evidenced by the essentially nonlinear character of the change in the curves shown in Figure 10.21a.

Figure 10.21b shows the dependence of the stresses on time for two points of the solid. During the calculations, the different models of modeling of the vaporization process and different volumes of the critical clarification are used. The letter A corresponds to a cell adjacent to the vaporization surface, whereas the letter B corresponds to the tenth cell beyond the vaporization front. It should be noted that while curves 2–4 of series A approach curve 1 with a reduction in V_0, curves 1–4 of series B correlate only at $t \leq 9 \times 10^{-11}$ second and subsequently differ significantly. This difference is minimal at $V_0 = 1.2$. At the same time, the agreement of the stresses

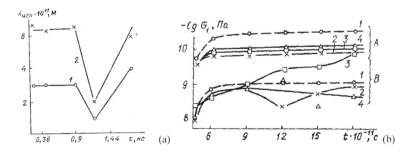

FIGURE 10.21 Dependence of the thickness of the vaporized material (L_{ucp}) on the duration of the pulse: 1) $V_0 = 1.3$; 2) $V_0 = 1.2$ (a). The stress evolution at points A and B of the solid: 1) Knudsen model, 2) $V_0 = 1.5$, 3) $V_0 = 1.3$, and 4) $V_0 = 1.2$(b).

Extreme Waves Excited by Radiation Impact

cannot be considered good. Even in the best case, the stresses calculated by the second model are nearly twice as great as the stresses calculated by the first model.

10.3.3 Calculations: Effect of Vaporization on Spalling

The reduction in the volume of the solid body is associated with both external spalling and evaporation.

Each of the two specified mechanisms predominates in different irradiation ranges; that is, with the growth of the radiation, the splitting in tension waves is replaced by an impulse evaporation of near-surface material. Let us study more attentively the evaporation effect.

As shown earlier, if the radiation energy is not very large, then due to the thermal expansion of the surface material, it can be destroyed by spallation. If the radiation energy is sufficient to evaporate the material, then the pressure of the resulting vapors will begin to impede the spallation, up to its complete disappearance. Of course, the above is only a rough outline. We will discuss the problem in more detail below.

Calculations were carried out for a single-layer plate of material A (iron) and a two-layer plate of materials A and B (aluminum). For calculations, it is necessary to specify the constants and functions included in the Eqs. (1.10) and (1.15)–(1.22). Parameters of the Eq. (10.27) determining the kinetic of the damages were for both materials the same: $\xi^* = 0.075$, $\tau_\xi = 5 \times 10^{-8}$ second, $\sigma_1 = 25 \times 10^8$ H/m², and $\sigma_{21} = 18 \times 10^8$ H/m² (see Section 10.2.2).

We now present the results of calculations performed with allowance for all basic thermomechanical processes occurring under a pulsed loading of the surface by a flow of energy. The attenuation coefficient of the radiation L in (10.44) is assumed to be equal to 5.5×10^{-7} m⁻¹. The loading was considered distributed. The energy flux Q_0 in (10.54) varied by step law. The time of action of Q_0 was 5×10^{-10} second, and the amplitude was changing. The cases $Q_0 = 1.2 \times 10^6$, 12×10^6, 36×10^6, and 12×10^7 W/cm² were considered. In these cases, the temperature reached on the surface of values 260°C, 2700°C, 8000°C, and 25,000°C.

Figure 10.22a–d shows the profiles of stress waves arising in a two-layer body under the specified loading cases. The numbers 1–3 are highlighted profiles at times $t = 0.5 \times 10^{-10}$ (1), 2.5×10^{-10} (2), and 4×10^{-10} seconds (3). The dot-dashed line defines the interface of the layers. The rupted lines are spalling surfaces. It is possible to trace the transition from a two-wave pattern (compression–rarefaction) to the compression wave. Changing of the destruction mechanism is clearly visible. Then, the destruction as a result of external splitting, characteristic of low radiation intensities, is replaced by evaporation. In the latter case (Figure 10.22d), the compression wave decays to the safe value from the point of view of the spalling strength at the depth of 10^{-5} m. It cannot cause the internal spalling. The regions of the evaporated $(z_{ev} = x_{ev})$ and the spalled $(z_{sp} = x_{sp})$ material are compared. In the figure, four cases are listed: a) $z_{ev} = 0$, $z_{sp} = 0$, b) $z_{ev} = 3 \times 10^{-8}$ m, $z_{sp} = 2.7 \times 10^{-7}$ m, c) $z_{ev} = 9 \times 10^{-8}$ m, $z_{sp} = 2.7 \times 10^{-7}$ m, d) $z_{ev} = 2.4 \cdot 10^{-7}$ m, and $z_{sp} = 0$.

Change of spalling into evaporation can be traced clearly. At the same time, comparing the results calculated on variants b and d in Figure 10.22, we conclude that

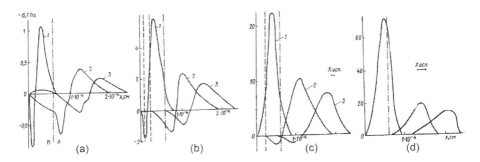

FIGURE 10.22 Stress profiles in two-layer material calculated for the instants of time $t = 0.5 \times 10^{-10}$ (1), 2.5×10^{-10} (2), 4×10^{-10} seconds (3), and four cases of thermal loading with an amplitude $T = 260°C$ (a), $2700°C$ (b), $8000°C$ (c), and $25{,}000°C$ (d) (here $x_{ucp} = z_{ev}$ and x corresponds z).

the region of destruction does not necessarily increase with increasing the radiation amplitude. It is connected with the fact that the destruction of the material by splitting requires much less energy than its evaporation. Only beginning with enough high temperatures, the destruction by evaporation approaches to the efficiency to spalling, which took place at more low temperatures.

As is well known, two different mechanisms form waves during impact radiation. The first is thermal. The radiation energy transforms into oscillations of the atoms in crystal lattice. When the energy increases, the atoms begin to exert each other a stronger impact, resulting in the sequence waves of compression and tension. This mechanism works mainly at low temperatures (Figure 10.22a). With increasing T up to the evaporation temperature of the material, another mechanism is associated with the impact on the surface high-temperature vapors erupted of the material surface. When the energy increases, the reaction of the vapors increases and the second mechanism suppresses the first. It can clearly be seen from Figure 10.22b–d. The rarefaction wave associated with the thermal mechanism is increasingly suppressed by the products of the evaporation. Process of the destruction in the tension wave is slowed down (see Figure 10.22b and c) and completely terminates at $T = 25{,}000°C$. Thus, the increasing of impact radiation (temperature) does not necessarily lead to an increase of the zones of the destruction.

Conclusions. Thus, it was shown on the basis of the unidimensional approach that the results of calculations based on the two different models and the numerical algorithms are in satisfactory agreement. This provides a hope for a certain degree of objectivity in the calculated results. Thus, it shows, based on numerical experiments, the possibility of amplification of thermal waves in multilayer structures. The influence of the change in the time of action of the heat pulse from 10^{-6} to 10^{-9} ms on the amplitude of stress waves was studied. A significant influence of the process of evaporation of surface material on the dynamic destruction of near-surface layers is demonstrated as well. These results can be verified experimentally, which would allow us to judge the correctness of theoretical models that are difficult to verify directly.

Extreme Waves Excited by Radiation Impact

Specifically, for example, it was objectively shown that vaporization of a substance with an impulsive energy input is basically a nonlinear process during the initial moments of time. Finally, despite the fact that the results calculated from the above two models are largely identical, the first model is more algorithmic and does not require additional restrictive conditions; therefore, it is more applicable.

10.4 TWO-DIMENSIONAL FRACTURE AND EVAPORATION

The destruction of a solid acted upon by an energy pulse was examined in [6,32] using a two-dimensional formulation. Here, we focused on studying the effect of the depth of energy absorption ($1/L$) on the character of the destruction. We will use an approach based on wide-range state equations of the material.

A circular aluminum plate of thickness h and radius R is fixed at its edge. Intensive energy release takes place within the plate in accordance with the following law:

$$Q = LQ_0 \exp(-r^2 r_0^{-2})\exp(-Lz)[H(t) - H(t+\tau)], \tag{10.60}$$

where Q_0 is the heating rate, r_0 is the radius of the characteristic loading area, and H is the Heaviside function. In the general case, L depends on the density and temperature of the material. The time of energy input τ is assumed to be small enough so that the processes of heat conduction and expansion of the cells cannot influence the energy absorption.

We used relations which serve as the basis of the Wilkins numerical algorithm [38]. We present the equations of continuity and motion:

$$\dot{V}/V = \dot{e}_{zz} + \dot{e}_{rr} + \dot{e}_{\varphi\varphi}, \tag{10.61}$$

$$\rho \frac{\partial v_z}{\partial t} = \frac{\partial \sigma_z}{\partial z} + \frac{\partial \tau_{zr}}{\partial r} + \frac{\tau_{zr}}{r}, \tag{10.62}$$

$$\rho \frac{\partial v_r}{\partial t} = \frac{\partial \tau_{zr}}{\partial z} + \frac{\partial \sigma_r}{\partial r} + \frac{\sigma_r - \sigma_\phi}{r}. \tag{10.63}$$

The geometrical relations are

$$\dot{e}_{zz} = \partial v_z/\partial z, \ \dot{e}_{rr} = \partial v_r/\partial r, \ \dot{e}_{\varphi\varphi} = v_r/r, \ \dot{e}_{zr} = (\partial v_z/\partial r + \partial v_r/\partial z)/2. \tag{10.64}$$

The governing relations are

$$\sigma_i = -p - \omega + s_i, \ \dot{s}_i = 2\mu(\dot{e}_{ii} - \dot{V}/3V), \ \dot{\tau}_{zr} = 2\mu\dot{e}_{zr}. \tag{10.65}$$

Here, $i = z$, r and ϕ. The von Mises ideal plasticity condition is

$$s_z^2 + s_r^2 + s_\varphi^2 + 2s_{zr}^2 \leq 2\sigma_0^2/3. \tag{10.66}$$

The above system of equations is closed by the state equations:

$$p = p_\rho + p_T, \quad p_\rho = KA\frac{mn}{3(m-n)}\left(V^{-m/3-1} - V^{-n/3-1}\right), \quad p_T = \gamma Q \tau V^{-1}. \quad (10.67)$$

Thus, the expression for p_ρ corresponds to (1.15), whereas that for p_T corresponds to (1.16) (see, also, Eq. (10.36)). The temperature impact p_T is determined using (10.60) and τ. The formulation of the problem must be augmented by certain fracture criterion. Here, we used the simplest criterion. We assumed that if the maximum normal stress in a numerical cell of the matter exceeds a certain constant value σ_{cr}, which represents the strength of the matter at rupture, then a crack which is perpendicular to the direction of this normal stress develops instantaneously in the given cell.

The algorithm for solving the above-formulated problem is similar to that described in detail in Ref. [1,7,32]. Calculations were performed with the following parameters: $R = 0.001$ m, $h = 0.0005$ m, and $r_0 = 0.0006$ m.

We examined three cases of loading: a) energy absorption occurs over roughly 1/3 of the plate thickness, b) over 1/2, and c) over 2/3.

In the first case (Figure 10.23a), it can be assumed that the energy absorption occurs near the surface. The high-temperature vapors, that are formed, form of a pressure pulse acting on the unvaporized material. The compression wave which is formed is reflected by the tensile pulse when it reaches the free surface. A disk-shaped fracture region is formed near the surface. With an increase in the depth of heating of the material (Figure 10.23b), a relatively long compression wave is formed, and the fracture is not formed explicitly. The calculations show that the volume of solid material adjacent to the free surface acquires a velocity that is nearly constant through its thickness. The vapors dislodge this material from the surrounding material, so to speak.

Further increase of the thickness of the absorbing layer leads to a situation, in which the mass of heated substance over the thickness of the plate begins to exceed the mass of unheated material. At the same time, the quantity of vaporized substance does not increase as in the preceding case (Figure 10.23b). The heated cells begin to exert pressure against the surrounding cells and fracture them. This ultimately results in through fracture of the plate (Figure 10.23c).

Figure 10.24 shows the dependence of the processes of the expansion of the evaporated matter and plate destruction (cells noted by points) from L: $L = 8000$ m^{-1} (a), $L = 4000$ m^{-1} (b), $L = 2000$ m^{-1} (c and d), and instants of time 0.116×10^{-6}, 0.118×10^{-6}, 0.0605×10^{-6}, and 0.155×10^{-6} seconds, respectively.

In the case a, the energy release can be considered as a near-surface one. The expansion of the high-temperature vapor leads to the formation of a pressure pulse on the condensed material. The resulting pressure waves, when reflected on a free surface, are reflected by a tensile pulse.

Near the surface appears disc-shaped area of destruction. The rates of expansion of evaporated matter and solid cells of the rear side of the plate in the considered cases reach 14,000 and 1600 m/s, respectively.

When the depth of the material's heating is increased (see Figure 10.24b), a longer wave of pressure is formed. In this case, the destruction is not connected with the spalling away from the solid material of the plate. Cells of the rear surface and the zone of intensive heating acquire, according to the calculations, almost the

Extreme Waves Excited by Radiation Impact 281

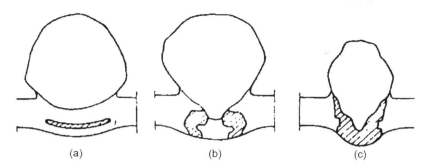

FIGURE 10.23 Zones of vaporization (points) and fracture (hatched region) of a plate after completion of damage development: (a) $L = 8000\,\text{m}^{-1}$, $t = 0.13 \times 10^{-6}$ second; (b) $L = 4000\,\text{m}^{-1}$, $t = 0.18 \times 10^{-6}$ second; and (c) $L = 2000\,\text{m}^{-1}$, $t = 0.22 \times 10^{-6}$ second (from the left to the right).

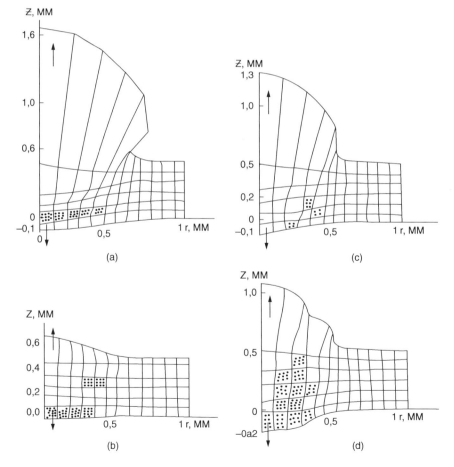

FIGURE 10.24 (a–d) Dependence of the dynamics of expansion and destruction of cells of matter on the depth of energy absorption.

same speed. As a result, the pairs seem to knock out the loaded cells from the surrounding material. The velocities of the cells indicated in Figure 10.24b are equal to 9500 and 1400 m/s. Further growth of the absorption layer leads to the fact that in thickness, the mass of heated material begins to exceed the mass of the unheated material. Together with this, the amount of evaporated substance compared to the previous case does not increase, and the temperature the pair does. Heated cells start to press on the surrounding cells and destroy them (see Figure 10.24c).

Over time, the number of the destroyed cells increases, and we have the full local destruction of the plate (destruction of all thickness of the plate). Note that the speed of the vapor cells in the case of $L = 2000 \, m^{-1}$ increased from 3400 m/s (B) up to 5500 m/s (г). At the same time, the speed of the rear side of the plate practically did not change and was equal, i.e., approximately 1500 m/s. Thus, an increase in absorbing capacity of the energy of a substance causes a drop in speed expiration of vapor and leads to a change in the nature of destruction with a slight change in the speed of motion of the rear surface of the plate.

Experimental data. In general, the described destruction is similar to the destructions of the plates by various strikers (impactors), carrying the same energy. With increasing of the striker mass is broken mechanisms of knocking out the cork and crushing of the material. Thus, the action of the radiation energy can be simulated by high-speed striker impact.

10.5 FRACTURE OF SOLID BY RADIATION PULSES AS A METHOD OF ENSURING SAFETY IN SPACE

To determine the optimum conditions of fracturing or altering the trajectory of dangerous space objects (DSOs) (large iron or rock space bodies, threatening to collide with the Earth), and also destroy space debris in the space around the Earth, we present a mathematical formulation for calculating the dynamic strength of solids under the action of high-energy loading pulses. The results of numerical modeling are compared with the experimental cupola-shaped spalling areas on the rear surface of a metallic plate subjected to laser radiation. The change in the fracture mechanisms (front and rear spallation) with increasing energy of the action has been detected. Recommendations are given for optimizing the pulsed laser action on DSOs to ensure the fragmentation of them or changing of their orbits.

10.5.1 INTRODUCTION

High-energy radiation sources have made it possible to develop new effective methods of actions on materials to change their properties and aggregate states [21,23,24,29]. The most accessible sources of intensive coherent electromagnetic irradiation are lasers. The continuous improvement of the technology of pulsed lasers is their increased power [46]. Now, they may be used in new unconventional areas. In particular, it is possible to examine the mechanical properties of various matter and materials under the conditions of energy, pressure, and loading rates which have been previously inaccessible for examinations [48–52].

At the same time, these properties are absolutely essential in developing the new technologies associated with the treatment of materials under the extreme temperature and force conditions. It is very important to take into account the strength properties of materials especially for the extreme radiation pulses. In these cases, the energy absorption by the material and very rapid evaporation of it may lead to the formation of stress waves whose amplitude is sufficient for fracture.

As an important example of an unconventional use, one can mention the possibility of utilizing high-energy lasers for evaluating the resistance of meteorite screens of space systems [53].

One of the most promising areas of application of laser emitters is space ecology. It is well known that a planet may collide with DSOs which include asteroids and large comets [54]. Regardless of the low probability of such an event, the potential damage caused by it in the case of the Earth is so large that preventive protective measures cannot be regarded as unnecessary.

Using for the protection a rocket carrying a powerful hydrogen bomb and its explosion near the DSO is a very complicated problem. In addition to this, it requires considerable investment. To rectify these problems, an important alternative is to use remote emitters which produce high-energy radiation of various natures. It should be taken into account that the attenuation of pulses on space distances leads to the situation in which the most obvious method – the variation of the trajectory of the DSO by generating the reactive jet as a result of vaporization of the DSOs – is possible only on small space distances from the Earth.

There is a possibility of efficiently using the laser-pulsed radiation for such space distances where the energy of the pulse is still insufficient to the vaporization of the material but is already sufficient for fracture of it within surface rarefaction wave. This wave is formed as a result of unloading the almost instantaneously heated material toward the side of the source of pulse radiation. With other conditions being equal, this effect may be applied over considerably longer distances than the effect whose task is to vaporize the DSOs. This greatly simplifies the method of achieving the given goal – to deflect sufficiently the DSO from its initial orbit along which the collision with the Earth may occur (Figure 10.25).

While solving the problems of laser thermonuclear synthesis, it has been indicated that it is possible to select the optimum nontrivial profile of the radiation power causing the maximum utilization of the pulse energy for accelerating the particle of the material medium [53]. It is assumed that this method is carried out not by a single but by a series of consecutive pulses graded with respect to duration and amplitude. However, here we examine the effect of only single pulse.

Liquidation of "space debris" – antropogeneous objects traveling on the orbits close to the Earth – is another problem of space ecology. Powerful lasers can be used for this problem. The importance of the problem was indicated by cases of collision of these objects with the space stations and also with other artificial satellites of the Earth. As there are vast quantities of the space debris, they cannot be removed by direct collection. The only realistic method of liquidating them is long-range vaporization or reduction to safe dimensions followed by combustion in the atmosphere. However, there is a risk that the degree of fragmentation of the individual objects will be insufficient and this will only increase the total number of the

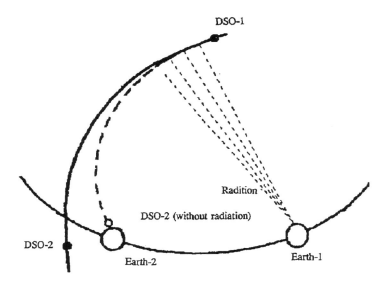

FIGURE 10.25 Diagram of deflection of a DSO as a result of series of radiation pulses (dashed line is a trajectory of the DSO in the absence of radiation up to the collision. Numbers 1 and 2 denote the positions of space bodies at different moments).

dangerous objects. The selection of the efficient and, at the same time, economical method of affecting on these satellites of the Earth requires careful analysis of fracture of certain solid structures by radiation pulses.

Here, a mathematical formulation of the problem of evaporation and fracture of condensed elastoplastic medium loaded with a radiation pulse is presented. Wide-range equations of state are used for a description of the thermodynamic behavior of matter in different aggregate states including melting, evaporation, and plasma formation. In addition, the model describes the dynamic failure of the material in tension waves by the way of the formation of microcracks.

The examined problems in the two-dimensional axisymmetric formulation were solved numerically using an explicit Lagrange algorithm. Comparison of the calculated results with the data obtained in experiments made it possible to verify the numerical method used. As a result of automatic rearrangement of the numerical mesh all types of fracture formed in dynamics could be examined.

As the presented investigations continue, we paid particular attention to the special features of the fracture of solids by external spallation. This type of fracture is regarded as a method which, if it will be sufficiently developed, could become a basis of a technology for the space protection of the Earth with the minimum losses. At the same time, a large number of calculations have been made to examine the rear spallation (free surface fracture). This phenomenon results naturally from presented above investigations and the general formulation of the problem. In particular, this may be important for solving the problem of liquidating of the space debris.

10.5.2 MATHEMATICAL FORMULATION OF THE PROBLEM

The mathematical model of the dynamic behavior of a solid in two-dimensional axisymmetrical formulation is constructed on the basis of a system of equations written in general Lagrange formulation [55–60]. The use of three cylindrical coordinate systems is assumed: Euler (or spatial) with the actual configuration of the examined solids, initial Lagrangian (material) which coincides at the initial moment with the coordinates of Euler, and finally, computed Lagrangian system (also material) in which the regular initial calculation configuration of the examined solid, obtained from its real initial configuration by a means of some homeomorphous transformation is specified. The coordinates of the points in these systems are denoted $(\tilde{z}, \tilde{r},$ and $\tilde{\varphi})$ and $(z, r,$ and $\varphi)$, respectively. The equations of motion used in this case have the form [32,59,60]:

$$\rho_0 \frac{\partial v_i}{\partial t} = \frac{\partial D_{ji}}{\partial i} + VQ_i, i,j = \{z,r\} \qquad (10.68)$$

where ρ_0 is the initial density of the material, v_i are the components of the velocity of the material point; D_{ji} are components of the first Piola–Kirchhoff stress tensor, and

$$Q_z = \sigma_{zr}/\tilde{r}, Q_r = (\sigma_{rr} - \sigma_{\varphi\varphi})/\tilde{r}. \qquad (10.69)$$

The accepted model of the elastoplastic solid (Mises's model) includes geometrical relations:

$$e_{ij} = \tfrac{1}{2}(\partial v_i/\partial j + \partial v_j/\partial i),$$

$$e_{\varphi\varphi} = v_r/r, e_{i\varphi} = 0, i,j = \{z,r\}, \qquad (10.70)$$

where $\{e_{ij}\}$ are the components of the strain rate tensor. Equations for the deviator component of the Cauchy stress tensor

$$s_{ij} = 2\mu(e_{ij} - \tfrac{1}{3}\delta_{ij}\dot{V}/V), i,j = \{z,r,\varphi\} \qquad (10.71)$$

and also the Mises condition of ideal plasticity are used (10.66). In the above equation, s_{ij} are the components of the stress deviator, and δ_{ij} is the Kronecker symbol.

The components of the Cauchy stress tensor σ_{ij} are computed according to (1.5). The hydrodynamic pressure p and temperature T are computed using the wide-range equations of state (1.10) [44,56]:

$$p = p_e(\rho) + p_T(\rho,T), E = E_e(\rho) + E_T(\rho,T). \qquad (10.72)$$

Generally speaking, described equations resemble the equations of Section 10.2. Some results of calculations are presented in Figure 10.26. The evolution of cold, thermal, and total pressures in one of the evaporated cells are shown. To provide more comprehensive information, the evolution is shown in relation to a monotonically

286   Extreme Waves and Shock-Excited Processes

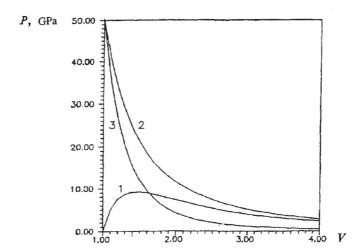

FIGURE 10.26 Dependence of the cold (1), thermal (2), and total (3) pressure on the relative volume.

decreasing quantity, i.e., relative volume, and not in relation to time. At the initial moment, the thermal compression pressure forms as a result of irradiation, where the material is undeformed and, consequently, cold pressure is equal to zero. The total positive pressure causes an expansion of the material with an increasing of its rate. The expansion is accompanied by an increase of the cold tensile pressure which partially compensates the gradually decreasing thermal pressure. However, the cold pressure has the maximum value $P_{c,\max}$ at some relative volume. When the latter exceeds certain critical level $V_{\max}$, the cold pressure starts to decrease and the material continues to expand without obstacles; that is, it is evaporated.

It should be mentioned that if the initial thermal pressure was lower than $P_{c,\max}$, then the tendency for expansion of the material would be retained only up to the moment of reaching the relative volume at which the cold and thermal pressures are equal to each other. In this case, the matter would be in the state of thermal equilibrium characterized by the absence of stresses in the material deformed in relation to its initial state.

To examine the material fracture, we used the criterion of instantaneous local fracture (Section 10.4). It was assumed that the distribution of energy as a result of instantaneous absorption of the emission pulse by the plate material can be represented in the following form:

$$E = E_s f(z) g(r), \qquad (10.73)$$

where E is the specific internal energy, which is absorbed symmetrically relatively in the center of the loaded spot; E_s is the specific energy on the surface of the plate in the center of the loading spot; $f(z)$ and $g(r)$ are functions which depend only on z and r, respectively.

Extreme Waves Excited by Radiation Impact

The formulated problem is solved numerically using the explicit Lagrangian finite-difference schema [60].

There were two aims for calculations: the first was to verify the numerical algorithms, and the second was to examine the dynamics of failure of solids by pulsed radiation when the absorption of energy increases. It was attempted to examine the possibility of changing the fracture mechanisms of DSO during its approach to the Earth where the power of effect on the DSO of every pulse in a series increases. In particular, it was attempted to examine the fracture of DSO by pulses when the power of every pulse in a series is increased.

10.5.3 Calculation Results and Comparison with Experiments

Verification of the model. The absorption of radiation in the depth of the material is described by the Bouguer–Lambert law (10.44):

$$f(z) = \exp(-Lz), \tag{10.74}$$

where L is the coefficient of attenuation of radiation in the material which depends on the mechanical properties of the absorbed material and on the wavelength of radiation.

However, the calculations based on Eq. (10.74) require a highly detailed numerical mesh. It is necessary to use from seven to ten cells to describe the exponent. Therefore, a simplified approach was used in which the absorption of radiation along the depth reduces linearly.

$$f(z) = 1 - Lz/2, \quad z \le 2/L. \tag{10.75}$$

The variation of the density of energy release along the radius was described using the Gaussian distribution [49]:

$$g(r) = \exp\left(-r^2 R_i^{-2}\right), \tag{10.76}$$

where R_i is the effective radius of the irradiation spot. Quantity $R_i = 1.0$ mm was taken from Ref. [61]. The pulse energy was founded using the following relationship:

$$\pi R_i^2 Q = \int_0^{R_i} \int_0^{\infty} E(z, r) \, dz \, dr, \tag{10.77}$$

where the initial distribution of specific energy $E(z, r)$ is taken from (10.73). Since the law of attenuation of radiation in the material was not specified in the calculations carried out in Ref. [61], here the attenuation coefficient L value typical for all metals is 10^5 m^{-1}.

To fulfill the first task, numerical modeling was carried out of the formation and development of experimentally observed [61] thin-wall cupolas on the rear side of an aluminum barrier under pulsed laser radiation. The typical forms of the cross section of the specimen are shown in Figures 10.27 and 10.28. The form of the cupola

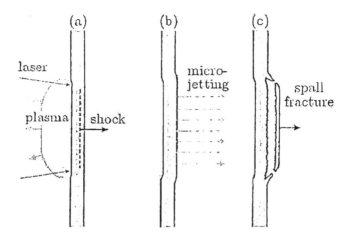

FIGURE 10.27 (a–c) A scheme of laser shock-induced fragmentation of aluminum target.

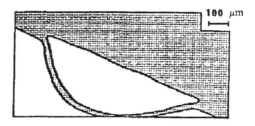

FIGURE 10.28 Form of the cupola on the rear (free) surface of the plate loaded by laser radiation.

shown in Figure 10.28 corresponds to the variant of irradiation of an aluminum round plate constrained at the edges. The thickness of the plate was $h = 0.65$ mm, radius $R = 1.0$ mm.

Formation of the cupola-like forms on the rear surface of the irradiated plates is a phenomenon which is utilized here, on the one hand, as a test but, on the other hand, presents a special subject for theoretical examination.

For the calculations, the material constants for aluminum were taken mainly from [44,56,48,61,62]. We will call this version of the calculations as the version C.

The number of calculation cells in the plate along the axis of symmetry was selected equal to 20 and in the radial directions 16.

Analysis of the formation of cupola-shaped convex areas on the rear surface of the irradiated plate indicates that the process consists of two stages. In the first short-time stage (the wave stage), a disk-shaped crack forms in the vicinity of the rear surface of the plate. The crack separates the sheet of material from the main volume. The thickness of this sheet is considerably smaller than that of the plate. The second long-time stage (conventionally referred to as the deformation) consists of the inertia movement of the disk-shaped sheet with its deceleration as a result of energy

Extreme Waves Excited by Radiation Impact   289

losses for plastic deformation. It should be mentioned that during the development of dynamic fracture of sheet, there is a time period characterized by simultaneous rapid occurrence of both wave movements and plastic deformations.

Let us examine these stages of fracture. Comparison of the spallation depth (0.065 mm in calculations and 0.064 mm in experiments) and of the radius of the disk-shaped crack (0.5 mm in calculations and 0.4 mm in experiments) indicate that they are in satisfactory agreement.

Figure 10.29a shows the configuration of the difference mesh 0.2×10^{-6} second after the start of irradiation. This moment corresponds to the initial stage of development of the spherical cupola.

Irradiation takes place in the direction from top to bottom. The upper part of the figure shows a depression formed in the area of the cells in which the material evaporated. As in the experiments, the thickness of the spallation cupola is heterogeneous along the radius.

To explain the mechanism of formation of rear crack (spallation), preceding the development of a spherical cupola, it is necessary to take into account the evolution of the stressed state in the target (see Figures 10.29b [32] and 10.31). During irradiation of the top surface of the target, a disk-shaped zone of heating and compression forms in the vicinity of the surface. A stress wave, consisting of two halfwaves, propagates from this zone into the bulk of the material. The first halfwave (compression) repeats the form of the initial temperature distribution in the solid. The second halfwave is the tensile part of the wave. These halfwaves determine the steep of the stresses from the maximum compression to the maximum tension.

Thus, according to Figure 10.29, the impact of radiation can lead to the formation of a strong compression wave in the surface layer of the material. The interaction of this wave with the free surface of the target causes the development of a stretching wave propagating into the material behind the compression wave. The tensile stress may exceed the dynamic tensile strength of the material, which will cause the spalling of the front layer of the target (Figure 10.29b). Over time, that layer can break up into fragments (drops, cavitation), and the compression wave front can reach the back side of the target and cause the spallation of the back surface of the target (see, also, Figures 10.27, 10.28, and 10.31(right)).

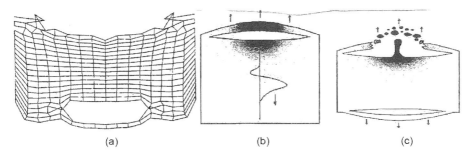

FIGURE 10.29 Configuration of the calculation mesh in the zone of formation of a cupola ($t = 0.203 \times 10^{-6}$ second) (a). A scheme of the processes occurring during irradiation of the target by a laser radiation (b and c).

However, if pulsed radiation is accompanied by intensive evaporation, the tensile stresses are restricted and the pulse loses its central symmetry relatively compression and tension. This was shown in [37] using unidimensional calculations (see Sections 10.2 and 10.3). The same result is in two-dimensional calculations (Figure 10.30). The restriction and suppression of the tensile stresses takes place as a result of the recoil of the vapors accelerating from the top surface.

Generally speaking, during propagation of the compression wave, this wave can attenuate quite rapidly. However, reaching the rear surface, the wave can be reflected as the tensile wave with the amplitude sufficient for formation of rear spallation (Figure 10.30). For comparison, the dotted and dashed line 5 in Figure 10.30 shows the profile stresses obtained when fracture of the material was not taken into account.

The propagating flat compression wave has the form of a circle constantly narrowing as a result of lateral unloading. Since the diameter of the radiation spot was greater than the thickness of the target, the stress state in the center of the narrowing circle remains to be planar. In accordance with intuitive expectation, the radius of the disk-shaped crack (region in which the stress remains planar) is approximately equal to the difference of the spot radius and the plate thickness (see, also, Figures 10.23a and 10.24a).

10.5.4 Special Features of Fracture by Spalling

The calculation results confirm the sufficiently high accuracy of the numerical methods used. Consequently, they can be used to examine the problem of controlling

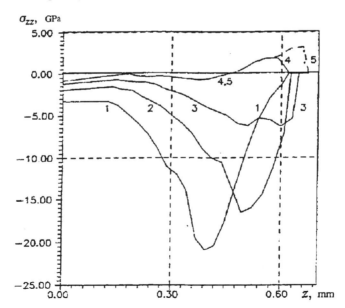

FIGURE 10.30 Profiles of stresses σ_{zz} on the axis of symmetry (modeling of experiments), 1) $t = 0.035 \times 10^{-6}$ second; 2) $t = 0.070 \times 10^{-6}$ second; and 3, 4) $t = 0.123 \times 10^{-6}$ second (line 5 is the calculation without taking the fracture into account).

Extreme Waves Excited by Radiation Impact

fracture and optimizing it during of spallation of the DSO by laser pulses and elimination of space debris. For it is necessary to construct a quite detailed picture of various variants of spallation fracture for different pulse energies. Let us consider two variants of pulsed irradiation which gives a sufficiently detailed picture of possible fracture mechanisms if they will be studied together with the experimentally recorded variant (variant **C**, see Section 10.5.3).

We shall examine two variants of calculating irradiation of an aluminum plate differing only in the surface density of the energy pulse: In variant **A**, $Q = 1.27 \times 10^{-2}$ J/m^2, and in variant **B**, $Q = 1.91 \times 10^{-2}$ J/m^2. The geometrical parameters of the calculations are as follows: $R = 0.6$ mm, $h = 1.0$ mm, and $R_i = 0.6$ mm. In addition, in comparison with variant **C**, we selected irradiation with a shorter wavelength, $L = 150,000$ m^{-1}. In these calculations, the distribution of the pulse energy in the depth was assumed as (10.75). The number of calculation cells in the plate along the axis of symmetry was 10 and along the diameter 12.

Variant A. Only the disk-shaped spallation was realized in the variant. Figure 10.31A shows the corresponding configuration of the target at time $t = 16$ ns.

The mechanism of formation of tension and compression zones, causing spallation at first moments, is shown in Figure 10.32 which illustrates the profiles of the axial stresses along the axis of symmetry of the plate (its cylindrical volume). The general form of the stress waves resembles that shown in Figure 10.30, but the level of the tensile stresses is not reduced by the recoil pulse of the jet of the vapors.

The amplitude of the tensile wave is almost equal to the amplitude of the compression wave. A tensile pulse appears during the departure of the stress wave from the free surface. This tension may be enough to form external spallation [35] (see Sections 10.1.4, 10.2, and 10.3). However, the stress wave attenuates while moving through the material. When this wave reaches the rear surface, its amplitude does not reach the critical tension of the material. Thus, there is no rear spallation. Front spallation takes place almost without evaporation.

Variant B. Figure 10.31B shows the configuration of the cross section of the loaded plate at time $t = 22$ ns for the second variant **B**. Two types of spallation are detected in this case: rear and front. Neither of them are strictly in disk-like shape. The front spallation has the form of a toroidal crack. In other words, the material at the axis of symmetry in the vicinity of the front surface is not fractured. It should be mentioned that the formation of a ring spallation cannot be detected in

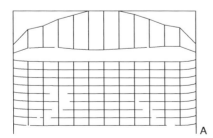 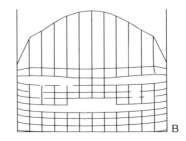

FIGURE 10.31 Configuration of the calculation mesh after spallation: variant A – only external spallation, $t = 16$ ns; variant B – external and rear spallation, $t = 22$ ns.

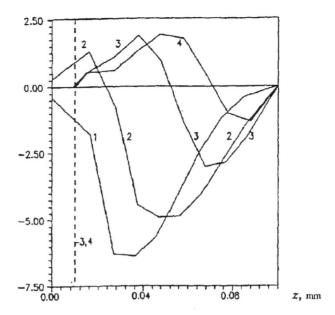

FIGURE 10.32 Profiles of stresses σ_{zz} (GPa) on the axis of symmetry for external spallation: 1) $t = 5.4$ ns, 2) $t = 8.1$ ns, 3) $t = 13.6$ ns, and 4) $t = 16.3$ ns (dashed line is the surface of external spallation).

unidimensional calculations. Observation of this type of fracture in experiments at present is associated with difficulties because laboratory lasers usually have a relatively long wavelength (in the vicinity of the optical range) and, correspondingly, relatively thin spallation sheets (of the order of tens of micrometers). In addition, the external spallation is formed in a relatively narrow range of the intensities.

The formation of the ring-like crack is caused by the stress state which is highly heterogeneous along the radius. In the vicinity of the center of the target, it is close to variant **C** as a result of extensive evaporation. External spallation does not take place in this case. At the periphery of the irradiation zone, the stress state is close to variant **A**. The external spallation can take place as a result of a low counter pressure of the vapors.

The rear spallation is not disk-shaped but rather semi-lens facing the rear side of the target by the convex center. This is also associated with the operation of two-dimensional effects: distortion of the edges of the flat wave as a result of lateral unloading.

The mechanism of formation of the tensile and compression zones for variant **B** for the cylindrical surface having the radius 0.3 mm is shown in Figure 10.33. This calculation was carried out in order to show the surfaces of both spallation fractures. The general form of the stress waves resembles that shown in Figure 10.32. The amplitude of the tensile wave is sufficient to initiate a front spallation although the magnitude of the tensile stresses is partially suppressed by the recoil pulse of the jet of the outgoing vapors. At the same time, attenuation of the wave in the material weakens this wave insufficiently so that rear spallation is also possible.

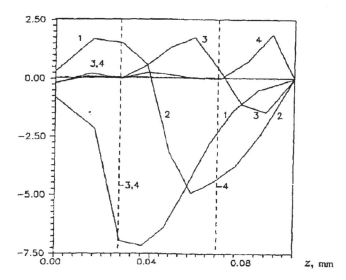

FIGURE 10.33 Profiles of stresses σ_{zz} (GPa) along the direction of the axial coordinate at the distance 0.3 mm from the axis of symmetry. Cases of the external and rear spallation: 1) $t = 5.4$ ns, 2) $t = 10.9$ ns, 3) $t = 16.3$ ns, and 4) $t = 21.7$ ns (dashed lines are surfaces of spallation).

It should be mentioned that there is no the spallation in the vicinity of the external surface if the initial pulse energy is lower than that in variant **A**.

10.5.5 Efficiency of Laser Fracture

The calculations show that the mechanism of external spallation may be utilized quite efficiently to fracture remote objects. However, in addition to principle feasibility of such a method of fracture it is necessary to optimize the method. The unidimensional calculations of pulsed irradiation of the plates, presented below, show the sufficient efficiency of fracture by a means of external spallation. In particular, in a certain range of the amplitudes the increase of the pulse energy may even reduce the mass of the removed material. It is due to the evaporation which suppresses the process of external spallation. In this calculation series, we examined only external spallation to apply it to the problems of the DSO failure.

Here are presented model calculations. The irradiation (instantaneous along the normal to the surface) of an aluminum sheet of infinite thickness coated with an external layer of iron 0.6 mm was examined. The physico-technical characteristics of the iron determined were given in Ref. [44].

Figure 10.34 shows the profiles of the stress waves and the spallation surfaces for four values of the radiation energy. The energy absorption was described by the Bouguer–Lambert law (10.44), and the surface temperature for various radiation variants was 260°C to 2700°C, 8000°C, and 25,000°C.

It may be seen that the mechanisms of the fracture and the vaporization of material are competing. The growth of the radiation energy increases the pressure inside the cloud of the evaporating matter and the stress in the subsurface material. As a result, the frontal region of the fracture decreases until it completely disappears at a specific threshold level of the absorbed energy. The volume of the removed material depends in a highly nonlinear manner on the energy of the irradiation pulse. This is associated with the fact that the amount of energy used by the fracture is considerably smaller than that used for evaporating the matter. Consequently, the relation of the mass of the fractured matter and pulse, acquired by the remaining unfractured part of the solid, has a local maximum (Figure 10.34). Similar maximum in the case of fracture caused only by evaporation of the matter can be obtained only at considerably higher radiation energies.

To confirm this, we shall compare the thickness of the evaporated z_{ev} and spallated z_{sp} material. For the four cases examined, we have the following values: A) $z_{ev} = 0$, $z_{sp} = 0$, B) $z_{ev} = 3 \times 10^{-8}$ m, $z_{sp} = 2.7 \times 10^{-7}$ m, C) $z_{ev} = 9 \times 10^{-8}$ m, $z_{sp} = 2.7 \times 10^{-7}$ m, D) $z_{ev} = 2.4 \times 10^{-7}$ m, and $z_{sp} = 0$. The change of the mechanism of fracture by external spallation to the mechanism of the evaporation of the material can be clearly seen when the energy grows. Calculations carried out using the variants B and D in

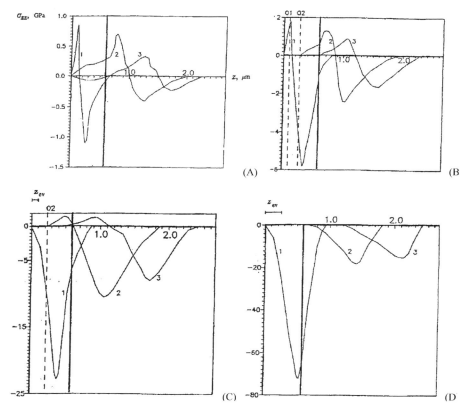

FIGURE 10.34 Profiles of stresses σ_{zz} caused by irradiation $T_s = 260$ (A), 2700 (B), 8000 (C), and 25,000°C (D): 1) $t = 0.05$ ns, 2) $t = 0.25$ ns, and 3) $t = 0.40$ ns (dashed lines show spallation surfaces 01 and 02 at moments 0.05 and 0.25 ns, respectively, and the double line is the boundary of the materials).

Figure 10.34 show that the volume of the material removed from the surface of the solid may not necessarily increase with the growth the energy. This associates with the fact that the fracture of the material by the spallation requires a considerably lower energy than the evaporation. There is the high efficiency of the methods of removing the mass by external spallation in comparison with evaporation in a specific range of the laser energy.

10.5.6 Discussion and Conclusion

The radiation variants (**C**, **A**, **B**) examined make it possible to construct an entire image of the change of the mechanisms of fracture with increasing energy of the radiation pulse.

At relatively low pulse energies, the surface evaporation is negligible. Stress waves are too weak to form the spallations. When some threshold value is exceeded, the fracture described by variant A starts to take place and the front crack can form. Thickness of the fractured material is proportional approximately to the coefficient of energy attenuation in the material. In particular, it depends on the radiation characteristics and only slightly on the pulse energy.

With increasing pulse energy, the front spallation begins to be suppressed by the recoil pulse of the vapor and the amplitude of the resultant compression wave increases at the same time. If the thickness of the irradiated solid is smaller than the width of the irradiation spot, the compression wave extends to the rear surface. The rear spallation can take place if the pulse power is above certain critical value. Both front and rear spallation can form at a specific combination of the characteristics of the load and the plate. The last case is the variant **B**.

With a further increase of the pulse energy, the halfwave of the tensile stresses, formed in the vicinity of the front surface of the irradiated solid, is suppressed. In this case, only the rear spallation can take place. This mechanism corresponds to the experimental data [48,61] – variant **C**. A further increase of the pulse energy causes the separation of the resultant cupola-shaped shell and its fragmentation [48,61]. However, the range of the energy in which rear spallation may occur is relatively narrow. The range in which double spallation may be observed is even smaller. Thus, the numerical experiments show that the pulsed irradiation, causing the front spallation, is an efficient means of influencing on the DSO. These pulses make it possible to affect on the DSO from large distances. It should be mentioned that the efficiency of the fracture is determined in the first instance not by the pulse energy but by the depth of its absorption. External spallation is proportional to depth of the absorption in the last case. In a specific range of distances to the target and the pulse energy, it is possible to find the best way for fracture of the DSO by external spallation and preventing premature suppression of this fracture mechanism by the evaporation. The development of powerful lasers in the X-ray and gamma ranges and the optimization of radiation pulses would make it possible to obtain and optimum combination of the front spallation and evaporation of material of the DSO.

To liquidate space debris – especially fine debris – the best results will be obtained by combining external and internal spallation. In this case, the depth of absorption should be sufficiently small to ensure fine fragmentation of the material.

Combination of the two types of spallation would enable highly efficient disintegration of small fragments of technogenous origin.

CONCLUSION

1. Numerical modeling of the experiments with the formation of cupola-shaped rear convex areas under laser radiation showed that the results are in agreement with the experimental data. The theoretical data describe rightly the depth and radius of the disk-shaped spallation crack observed in the experiment
2. A change of the mechanism of spallation fracture with increasing energy of the laser pulse was discovered: from the front spallation to a combination of front and rear spallations and then to the rear spallation only.
3. A highly two-dimensional effect of formation of the ring-shaped crack in the front spallation has been determined numerically.
4. We suggest to prevent the collision of the Earth with the DSO by deflecting the orbit and partial fragmentation of DSOs. It is proposed, as the most efficient procedure, to subject them to the effect of a series of pulses of laser radiation causing a frontal spallation and evaporation. For the maximum disintegration of fine space debris, it is recommended to utilize the laser pulses causing front and rear spallations at the same time.

REFERENCE

[32] Bulgakov AV et al. Synthesis of nanoscale materials under the influence of powerful energy flows on matter. Novosibirsk: Institute of Thermophysics SB RAS, 2009 (in Russian) (Булгаков АВ и другие. Синтез наноразмерных материалов при Воздействии мощных потоков энергии на вещество. Новосибирск: Институт теплофизики СО РАН, 2009).

11 Melting Waves in Front of a Massive Perforator

The penetration and perforation of metal plates by a cylindrical steel projectile are examined experimentally and simulated numerically. It is shown that conical melting wave is formed beneath the projectile [63–66] (Figure 11.1).

We emphasize that Mach cones have been well known for gas-like media.

This chapter is aimed at the study of all the stages of the penetration process right up to complete perforation. Particular attention is paid to investigate the formation of melting wave beneath the projectile. The influence of the impact velocity of the projectile on the temperature effects connected with the plastic shear deformation, the penetration depth, and the character of deformation and perforation of targets of various materials is also studied.

11.1 EXPERIMENTAL INVESTIGATION

Targets of two aluminum alloys, AMg6 (highly plastic) and Al-Zn-Mg (high strength) [63,64], were examined. The plates of the first alloy were 30 and 60 mm in thickness, and of the second, 30 mm. The test results are given in Figures 11.2 and 11.3.

We note that for the 30-mm plates the material failure occurs are approximately identical and are in the 220–240 m/s range. For the 60 mm AMg6 plate, the point of failure is located substantially further along the velocity scale in the 450–480 m/s interval.

The influence of the impact velocity on the penetration depth and the deformation and perforation modes was investigated. Particular attention was given to the cone formation in the targets, due to an increase in temperature and melting of material at the sites of intensive plastic shear deformations. Some results of this research are

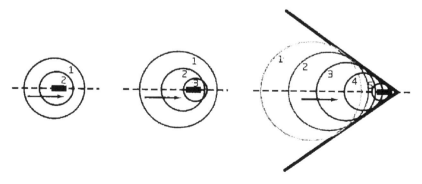

FIGURE 11.1 The Mach cone arising in gas when the speed of a body exceeds some critical speed.

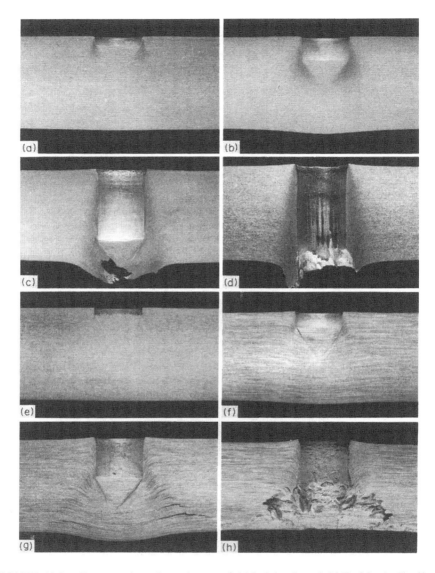

FIGURE 11.2 Cross section of specimens of AMg6 (a–d) and Al-Zn-Mg (e–h) alloys after tests (the surfaces were polished and pickled with hydrofluoric acid): (a) $v_0 = 138$ m/s; (b) $v_0 = 224$ m/s; (c) $v_0 = 390$ m/s; (d) $v_0 = 437$ m/s; (e) $v_0 = 157$ m/s; (f) $v_0 = 240$ m/s; (g) $v_0 = 340$ m/s; (h) $v_0 = 440$ m/s [66].

presented in Figure 11.2 for different moments of intrusion of a solid body (the projectile with a flat penetrating end) into softer material.

The deformation of the target is shown in Figure 11.2. A common feature for the examined alloys even for low impact velocities is the presence of zones in which structural changes in the material were noticed near the corners of the craters.

Melting Waves in Front of Massive Perforator

FIGURE 11.3 The typical plug and cone after complete perforation of 60 mm thick aluminum alloy target. Melting cone surfaces are formed by intensive plastic shear deformations. The projectile velocity $v_0 = 660$ m/s. Plug and cone heights are 15 and 8 mm, respectively [66].

These changes are most naturally explained by the intensive local plastic deformations of the material leading to the occurrence of adiabatic shear [63–66]. These zones increase dimensionally with an increase in the impact velocities, but remain localized in the areas, adjoining the impacting body. Here, in front of the projectile, a cone is finally formed of the strips that join at the axis of symmetry. With further increase in the velocity, this cone is particularly clearly pronounced in targets of the Al-Zn-Mg alloy and no longer changes its configuration.

The deformation patterns in plate sections that form during the penetration are shown. At velocities close to the ballistic limit (corresponding to complete perforation), the materials' response is different. The Al-Zn-Mg alloy is characterized by the initiation of internal spallation cracks (Figure 11.2g). Conversely, in the AMg6 alloy, such cracks are absent for all interaction velocities. There is also a notable difference between the failure modes in the cases of complete perforation (Figure 11.2(d, h)). The formation of the Mach cone is clearly shown in photos b, f, and g. In particular, an example of the Mach cone and cracks ahead of it is clearly shown in Figure 11.2g, an example of a Mach cone is shown in Figure 11.2f, and the generation of the Mach cone is shown in Fig. 11.2b.

At the same time, Figure 11.3 shows that the plug forced out after the complete perforation of the AMg6 plate is separated into two parts. The first part is the cone-shaped crater, whose surface is smooth and shiny. Such specific failure modes can naturally be associated with melting of the material in the sites of intensive plastic shear deformations. The second part is formed by the material taken from the crater (at the right).

It needs to be noted that in some works (see [63,64]), the initial formative stage of the cone of failure in various target materials was observed, whereas the separation of the cone from the plug with complete perforation was observed in others. Here, we considered the dynamics of the formation and development of such structures and the accompanying temperature effects in relation to the impact velocities. From our point of view, a comprehensive study of the complex combination of mutually interacting physicomechanical processes occurring in the target material, including elastic and plastic deformation, heating and melting of the materials, is possible only by using the method of numerical modeling based on experimental data.

11.2 NUMERICAL MODELING

The calculations in this chapter are made in accordance with the algorithm in [38]. The behavior of the projectile materials was assumed to be elastoplastic with ideal plasticity, which is defined by a hyperbolic system of laws of conservation in a differential form supplemented with corresponding determining relationships.

The mathematical model was constructed on the basis of a hyperbolic two-dimensional system of the continuum mechanics equations for the asymmetrical case. The equations for this case include (10.61)–(10.63), the equation of state (10.36):

$$p = a_1(V^{-1}-1) + a_2(V^{-1}-1)^2 + a_3(V^{-1}-1)^3, \tag{11.1}$$

the equations for the deviator components (10.71) and von Mises yield condition (10.66).

Since it was assumed that the experimentally obtained results are connected with the separation of the body, it was additionally assumed that all the work of plastic deformation in one volume of the body or the other is converted into internal energy and is used up to increase the temperature of this volume. Here, since the time taken for plastic deformation with the given type of loading is measured in microseconds, the heat-transfer processes were disregarded. In other words, the plastic deformation was assumed to be adiabatic.

The rate of increase in temperature was determined by a formula,

$$\dot{T} = (C_p \rho_0)^{-1} \sigma_{ij} \dot{\varepsilon}_{ij}, \tag{11.2}$$

where T is the temperature, C_p is the specific heat of the material, ρ_0 is its initial density, and σ_{ij} and $\dot{\varepsilon}_{ij}$ are the components of the stress and deformation rate tensors, respectively (the dot on top of the symbol signifies a time derivative along the trajectory).

Here, the increase in temperature was reckoned from the instant the plastic deformation started in a particular volume of the material up to the time the melting point is reached in the same volume. After this, the material was considered to be in a molten state and was treated as a fluid in subsequent calculations (with a zero second coefficient of Lame) by assuming that its temperature no longer increases. It was assumed that, for the AMg6 alloy, $C_p = 0.9 \times 10^3$ J/(kg·deg), $T_m = 450°C$; for the Al-Zn-Mg alloy, $C_p = 10^3$ J/(kg·deg), $T_m = 570°C$; for steel $C_p = 1.1 \times 10^3$ J/(kg·deg), $T_m = 1550°C$, where T_m is the melting point. The initial temperature of the interacting bodies was 20°C.

The present work uses the boundary conditions at the surface of contact between the projectile and the target. A condition of total slip was set at this boundary, which can introduce an additional error into the calculations. An algorithm [60] has been used in this work to calculate the position of the contact boundaries. This algorithm makes it possible to account for the sliding past of the elements of the surfaces of the interacting bodies and their partial or total separation from one another. The algorithm includes three stages at each spacing relative to the time.

The preliminary values of the velocities and coordinates of the boundary nodes of the calculation grid are calculated in the first stage, and here, it is assumed that for each body, there exists no other contacting body.

In the second stage, the nodes intersecting the boundary of another body are determined. The velocities of these nodes and the sections of the boundaries which they intersected are determined so that, as a result, the nodes coincide exactly with the corresponding sections of the boundaries. The magnitude of the corrected velocity here is assumed to be a set function of the velocities of the adjoining boundary nodes of the grid at the instant the boundary is intersected and of the stress–strain state of the volumes of the material that adjoin this boundary. The form of this function is determined in accordance with the law of conservation of momentum.

The location of the contact boundaries as a whole is established in the third stage from the local values of the corrected velocities of the nodes. The use of the above algorithm yielded a form of the formed crater.

11.3 RESULTS OF THE CALCULATION AND DISCUSSION

As is known, the accuracy in a numerical experiment depends on the values of the spacings used in the time and space coordinates.

In the case examined, the size of the cells should provide the possibility of obtaining, in the numerical calculations, temperatures that reach the melting point of the material. With a view to finding such sizes, calculations for impacts with 168 m/s velocity were made for various variants of splitting of the area adjoining the front face of the projectile. In particular, cells measuring 3.8×3.8, 1.9×1.5, and 1.0×0.7 mm were used. The maximum temperatures reached in the AMg6 alloy target in these cases were 272, 408, and 450, respectively. The increase in the maximum temperature "with condensation" of the grid was evidently due to the fact that, in using a coarse grid, the work of the plastic deformation is found to be distributed over a larger mass of the material than is actually the case, and this itself leads to a reduction in the calculated value of the temperature. With a finer grid, a better approximation of the area is obtained wherein the conversion of the work in the plastic deformation to internal energy is affected. It was this particular type of grid that was used to obtain the results presented below.

The picture of deformation of the Lagrangian grid in the target with the passage of time is illustrated in Figure 11.4. The grid in the projectile is not shown in view of its minor deformation. It can be seen that maximum grid distortion in the target is concentrated close to the corner of the penetrating projectile, and, here, these distortions decrease more slowly only in the vicinity of the line located at ~45° to the leading face of the projectile in the direction of the axis of symmetry. These relatively large distortions of the Lagrangian grid can be associated with a sufficient level of confidence and with the experimentally observed bands of adiabatic shear (Figure 11.2).

The above situation is further confirmed in the analysis of the temperature distribution in the AMg6 alloy target sections (see tables 1–3 in [64]).

The variation in the calculated temperature values in the Lagrange grid cells in relation to the impact velocity can be seen from tables 1–3 in [64]. For $v_0 = 112$ m/s, the melting point is attained in only one cell adjoining the corner of the projectile section. The number of such cells increases with an increase in the impact velocity, and these cells form the already-known band at ~45° angle to the leading face of the

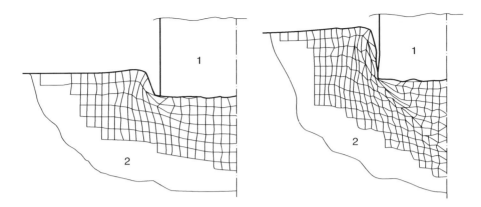

FIGURE 11.4 Lagrangian grid distortions near the contact boundary between projectile (1) and 30 mm thick AMg6 alloy plate (2) for $v_0 = 224$ m/s at instants 10.8×10^{-6} s (a) and 32.2×10^{-6} s (b).

projectile. The temperature gradient in the vicinity of the given band is very high. It should be noted that there is good agreement between the lengths of the calculated areas of melting and the experimentally observed structurally changed zones of the target material. The large width of the calculated zones is probably explained by the imperfection of the model used.

The temperature distribution in the Al-Zn-Mg alloy target was given in table 4 in [64]. It is notable for its raised shear modulus and creep limit. It is not difficult to see that the temperature gradient in the given case is considerably higher than in the AMg6 alloy. This evidently qualitatively corresponds to the experimentally observed reduction in the width of the zones of structural changes in the Al-Zn-Mg alloy as compared to the AMg6 alloy. A better agreement between the calculations and experiments (with respect to the size of the adiabatic shear bands) could have been obtained in the case of the Al-Zn-Mg alloy with a condensation of the grid.

However, even with the use of the breakup, the distortions of the individual cells are so large (Figure 11.4) that, due to the reasons noted, it was not possible to complete the calculations for the entire experimentally investigated range of velocities. Such calculations require further improvement of the numerical method used.

Thus, we have illustrated the possibility of the emergence in solids of waves similar to Mach waves in gas. It can be assumed that such a wave and melting can occur when a giant crack propagates along the junction of tectonic plates during catastrophic earthquake [65,66].

REFERENCES

1. Galiev ShU. *Nonlinear Waves in Bounded Continuous.* Naukova Dumka, Kiev (1988) (in Russian).
2. Vick B, Ozisik MN. Growth and decay of a thermal pulse predicted by the hyperbolic heat conduction equation. *J Heat Transfer* 105 (4): 902 (1983).
3. Bubnov VA. Wave concepts in the theory of heat. *Int J Heat Mass Transfer* 19: 175–184 (1976).

4. Lutset MO, Tsai AN. Unsteady heat transfer in superfluid helium. In: *Heat Transfer During of Phase Transformations.* Kutateladze SS (ed.), pp. 70–75. Institute of Thermophysics SB AN USSR, Novosibirsk (1983) (in Russian).
5. Khalatnikov IM. *Theory of Superfluidity.* Nauka, Moscow (1971) (in Russian).
6. Galiev ShU, Zhurakhovskii SV, Titarenko SI. *Mathematical Models of the Impact of a Thermal Pulse on Materials.* Preprint of the Academy of Sciences of the Ukrainian SSR. Institutet Problems of Strength, Kiev (1985).
7. Galiev ShU, Babish UN, Romashchenko VA, Zhurakhovskii SV, Nechitailo NB. *Numerical Modeling of Processing in Confined Media.* Naukova Dumka, Kiev (1989) (in Russian).
8. Esel'son VI, Grigoryev VI, Ivantsev VG, Rudavsky EJ. *Properties of Liquid and Solid Helium.* Standartov, Moscow (1978) (in Russian).
9. Galiev ShU, Zhurakhovskii SV. Fracture of multilayer plates in thermoviscoelastoplastic waves. *Strength Mat* 11: 37–43 (1984).
10. Lutset MO, Tsoi AN. *Experimental Study the Propagation of Thermal Pulses in Superfluid Helium. Boiling and Condensation. Hydrodynamics and Heat Transfer,* pp. 91–101. Izdatelstvo Institute of Thermophysics SB AN USSR, Novosibirsk (1986).
11. Galiev ShU, Zhurakhovskii SV, Ivashchenko KB. Effect of the HE II phase transition in He I on the formation of second-sound waves. *Phys Low Temp* 17 (8): 1034–1038 (1991).
12. Rakhmatullin XA, Shemyakin EI, Demyanov YuA, Zvyagin AV. *Strength and Destruction under Short-Term Loads.* Moscow (2008).
13. Peshkov V. Second sound' in Helium II. *J Phys* 8: 381 (1944).
14. Maxwell JC. On the dynamical theory of gases. *Philos Trans Roy Soc* 157: 49 (1867).
15. Frankel JI, Vick B, Özisik MN. Flux formulation of hyperbolic heat conduction. *J Appl Phys* 58: 3340 (1985).
16. Kim WS, Hector LG, Ozisik MN. Hyperbolic heat conduction due to axisymmetric continuous or pulsed surface heat sources. *J Appl Phys* 68 (11): 5478 (1990).
17. Zel' dovich YaB. On the possibility of rarefaction shock waves, *Zh Eksp Teor Fiz* 4: 363–364 (1946); Borisov AA, Borisov AL, Kutateladze SS, Nakoryakov VE. Rarefaction shock wave near the critical liquid vapour point. *J Fluid Mech* 126: 59–73 (1983).
18. Samarskii AA, Popov YuP. *Finite-Difference Methods for the Solution of Gas Dynamics Problems.* Nauka, Moscow (1980).
19. Torczynski JR. On the interaction of second sound shock waves and vorticity in superfluid helium. *Phys Fluids* 27 (11): 2636 (1984).
20. Galiev ShU, Zhurakhovskii SV, Ivashchenko KB. *Strong Nonlinearity of Thermodynamic Functions as a Cause of Cooling Shock Formation.* Preprint. Institute Problems of Strength, Kiev (1995).
21. Fortov VE, Lomonosov IV. Ya B Zeldovich and equation of state problems for matter under extreme conditions. *Physics-Uspekhi* 57: 219–233 (2014).
22. Mayer PN, Mayer AE. Cavitation and evaporation in metals under the action of ultrashort intensive irradiation. *11th World Congress on Computational Mechanics (WCCM XI),* Barcelona, Spain, 20–25 July (2014).
23. Fortov VE. Extreme states of matter on earth and in space. *Physics-Uspekhi* 179: 653–687 (2009).
24. Fortov VE. Powerful shock waves and extreme states of matter. *Physics-Uspekhi* 177: 347–368 (2007).
25. Leveugle E, Ivanov DS, Zhigilei LV. Photomechanical spallation of molecular and metal targets: Molecular dynamics study. *App Phys A* 79: 1643–1655 (2004).
26. Zhigilei LV, Lin Z, Ivanov DS. Atomistic modelling of short pulse laser ablation of metal: Connections between melting, spallation and phase explosion. *J Phys Chem C* 113: 11892–11906 (2009).

27. Fortov VE, Kostin VV, Eliezer S. Spallation if metals under laser irradiation. *J Appl Phys* 70 (8): 4524–4531 (1991).
28. Anisimov SI, Prokhorov AM, Fortov VE. Application of high-power lasers for studying substances at ultrahigh pressures. *Physics-Uspekhi* 142: 395–434 (1984).
29. Fortov VE, Yakubov IT. *Nonideal Plasma*. Énergoatomizdat, Moscow (1994) (in Russian).
30. Delone NB. *Interaction of Laser Radiation with the Matter*. Nauka, Moscow (1989).
31. Novikov SA, Ruzanov AI, Trunin IR, Uchaev AYA. Stress wave propagation and fracture processes in metal during rapid bulk heating. *Strength Mat* 26 (2): 137–140 (1994).
32. Bulgakov AV et al. *Synthesis of nanoscale materials under the influence of powerful energy flows on matter*. Novosibirsk: Institute of Thermophysics SB RAS, 2009, 462 pages (in Russian: Булгаков АВ et al. Синтез наноразмерных материалов при воздействии мощных потоков энергии на вещество. Новосибирск: Институт теплофизики СО РАН, 2009. 462 с.); Ivanov DS et al. The mechanism of nanobump formation in femtosecond pulse laser nanostructuring of thin metal films. Applied Physics A 92 (4): 791–796 (2008); Rethfeld B, Ivanov DS, Garcia ME, Anisimov SI. Modelling ultrafast laser ablation. *Journal of Physics D: Applied Physics* 50 (19): 193001 (2017); Zhigilei LV, Ivanov D, Leveugle E, Sadigh B. Computer modeling of laser melting and spallation of metal targets. *Proceedings of SPIE (The International Society for Optical Engineering)* 5448, 505–519 (2004).
33. Жиряков БМ, Обеснюк ВД. Влияние прозрачных покрытий на генерацию лазериноиндуцированных ударных волн в металлах. *Физика и химия обработки металлов* 5: 29–33 (1984) (Zhiryakov BM, Obesnyuk VD. Effect of transparent coatings on the generation of laser-induced shock waves in metals. *Phys Chemistry Met Process* 5: 29–33 (1984)).
34. Харитонов ВВ et al. Динамические эффекты при импульсном нагреве лазерных зеркал. *Теплофизика высоких температур* 21 (3): 1127–1133 (1983) (Kharitonov VV, Subbotin VI, Grishunin PA, Timonin AS. Dynamic effects during pulsed heating of laser mirrors. *Thermal Phys High Temp* 21 (3): 1127–1133 (1983)).
35. Zhurakhovskii SV. Optimization of the outer laser-induced spalling. *Strength Mat* 35 (1): 111–116 (2003).
36. Bouguer P. *Essai d'optique sur la gradation de la lumière*. Gauthier-Villars et Cie, Paris (1729) (reprinted in 1929 with a Biographic note and a Preface).
37. Galiev ShU, Zhurakhovskii SV. Fracture of multilayer plates in thermoviscoelastoplastic waves. *Strength Mat* 11: 37–43 (1984).
38. Wilkins ML. *Calculation of Elastic-Plastic Flow*. University of California, Lawrence Radiation Laboratory, Livermore (1963).
39. Zel'dovich YaB, Raizer YuP. *Physics of Shock Waves and High-Temperature Hydrodynamic Phenomena*. Dover Publications, Moscow (2002).
40. Orlenko LP. *Behavior of Materials under Intensive Dynamic Loads*. Nauka, Moscow (1966) (in Russian).
41. Филипп Баум, Леонид Орленко, Кирилл Станюкович Физика взрыва (1975).
41. Baum F, Orlenko L, Stanyukovich K. *Physics of the Explosion*. Nauka, Moscow (1975).
42. Nigmatulin RI, Akhmadeev NKh, Akhmetova NA. Fracture of plates of Armco iron in unloading waves. *Detonatsiya* 2: 145–149 (1981).
43. Akhmadeev MKh, Nigmatulin RI. Dynamic cleavage fracture in unloading waves. *Dokl Akad Nauk SSSR* 226 (5): 1131–1134 (1982).
44. Kolgatin SN, Khachatur'yants AV. Interpolational equations of state of metals. *Thermal Phys High Temp* 20 (3): 447–451 (1984).
45. Knight CJ. Evaporation from a cylindrical surface into vacuum. *J Fluid Mech* 75(3): 469–486 (1976).

46. Anisimov SI. Vaporization of metal absorbing laser radiation. *Zh Eksp Teor Fiz* 54: 339–342 (1968).
47. Mazhukin VI, Pestryakova GA. Mathematical modeling of surface vaporization by laser radiation. *Dokl Akad Nauk SSSR* 278(4): 843–848 (1984).
48. Elieser Sh, Gazit Y, Gilath I. Shock wave decay and spall strength in laser-matter interaction. *J Appl Phys* 68 (1): 356–358 (1990).
49. Reddy G. *Effect of Powerful Laser Radiation.* Moscow (1974) (in Russian).
50. Boyko VI et al. A model of interaction of a powerful ion beam with a metallic absorber. *Zh Prikl Mekh Tekh Fiz* 4: 26–31 (1990).
51. Boyko VI, Shamanim IV, Yushcbltsin KV. Thermal shock loading of metal with a pulsed proton beam. *Fiz Khim Obrab Mater* 1: 29–33 (1992).
52. Dobkin AV, Son EN. Thermoelastic fracture of materials under the effect of pulsed beams of relativistic electrons. In: *Extreme States of Matter,* pp. 196–201. Moscow (1991) (in Russian).
53. Agureikin VA, Anisimov SI, Bushman AV et al. Thermophysical and gas-dynamic problems of antimeteorite protection of Vega space system. *Thermal Phys High Temp* 22 (5): 964-983 (1984).
54. Harrington J, LeBea RP, Backe KA, Dowling TE. Dynamic response of Jupiter's atmosphere to the impact of comet Shoemaker-Levy 9. *Nature 367*: 525-527 (1994).
55. Galiev ShU, Zhurakhovskii SV, Ivashchenko KB. Variation of the orbit and fracture spallation of dangerous space objects by repeating high-energy pulses. *Proceedings of an International Conference SPE'94*, Snezhinsk, Chelyabisnk, Russia, 26–30 September, pp. 48–49 (1994).
56. Galiev ShU, Zhurakhovskii SV, Ivashchenko KB. Fracture of bodies caused by pulses of radiation as a method for guaranteeing safety in the outer space. *Strength Mat* 3: 31–51 (1996).
57. Bunatyan AA et al. Numerical examination of development of perturbation in compressing a target with a high-intensity pulse. In: *Numerical Methods in Plasma Physics,* pp. 83–89. Moscow (1977) (in Russian).
58. Bathe K-J. *Finite Element Procedures in Engineering Analysis.* Prentice-Hall, Englewood Cliffs, NJ (1982).
59. Pozdeev AA, Trusov PV, Nyashin YUI. *High Elastoplastic Strains.* Nauka, Moscow (1986) (in Russian).
60. Ivashchenko KB. *A Method of Applying Boundary Conditions at Contact Boundaries in Numerical Examination of the Interaction of Deformed Solids.* Preprint of the Institute of Problems of Strength, Academy of Sciences of the Ukraine, USSR, Kiev (1990) (in Russian).
61. Eliezer Sh, Gilath I, Bar-Noy T. Laser-induced spall in metal: Experiment and simulation. *J Appl Phys* 67 (2): 715–724 (1990).
62. *Tables of Physical Quantities (Handbook).* Ed. IK Kikoin. AtomIzdat, Moscow (1976) (in Russian).
63. Astanin VV, Galiev ShU, Ivashchenko KB. Pecularities of the deformation and failure of aluminum targets in interaction with a steel projectile along the normal. *Strength Mat* 12: 1605–1611 (1988).
64. Astanin VV, Galiev ShU, Ivashchenko KB. Cone formation in targets beneath a penetration projectile. *Int J Impact Eng* 11(4): 515–525 (1991).
65. Galiev ShU. *Darwin, Geodynamics and Extreme Waves.* Springer, Cham (2015).
66. Galiev ShU. *Charles Darwin's Geophysical Reports as Models of the Theory of Catastrophic Waves.* Center of Modern Education, Moscow (2011) (in Russian).

Index

A

Absorption of radiation, 287
Akhmadeev–Nigmatulin model, 35, 47
Annular stiffeners, 171–173
Anomalous behavior, of plate, 218, 225–229
Atmosphere, on counterintuitive behavior, 217–218
Attenuation coefficient, 287
Axisymmetric momentless (flexible) shells, 30–31

B

Bending stress, 195–200
Bifurcation, 211–213, 217
Bilinear (constant pressure) model, 220, 226, 227
Bilinear (constant sound velocity) model, 220–221, 226
Bleich–Sandler plate problem, 147, 148
Bouguer–Lambert law, 236, 257, 267, 287, 293
Boundary conditions, melting waves, 300
Boussinesq's equation, 80
BrKMts alloy plate, 113–114
Bubble-hydroelastic systems, 123
Bubble media
    linear acoustics, 69–73
        one-speed wave equations, 72–73
        three-speed wave equations, 70
        two-speed wave equations, 71–72
    nonlinear acoustic, 79–82
        Boussinesq's equation, 80
        Klein–Gordon equation, 81–82
        long-wave equation, 80–81
        Schrödinger equation, 82
    one dimensional governing equations, strongly nonlinear wave equation, 67–69
    volcanic activities, 73–75
Bubbly liquid model, 126, 176, 177, 221–223, 226
Bulk cavitation, 93, 98
Butterfly effect, 229

C

Caloric equations, 5–8
Cap/permeable membrane system, 200–202
Cap pole, deformation for, 199–200
Cartesian coordinate system, 42–43
Cauchy–Lagrange integral, 107
Cavitation
    on counterintuitive behavior, 217–219
        calculation details, 224–225
        gas models, 223
        liquid models, 219–224 (*see also* Liquid models)
        numerical investigations, 225–230
        Timoshenko's model, 219
    waves
        generation of, 163–166
        in plate, 159–163
        in thin plate, 155–159
Circular plates, counterintuitive behavior of, 187–188
    elastic–plastic elements, 192
    perforated plates, 191, 192
    plastic deformations, 190–191
    remarks, 192–193
    results, 188–190
    thick two-layer plate, 192
Compression shock wave, 185–186
Conical melting wave, 297
Conical panels, counterintuitive behavior of, 203–205
Conservation equation, 164
Continuum mechanics
    Caloric equations, 5–8
    classical models, 3–4
    conservation laws of, 260–261
    equations, 300
    kinematic relations, 4
    state equations, 4–5
    thermal equations, 5–8
Continuum model, 267–268
Cooling shock wave
    formation of, 242–248 (*see also* Thermodynamic function)
Counterintuitive behavior (CIB), 181
    atmosphere and cavitation on, 217–219
        calculation details, 224–225
        gas models, 223
        liquid models, 219–224 (*see also* Liquid models)
        numerical investigations, 225–230
        Timoshenko's model, 219
    cap/permeable membrane system, 200–202
    of circular plates, 187–193
    numerical calculations
        cylindrical panel, 215–217

Counterintuitive behavior (*Cont.*)
   plates, 210–214
      spherical cap, 214–215
   of panels, 203–205
   of plates and shallow shells, 207
      investigation techniques, 207–210
      results, 210–217
   of rectangular plates, 193
   of shallow caps, 194–200
   of spherical shallow shells, 193–194
Counterintuitive displacement
   of center of specimens, 186, 187
   of double-layer perforated plate., 189
Cryogenic liquids, 140–142
Cylindrical coordinate system, 144
Cylindrical elastic container, 129–131
   bubble dynamics features, 133–135
      distributed bubbles, 133–134
      influence of bubble screens, 134–135
   cool boiling effects, 130–133
      governed equations, 130–131
      ideal fluid model, 131
      influence of, 131–133
      liquid model, 131
Cylindrical panels
   counterintuitive behavior of, 203
      numerical calculations, 215–217
      one-dimensional equations, 208
Cylindrical shell
   basic equations, 127–128
   calculations results, 128–129
   cylindrical elastic container, 129–131
      bubble dynamics features, 133–135
      cool boiling effects, 130–133
   dynamic cavitation hydroelasticity problems
      governing equations, 143–145
      numerical–analytical method, 145–147
      results, 147–153
   hydroelastic systems
      dynamic phenomena in cryogenic liquids, 140–142
      pressure waves sources, 139–140
   hydro-gas-elastic systems, 135–139
      calculations results, 136–139

D

Dangerous space objects (DSOs), 282–284, 287, 291
   deflection of, 283, 284
   fracture mechanisms of, 287
   pulsed laser action on, 282
   spallation of, 290–291, 293, 295
   vaporization of, 283
D16AT duralumin plate, 103
Depression wave, 184–186, 190–194, 229
Distributed load, 270

Double-layer membrane, 201, 202
DSOs, *see* Dangerous space objects (DSOs)
Duralumin plate, 103, 108, 111–113
Dynamical buckling, 205
Dynamic cavitation hydroelasticity problems
   governing equations, 143–145
   numerical–analytical method, 145–147
   results, 147–153
Dynamic fracture processes
   interacting voids models, 39–43
      Cartesian coordinate system, 42–43
      spatial model, 43
      two interacting pores (bubbles), 40–41
      unidimensional model, 41–42
   microstructural models, 36–39
      numerical study, 38–39
      particular cases, 38–39
      remark, 39
   phenomenological models, 34–36
   pore dynamics
      coalescence of pores, 46
      compressible porous material, 43–46
      macro-cracks, 33–34
      mathematical model, 47–49

E

Elasticity theory, 174
Elastic liquid model, 120
Elastic–plastic model, 118
Elastic–plastic problems, 11
Elastoplastic with ideal plasticity, 300
Electromagnetic sensors, 186
Energy absorption, 280, 281, 283
Energy conservation law, 235
Equations of motion, 207, 217, 224, 285
Evaporation surface
   continuum model, 267–268
   Knudsen model, 268–269
Experimental data
   counterintuitive behavior
      cap/permeable membrane system, 200–202
      of circular plates, 187–193
      of elastic–plastic structures, 183–184
      of panels, 203–205
      of rectangular plates, 193
      of shallow caps, 194–200
      of spherical shallow shells, 193–194
   impact loading
      introduction of, 183–184
      method, 185–187
Explicit difference scheme, 210
Explicit numerical scheme, 264
Explosion system, underwater waves
   fracture and cavitation waves in
      plate, 159–163
      thin plate, 155–159

Index

External spallation, 284, 291–295
Extreme waves excitation
  impulse loading
    algorithm and results, 263–267
    fracture, mathematical model of, 260–263
  impulsive deformation and destruction
    outer laser-induced spalling, 256–260
    stress and fracture waves in metals, 253–256
    thermoelastic waves excitation, 250–253
  modeling process
    boundary conditions, 269–270
    continuum model, 267–268
    fracture, 274–277
    initial conditions, 270
    Knudsen model, 268–269
    spalling, vaporization on, 277–279
    temperature effects, 270–273
    vaporization effects, 273–277
  overview of, 249–250
  radiation pulses, solid fracture by
    calculation results, 287–290
    introduction of, 282–284
    laser fracture efficiency, 293–295
    mathematical model, 285–287
    spallation, 290–293
    two-dimensional fracture and evaporation, 279–282

F

Finite-difference method/scheme, 164, 169, 254
  two-dimensional dynamic boundary problems, 146
Flow plasticity theory, 10–12
Fourier equation
  linear analysis, 235, 236
  nonlinear heat waves, 238
Fracture and cavitation waves
  underwater explosion system
    plate, 159–163
    thin-plate, 155–159
Fracture mechanisms
  dangerous space objects, 287
  mathematical model of
    impulse loading, 260–263
  modeling process, 274–277
  of solid by radiation pulses
    calculation results, 287–290
    cupola formation, 287–289
    introduction of, 282–284
    laser fracture efficiency, 293–295
    mathematical model, 285–287
    spallation, 290–293, 295, 296
  and stress waves in metals, 253–256

G

Galiev's method, 120, 148, 149
Gas-hydroelastic system
  liquid in a tank, 124–126
    bubble screen effect, 126
  Rayleigh equation, 122
  tank impact loading
    concentration of bubbles, 123–124
    ideal liquid model, 122–123
    influence of dynamics, 123–124
Gas–liquid system, 224–225
Gassy media modeling
  state equation
    condensed matter–gas mixture, 58–60
    strongly nonlinear model, 60–62
    tait-like form, 62–65
  wave equations, 65–66
    longitudinal waves, 66
    spherical waves, 66
Gaussian distribution, 287
Generalized thermodynamics, 239

H

Heat conduction, 235
Heat flux equation, 235, 261
Heat waves, 235
  interaction of, 273
  linear analysis, 235–238
  nonlinear analysis, two-stage waveform
    in liquid helium, 238–239
    reflection and amplification, 241–242
    of second sound, 239–241
  profiles, 237–238
  thermodynamic function
    introduction of, 243
    numerical results, 244–248
    relations and equation, 243–244
Hooke's law, 8, 9, 11, 47, 160, 261
Hook's law, 115, 208–209
Hugoniot relations, 269
Hull cavitation, 92, 93
Hydrodeformable systems
  experimental approach, 96–101
    elastic plate–underwater wave interaction, 97–98
    elastoplastic plate–underwater wave interaction, 98–101
  extreme waves, 93–96
  gas–water vapor appears, 93
  metal work pieces, 91–92
  plasticity effects
    comments, 120
    elastic plate, 117–118
    elastoplastic plate, 118–120
    problem statements, 114–117

Hydrodeformable systems (*cont.*)
    results, 117
    solution method, 117
    plate interaction
        BrKMts alloy plate, 113–114
        cavitation effects, 103–106
        deformability effects, 101–103
        duralumin plate, 111–113
        qualitative analysis, 106–110
        results, 111
        solution method, 111
    ships and offshore structures, 92–93
    underwater explosions, 93–96
Hydroelastic systems
    dynamic phenomena in cryogenic liquids, 140–142
    pressure waves sources, 139–140
Hydrogasdynamics equations, 156
Hydro-gas-elastic systems, 135–139
    calculations results, 136–139
Hyperbolicity, influence of, 235–238
Hyperbolic two-dimensional system, 300

I

Ideal elastic liquid model, 219, 226
Ideal fluid model, 126
Ideal gas model, 223
Ideal liquid model, 122, 131, 133
Ideal plasticity, 10
Impact loading
    introduction of, 183–184
    method, 185–187
Impact velocity, 297, 299
Implicit iterative procedure, 264
Impulse loading, melting effects
    algorithm
        of calculation, 263–264
        and results, 264–267
    fracture, mathematical model of, 260–261
        boundary conditions, 263
        initial conditions, 263
        material damage, 261–262
        melting material, 262–263
Impulsive deformation and destruction
    outer laser-induced spalling, 256–260
    stress and fracture waves in metals
        rapid bulk heating, 253–256
    thermoelastic waves excitation
        by long-wave radiation, 250, 251
        by short-wave radiation, 250–253
Initial conditions, counterintuitive behavior, 225
Instantaneous wave detonation, 156
Instant load, 270
Interacting voids models, 39–43
    Cartesian coordinate system, 42–43
    spatial model, 43
    two interacting pores (bubbles), 40–41
    unidimensional model, 41–42

K

Kirchhoff–Love hypotheses, 24–28
Klein–Gordon equation, 81–82
Knudsen model, 268–269
Kronecker delta, 4
Kuznetsov's equations, 140

L

Lagrange formulation, 284, 285
Lagrangian grid distortions, 301, 302
Lame equation, 8
Laser fracture efficiency, 293–295
Laser thermonuclear synthesis, 283
Linear elastic analysis, shallow caps, 195–196
Liquid models
    bubbly liquid, 221–223, 226
    into cavitation phenomenon, 220–221
    displacement–time curves, 226, 227
    ideal elastic liquid, 219
    wide-range equation, state of water, 223, 224
Liquid pressure variation, 228
Long-wave equation, 80–81
Long wave, extreme amplification of, 175, 176

M

Mach cone, 297, 299, 302
Manganin transducer, 256
Mass conservation law, 269
Mass velocities, 258, 259
Mechanical criterion of failure, 161
Melting effects of materials
    impulse loading
        algorithm and results, 263–267
        fracture, mathematical model of, 260–263
Melting waves
    boundary conditions, 300
    calculation results, 301–302
    experimental investigation, 297–299
    numerical modeling, 300–301
Membrane stress, 195, 196
Mesh-analysis method, 224
Microdamage wave propagation, 265, 266
Microscopic damage, of material, 261, 262
Microstructural models, 36–39
    numerical study, 38–39
    particular cases, 38–39
    remark, 39
Mirror destruction, 252, 253
Modeling process
    boundary conditions, 269–270
    calculations

Index

temperature effects, 270–273
vaporization effects, 273–277
continuum model, 267–268
fracture, 274–277
initial conditions, 270
Knudsen model, 268–269
spalling, vaporization on, 277–279
Modified Fourier law, 5
Momentless shells
   Kirchhoff–Love hypotheses, 24–28
   shallow shells, 24–28
Mount St Helens eruption, 75–77

N

Natanzon's experiments, 21
Navier–Stokes law, 12
Newton's method, 148
Non-cavitation model, 104
Nonlinear acoustic
   bubble media, 79–82
      Boussinesq's equation, 80
      Klein–Gordon equation, 81–82
      long-wave equation, 80–81
      Schrödinger equation, 82
Nonlinear airy-type equations, 82–84
Nonlinear elastic analysis, shallow caps, 195–197
Nonlinear elastic–plastic analysis
   shallow caps, 197–198
Nonlinear equations of motion, 184, 217, 229
Nonlinear heat waves, 238–242
   reflection and amplification, 241–242
   thermodynamic function
      introduction of, 243
      numerical results, 244–248
      relations and equation, 243–244
   two-speed wave front, liquid helium, 238–239
   two-stage waveform, second sound, 239–241
Numerical modeling, melting waves, 300–301

O

Observable extreme waves examples
   Mount St Helens eruption, 75–77
   volcano Santiaguito eruptions, 77–79
One-dimensional cavitation interaction, qualitative picture, 106, 107
One dimensional governing equations
   bubbly media, strongly nonlinear wave equation, 67–69
   state equation
      differential form, 66–67
      linear bubble oscillations, 67
One-dimensional nonlinear wave equation, 242
Oscilloscope screen, 97
Outer laser-induced spalling, optimization, 256–260

P

Panels, counterintuitive behavior of, 203–205
Phenomenological models, 34–36
Piezoelectric pressure sensors, 97
Plastic deformations, 190–191, 203, 213, 214, 300, 301
Plastic flow theory, 9–10
Plate/liquid contact condition, 225
Plate–liquid interaction
   cavitation interaction process
      BrKMts alloy plate, 113–14
      duralumin plate, 111–113
      qualitative analysis, 106–110
      results, 111
      solution method, 111
   plasticity effects
      comments, 120
      elastic plate, 117–118
      elastoplastic plate, 118–120
      problem statements, 114–117
      results, 117
      solution method, 117
Plates
   counterintuitive behavior of
      anomalous behavior, 218, 225–229
      numerical calculations, 210–214, 224, 225
Pore dynamics
   coalescence of pores, 46
   compressible porous material, 43–46
   macro-cracks, 33–34
   mathematical model, 47–49
Prandtl–Reuss constitutive equations, 210
Pulsed laser radiation, 287, 290
Pulse energy, 287

R

Radiation power density, 251, 253
Radiation pulses
   solid fracture by
      calculation results, 287–290
      introduction of, 282–284
      laser fracture efficiency, 293–295
      mathematical model, 285–287
      spallation, 290–293, 295, 296
Rapid bulk heating, stress and fracture waves, 253–256
Rarefaction zones
   bubbly liquid
      fractured liquid, 22–23
      fracture of, 23–24
      gas–liquid mixture formation, 17
      plane waves, 21–22
      Rayleigh equation, 17–18

Rarefaction zones (*cont.*)
　remarks, 24
　state equation, 18–19
　fracture (cold boiling), 19–21
Rayleigh equation, 121, 122, 143
Rayleigh–Plesset equation, 66
Rayleigh's equation, 221
Rectangular plates, counterintuitive behavior of, 193
Relaxation heat transfer, 242, 243
Remote emitters, 283
Reverse snap-buckling, 184, 186, 187, 189, 196, 228, 229
Runge–Kutta method, 38, 122, 132

S

Saha equation, 6
Schrödinger equation, 82
Shallow caps, counterintuitive behavior of, 194–195
　deformation for pole, 199–200
　failure of cap pole
　　linear elastic analysis, 195–196
　　nonlinear elastic analysis, 195–197
　　nonlinear elastic–plastic analysis, 197–198
　　reasons for, 198–200
Shallow shells, 24–28
　nonlinear theory of, 208
Shauer's model, 120
Shock loading, 233
Shock-wave pressure, 250
Short pulse radiation, 250–253
Short wave, extreme amplification of, 175, 176
Single-layer membrane, 201, 202
Smooth shell, 170–171
Snap-buckling, 196, 205
Space debris, 283, 284, 291, 295, 296
Space safety
　solid fracture, radiation pulses
　　calculation results, 287–290
　　cupola formation, 287–289
　　high-energy lasers in, 283
　　introduction of, 282–284
　　laser fracture efficiency, 293–295
　　mathematical model, 285–287
　　spallation, 290–293, 295, 296
Spallation mechanism, 290–293, 295, 296
Spherical cap, 209
　counterintuitive behavior of, 214–215
Spherical shallow shells, CIB of, 193–194
SSS, *see* Stress-strain state (SSS)
State equation
　gassy media modeling
　　condensed matter–gas mixture, 58–60
　　strongly nonlinear model, 60–62

　　tait-like form, 62–65
　one dimensional governing equations
　　differential form, 66–67
　　linear bubble oscillations, 67
State (constitutive) equations
　elastic body, 8–9
　flow plasticity theory, 10–12
　ideal plasticity, 10
　plastic flow theory, 9–10
Stiffened spherical shells
　in crossing system, 173
　extreme amplification, 174–177
　hydrodynamic equations, 169
　motion equations, 167–168
　shell models, 168–169
　transient interaction of
　　numeric method, 169–170
　　problem and method, 166–169
　　results, 170–174
Stress profiles, 258, 259, 270–272, 277, 278
Stress–strain behavior, 209
Stress–strain state (SSS), 161, 162, 171
Stress wave propagation, 252–254, 265, 266
Structural orthotropy, 168
Surface rarefaction wave, 283

T

Tait's equation, 13–16, 143, 164
Temperature-dependent functions, 243–244
Temperature profile, 244–246
Thermal conductivity, 235–237, 243
Thermal equations, 5–8
Thermal force loading
　fracture, mathematical model of, 260–261
　　boundary conditions, 263
　　material damage, 261–262
　　melting material, 262–263
Thermal stresses, 257, 258
Thermal waves, amplification of, 237–238
Thermodynamic function
　nonlinearity of
　　introduction of, 243
　　numerical results, 244–248
　　relations and equation, 243–244
Thermoelastic waves
　equations for, 251
　excitation
　　by long-wave radiation, 250, 251
　　by short-wave radiation, 250–253
Thermomechanical behavior, 233
Thermomechanical process, 277
Threshold of destruction, 252
Timoshenko's model, 219
The Timoshenko theory, 29–30
Timoshenko type, 163
Transient cavitation zone, 94

Index

Tuler–Butcher formulas, 36
Two-dimensional axisymmetrical formulation, 284, 285
Two-dimensional fracture and evaporation, 279–282
2D theory, 112

U

Underwater explosion, 183, 184, 188
   system, 218
Underwater waves
   explosion system, fracture and cavitation waves, 155–159
   extreme amplification, 174–177
   stiffened spherical shells (*see also* Stiffened spherical shells)
   transient interaction of, 166–174

V

Vaporization effects, 273–277
Vaporization modeling
   continuum model, 267–268
   Knudsen model, 268–269
Viscous liquid
   simplifying the mathematical formulation, 13–15
   wide range of parameters, 15–17
Volcano Santiaguito eruptions, 77–79

W

Wave propagation processes, 253, 254, 256
Weakly cohesived media (WCM)
   bubbly media, volcanic activities, 73–75

gassy material properties, 53–54
gassy media modeling
   state equation
      condensed matter–gas mixture, 58–60
      strongly nonlinear model, 60–62
      tait-like form, 62–65
      wave equations, 65–66
      longitudinal waves, 66
      spherical waves, 66
geomaterials within extreme waves, 54–57
linear acoustics
   bubble media, 69–73
   one-speed wave equations, 72–73
   three-speed wave equations, 70
   two-speed wave equations, 71–72
nonlinear acoustic, bubble media, 79–82
   Boussinesq's equation, 80
   Klein–Gordon equation, 81–82
   long-wave equation, 80–81
   Schrödinger equation, 82
nonlinear airy-type equations, 82–84
observable extreme waves examples
   Mount St Helens eruption, 75–77
   volcano Santiaguito eruptions, 77–79
one dimensional governing equations
   state equation
      differential form, 66–67
      linear bubble oscillations, 67
Wide-range equation, state of water, 223, 224
Wilkins method, 157, 160, 164, 169
Wilkins numerical algorithm, 279

Z

Zhurkov model, 35

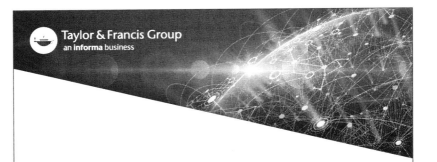